T0231098

Advanced Routing of Electronic Modules

Electronic Packages, Interconnects, and Product Series

Published

Long-Term Non-Operating Reliability of Electronic Products
Judy Pecht and Michael Pecht

Product Reliability, Maintainability, and Supportability Handbook
Edited by Michael Pecht

Forthcoming

Estimating the Influence of Temperature on Microelectronic Device Reliability
Pradeep Lall, Michael Pecht, and Edward Hakim

Michael Pecht
Yeun Tsun Wong

Advanced Routing of Electronic Modules

CRC Press
Taylor & Francis Group
Boca Raton London New York

CRC Press is an imprint of the
Taylor & Francis Group, an **informa** business

CRC Press
Taylor & Francis Group
6000 Broken Sound Parkway NW, Suite 300
Boca Raton, FL 33487-2742

© 1996 by Taylor & Francis Group, LLC
CRC Press is an imprint of Taylor & Francis Group, an Informa business

Visit the Taylor & Francis Web site at
http://www.taylorandfrancis.com

and the CRC Press Web site at
http://www.crcpress.com

PREFACE

The rapid growth of the electronic products market has created an increasing need for affordable, reliable, high-speed, and high-density multilayer printed circuit boards (PCBs). The complexity of a PCB is measured by the board size; the metallization geometry and density; the number of signal, ground, and voltage layers; the size and number of vias; and the electrical and mechanical characteristics.

Routing is a design step that determines the locations and dimensions of the metallization, clearances, and vias in each layer, described as a function of electronic and mechanical characteristics, testability, reliability, manufacturability, and assembly. Advanced routing includes detailed, global, and channel routing, via minimization, compaction, rip-up rerouting, and parallel and high-speed PCB and multichip module (MCM) routing. In high-density and high-speed PCB/MCM routing, signal integrity, thermal constraints, reliability, test access, and manufacturability must all be considered. Providing both fundamental theory and advanced technologies for improving routing, this book's eleven chapters cover the major features of advanced routing.

Chapter 1 discusses refinement approaches to approximate a minimum rectilinear Steiner tree from a minimum spanning tree. This approach is one of the traditional methodologies for approximating a minimum Steiner tree.

Chapter 2 surveys the techniques in detailed routing and introduces the obstacle-avoiding rectilinear Steiner tree problem for routing simple multi-terminal nets sequentially in a workspace with obstacles.

Chapter 3 discusses a theory for the general solution of the rectilinear Steiner's problem. Applying the theory, linear algorithms are developed for solving the switchbox problem and the case when terminals are lying on the boundary of a rectilinear polygon. An algorithm expansion is present to compute an accuracy-estimable solution efficiently, using a series of subalgorithms expanded from an inefficient algorithm for an exact solution. Based on the solution, a parallel routing is presented that selects solutions to minimize local congestion, measured by a figure of merit.

Chapter 4 describes timing delay, clock skew, and noise control requirements in signal integrity and introduces computer-aided approaches to manage these requirements in high-speed PCB/MCM routing. This chapter also describes an approach that uses a global router to efficiently determine routability, while using a predictor function based on precharacterized simulation results to ensure that time delay and noise requirements are met.

Chapter 5 introduces high-speed PCB routing for analog and mixed circuits when a large variety of simultaneous constraints must be satisfied. Several routing approaches and features are described in this chapter, based on the sensitivity analysis that transforms the performance specifications into constraints on parasitics.

Chapter 6 describes a bounded search algorithm and channel segmentation to improve the quality of segmented channel routing in field-programmable gate arrays (FPGAs).

Chapter 7 overviews global, contact, and main-channel routing and describes their applications in the restricted channel routing for CMOS gate arrays.

Chapter 8 discusses the two-layer wiring problem based on a knock-knee mode layout, its corresponding dual graph, and the constrained via minimization in the wiring and the macro cell resulting from preassigning pins. This chapter also discusses interconnection delay minimization in two layers, whether or not they have the same conductivities.

Chapter 9 presents compaction techniques, especially hierarchical and performance-oriented compaction. It also discusses performance-driven compaction as it relates to delay and crosstalk minimization.

Chapter 10 introduces rip-up and reroute approaches. Particularly, it discusses a hypergraph model and its algorithm.

Chapter 11 presents parallel routing, including global routing, boundary crossing placement, and detailed maze routing in hardware acceleration. This chapter also describes data structures, data management, and algorithms for parallel routing in a multiple-processor hardware system.

In sum, this book presents the technologies, algorithms, and methodologies for those people who will be developing the next generation of electronic products. In addition, the book describes some of the as yet unsolved technical problems and challenges.

Support for this book came from the Technology Reinvestment Program (EEC-9415445) and the National Science Foundation (EEC-9108844).

Michael Pecht and Yeun T. Wong

CONTENTS

CONTRIBUTORS

Ting-Hai Chao received his B.S. from Cheng Kung University, Tainan, Taiwan in 1971, and a M.S. in mathematics and a Ph.D. in computer science from Tsing Hua University, Hsinchu, Taiwan, in 1975 and 1992, respectively. Since 1980, he joined the Industrial Technology Research Institute(ITRI), Hsinchu, Taiwan, as a senior engineer. Currently, he serves as manager of the Quality Assurance Department at Computer and Communication Research Laboratories of ITRI.
Address: Q000, Bldg. 14 195 Sec. 4, Chung Hsing Rd., Chutung, Taiwan 310, R.O.C.

Tae Won Cho graduated from Seoul National University in Electronic Engineering, Korea, in 1973, received a M.S. from the University of Louisville, KY, in 1986, and a Ph.D. from University of Kentucky, in 1992, both in electrical engineering. He is a professor of electronic engineering at Chungbuk National University, Korea. He serves as a Chairman of the Study Group of "Low Power High Speed VLSI Design," a group organized by him with professors and experts in that field since 1993. From 1973 to 1984 he was a manager of the Gold Star Cable Co., Korea. He serves as consultant to a number of industrial laboratories in the area of processor design and CAD of VLSI circuits. His research interests include VLSI design, parallel processing, computer architecture, and placement and routing in the VLSI circuits.
Address: Department of Electronic Engineering, Chungbuk National University, Cheongju, Chungbuk, Korea 360-763.

James P. Cohoon received his Ph.D. in 1983 from the University of Minnesota. He then joined the Department of Computer Science at the University of Virginia where he is currently an associate professor. His department has twice nominated him for the university's best teacher award. His primary research interests lie in VLSI circuit layout with particular emphasis on algorithmic aspects of routing and placement. He is the author or co-author of more than fifty papers. His current work is developing a new layout methodology that treats routing and placement in an integrated manner. The methodology is applicable to both standard cell and FPGA layout. Other research interests include computational geometry, parallel algorithms, heuristic search, testing, and visualization. He has served on the programming committees for such conferences as DAC, ICCAD, and ICCD and was co-organizer of the first ACM Design Automation Workshop in Russia. He is Chair of ACM-SIGDA and a member of the ACM and IEEE Circuits and Systems professional societies. His honors include a Fulbright Award and the SIGDA Leadership award.

Address: The Department of Computer Science, University of Virginia, Charlottesville, Virginia 22903.

Paul D. Franzon is an associate professor in the Department of Electrical and Computer Engineering at North Carolina State University. He has more than 10 years experience in electronic systems design and design methodology research and development. Has worked at AT&T Bell Laboratories in Holmdel, NJ, and at the Australian Defense Science and Technology Organization.

His current research interests include design sciences/methodology for high speed packaging and interconnect, for high speed and low power chip design, and the application of micro electro mechanical machines to electronic systems. He has published more than 60 articles and reports. He is also the co-editor and author of a book about multichip module technologies published in October, 1992.

He is a member of IEEE, ACM, and ISHM. He served as the chairman of the Education Committee for the National IEEE-CHMT Society. In 1993, he received an NSF Young Investigator's Award.

He received a B.S. in physics and mathematics, a M.E. with First Class Honors in Electrical Engineering, and a Ph.D. in electrical engineering, all from the University of Adelaide, Adelaide, Australia.
Address: Department of Electrical and Computer Engineering, North Carolina State University, Raleigh, NC 27695-7911.

Joseph L. Ganley received his Ph.D. in 1995 from the University of Virginia. He is currently employed by Cadence Design Systems in San Jose, California. His primary research interests are in VLSI physical design automation, geometric and graph algorithms, scientific computing, and parallel algorithms. He is a member of ACM, SIGACT, SIGDA, SIAM, and Tau Beta Pi. His honors include a University of Virginia Dean's Fellowship and a Virginia Space Grant Fellowship.
Address: Cadence Design Systems, Inc., 555 River Oaks Parkway, Building 2, MS 2A2, San Jose, California 95134.

Yu-Chin Hsu received his B.S. in computer science from National Taiwan University, Taipei, Taiwan, in 1981, and a M.S. and a Ph.D. in computer science from the University of Illinois at Urbana-Champaign in 1986 and 1987, respectively. He is currently serving as an associate professor of the Computer Science Department at University of California, Riverside. Previously, Dr. Hsu served on the faculty of Tsing Hua University, Hsinchu, Taiwan.
Address: Department of Computer Science, University of California, Riverside, CA 92521.

Aiguo Lu received his B.S. (with honors) in electronic engineering from Xidian University, Xian, in 1983, and a M.S. degree in Nanjing University of Posts and Telecommunications, Nanjing, in 1986, both in China. Currently, he is studying for a Ph.D. at the Department of Electrical and Electronic Engineering, University of Bristol, United Kingdom.

His research interests include logic synthesis, layout for CMOS gate arrays, ASIC design using FPGAs, and network optimization.
Address: Room 0.38a Queen's Building, University Walk, University of Bristol, BS8 1TR, UK.

Enrico Malavasi received the "laurea" in electrical engineering from the University of Bologna (Italy) in 1984, and a M.S. in electrical engineering from the University of California at Berkeley in 1993. Between 1986 and 1989 he worked at the Department of Electrical Engineering and Computer Science (DEIS) of the University of Bologna on research topics related to CAD for analog circuits. In 1989 he joined the *Dipartimento di Elettronica ed Informatica* of the University of Padova, Italy, as assistant professor. Between 1990 and 1994 he has collaborated with the CAD group of the Department of EECS of the University of California at Berkeley, where he has carried out research on performance-driven CAD methodologies for analog design. In 1994 he became a consultant for Cadence Design Systems, Inc., in the area of constraint-driven physical automation for analog circuits.

His research interests are in the area of design automation and methodology for analog and mixed circuits.
Address: DEI - Universita' di Padova, via Gradenigo, 6/A 35131 Padova, PD - ITALY

Sharad Mehrotra received his B.Tech. from Indian Institute of Technology, Kanpur, India, in 1989, a M.S. from Vanderbilt University in 1991, and a Ph.D. from North Carolina State University in 1994, all in electrical engineering. He is currently employed as a Development Staff Member at IBM, Austin, Texas, where he works in the High Performance Circuit Design Tools group.

His research interests include CAD for performance driven layout synthesis, timing verification, and performance optimization of CMOS circuits.
Address: IBM Corp., 11400 Burnet Road, M/S 4358, Austin, TX 78758.

Paul Molitor is a professor and a chairman of the Computer Science Department of Martine-Luther University of Halle. Dr. Molitor has published a number of papers on wiring and via minimization.
Address: Institute fur informatik, Martin-Lurther Universitat Halle, Postfach 06099, D-06099 Halle (Saale), Germany.

Michael G. Pecht is a professor and director of the CALCE Electronic Packaging Research Center at University of Maryland. He has a B.S. in Acoustics, a M.S. in electrical engineering, and a M.S. and a Ph.D. in engineering mechanics from the University of Wisconsin. He is a professional engineer, an IEEE Fellow, and an ASME Fellow. He serves on the board of advisors for various companies and consults for the U.S. government, providing expertise in strategic planning in the area of electronics packaging. He is the chief editor of the *IEEE Transactions on Reliability*, a section editor for the Society of Automative Engineering, and on the advisory board of IEEE Spectrum.
Address: *CALCE* Electronics Packaging Research Center of University of Maryland, College Park, MD 20740.

Manuela B. Raith received her Dipl.-Math. degree in mathematics from the Technical University Braunschweig and the Dr. rer. nat. degree from the Ludwig-Maximilians-University Munich in 1987 and 1994, respectively. During a three-year scholarship she has been with the Department of Research and Development at Siemens AG. Since 1991 Dr. Raith has been working with Zuken-Redac, Munich. She is engaged in the topics of Routing, EMC, and High-Speed problems associated with printed circuit boards.
Address: Zuken-Redac, Muthmannstr. 4, D-80939 Munich, Germany.

Kaushik Roy received his Ph.D. (1990) from the University of Illinois at Urbana-Champaign in electrical and computer engineering, and his B.Tech. (1983) in electronics and electrical communications engineering from the Indian Institute of Technology, Kharagpur, India. From 1988 to 1993, he was employed with the Semiconductor Process and Design Center of Texas Instruments, Dallas, where he worked on Low-power VLSI and on Field Programmable Gate Arrays (FPGAs). From 1992 to 1993 he was also a Semiconductor Research Corporation (SRC) Mentor, and his responsibilities included monitoring research on VLSI testing at the University of Texas at Austin and the University of Illinois at Urbana-Champaign. In 1993 he accepted his current position as an assistant professor with the School of Electrical Engineering at Purdue University.

His principal areas of interest are in low-power VLSI, VLSI testing and reliability, and design of Field Programmable Gate Arrays. He is an associate editor of IEEE Design and Test of Computers and was on the program committee as guest editor for the special issue on low-power VLSI in the IEEE Design and Test of Computers which appeared in December of 1994.
Address: The School of Electrical Engineering, Purdue University, West Lafayette, IN 47907.

Hyunchul Shin received his B.S. in electronic engineering from Seoul National University, a M.S. in electrical engineering from the Korea Advanced Institute of Science and Technology in 1978 and 1980, respectively, and a Ph.D. in electrical engineering and computer sciences from the University of California, Berkeley, in 1987.

From 1980 to 1983, he was with the Department of Electronics Engineering, Kum-Oh institute of technology, Korea. In 1983, he received a Fulbright Scholarship. From 1987 to 1989, he was a member of the technical staff at AT&T Bell laboratories, Murray Hill, N.J. Since August 1989, he has been with the Department of Electronics Engineering, Han-Yang University, Korea, where he is the department chairman. He has written more than 50 technical papers.

His research interests include design and synthesis of integrated circuits and systems.

Address: Department of Electronics Engineering, Han Yang University, Ansan, KyungKi 425-791, Korea (R.O.K.)

Yeun Tsun Wong finished his undergraduate studies in Guangdong Polytechnic Institute (Polytechnic University of Guangdong) and advanced studies at the Iron and Steel Institute of Peking (Beijing Science and Technology University), China. He received a M.S. in mechanical engineering from the University of Maryland. He was with the Mechanical Engineering Department of GPI and is currently with the *CALCE* Electronics Packaging Research Center at the University of Maryland.

His current research focuses on Steiner's problem, algorithm expansion, software development, and their applications in placement and routing.

Address: *CALCE* Electronics Packaging Research Center of University of Maryland, College Park, MD 20740.

Xinyu Wu graduated from Harbin Institute of Engineering, Harbin, China, in 1961. He joined Nanjing University of Posts and Telecommunications in 1979. He is now a professor and the director of the Institute of Neural Networks and System Optimization. He is also an IEEE senior member. From June 1986 to June 1987, he was a visiting scholar, and from January 1994 to March 1994, he was a visiting professor, both in the Department of Electrical Engineering and Computer Science, University of Illinois at Chicago. From September 1993 to December 1993, he was a visiting professor in the Department of Electrical Engineering and Computer Science, University of California at Berkeley. He is the author of more than 70 technical papers and four books.

His current interests are in circuit theory, applied graph theory, control theory, VLSI circuit layout, and neural networks.

Address: Nanjing University of Posts and Telecommunications, Campus Box 150, 38 Guangdong Road, Nanjing 210003, P. R. China.

Bin Zhu received his B.S. in mathematics from Nanjing University of Posts and Telecommunications, Nanjing, China, in 1982, and a M.S. in electronic engineering from Xidian University, Xian, China, in 1987.

Since June 1993, he has been an associate professor in the Research Laboratory of Circuits and Systems, Nanjing University of Posts and Telecommunications. He is now a visiting scholar in the Department of Computer Science, University of Saarland, Germany.

His main research interests include layout algorithm, VLSI CAD system, synthesis and routing of FPGAs, and combinatorial optimization. Address: Fachbereich 14 - Informatik Sonderforschungsbereich 124: VLSI-Entwurfsmethoden und Parallelitaet, Universitaet des Saarlandes, 66041 Saarbruecken, Germany.

To
the students and the staff at the University of Maryland
CALCE Electronic Packaging Research Center
who helped in this effort, and to those who
will follow in their footsteps

Chapter 1

REFINEMENT APPROACHES FOR RECTILINEAR STEINER TREE CONSTRUCTION

Ting-Hai Chao and Yu-Chin Hsu

With the rapid evolution of VLSI technologies, design of integrated circuits is becoming increasingly complex and time consuming. To speed up the design work and improve the quality, Computer-Aided Design (CAD) techniques have become a necessity of design. Among the various design phase of VLSI, physical layout takes a major portion of the turn-around time, hence much CAD research works on automatic layout methodologies.

In general, there are three main stages in an auto-layout system, named placement, global routing, and detail routing. The primary objective of placement is to arrange the elements in a way that the interconnections can be routed and the overall chip area is minimal. After placement, all elements are placed on a fixed position, and the position of all pins of a net are determined. The main task of global routing is to define the routing region and decide which channels are to be used for each net. A routing tree is constructed for each net. Usually, there are two routing layers: one for horizontal segments and the other for vertical segments. Segments on different layers are connected through the vias. Each routing tree is composed of horizontal and vertical line segments in the proper routing channel. After global routing, detail routing are used to assign proper tracks for each line segment in the routing channel. The maximum tracks occurring in a channel is called the *channel density*. The routing area is determined by the summation of channel density of routing region. Since routing area occupies 30% or more chip area, minimizing the routing area has become a key issue in physical design.

A straightforward approach to reduce the routing area is to minimize the routing line segments used for each net. Hence, the classical global routing target is simply to minimize the total wire length of each routing tree. Total wire length is a critical parameter of the routing tree, since excess interconnect not only increases layout area but also results in greater capacitance, thus deforming the circuit speed. With this in mind, the rectilinear Steiner tree problem becomes a important issue in the classical global routing.

Given a set of points, P, on the plane, a Steiner tree for P is a tree that contains P as a subset of its vertex set. The vertices which are not in P are called the *Steiner points* of the Steiner tree. A *rectilinear*

1

Steiner tree (RST) is a Steiner tree which contains only horizontal and vertical line segments. The total edge length of an RST is referred to as the *cost* of the RST. The rectilinear Steiner tree problem is to find an RST with the minimum cost. This problem has been shown to be NP-complete [1] except for certain special cases [2 - 4]. M. Hanan [2] has shown the optimal RST of a point set S is a subgraph of the *underlying grid* G(S), where the underlying grid G(S) of point set S is the grid obtained by drawing horizontal and vertical lines through each point of S (see Figure 1.1a)

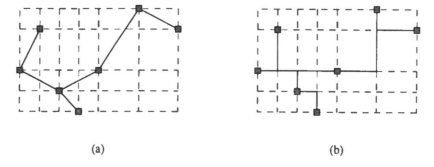

(a)   (b)

Figure 1.1 A spanning tree and a Steiner tree
(a) Underlying grid and an RMST; (b) An RST

RST's have many practical applications in VLSI physical design and computer network design. The problem has been studied extensively and many heuristic algorithms have been proposed [5 - 15]. F. Hwang [9] has shown that the ratio of the cost of a *rectilinear minimum spanning tree* (RMST) to that of an optimal RST is no greater than 3/2. Therefore, the cost improvement over RMST is usually used to judge the quality of an RST algorithm. The tree T of Figure 1.1a is an RMST of the point set. Figure 1.1b shows an RST of the point set.

1.1 EXISTING APPROACHES

M. R. Garey and D. S. Johnson have shown the RST problem is NP-complete in 1977 [1]. M. Hanan has shown the optimal solution can be found for a set of points with cardinalities under six [2]. J. P. Cohoon, D. S. Richard and J. S. Salowe have shown an optimal Steiner tree algorithm for a net whose terminals lie on the boundary of a rectangle in [3]. Except for the above special cases where an RST can be constructed optimally, there are many suboptimal algorithms for the general cases.

In general, we can classify the RST construction algorithms into two main categories. One is the construction approach, and the other is the refinement approach. Given a set of points, the construction approach builds an optimal RST for a subset of points (normally small enough to find an optimal solution in a short time). Then the RST is expanded to include the rest of the points. The LBH algorithm proposed by J. H. Lee, N. K. Bose and F. K. Hwang is a construction one [6]. It starts from a RST with three points, and iteratively construct a new RST by joining a new point which is the nearest point of the current RST. The average cost improvement of LBH is between 8% and 10%.

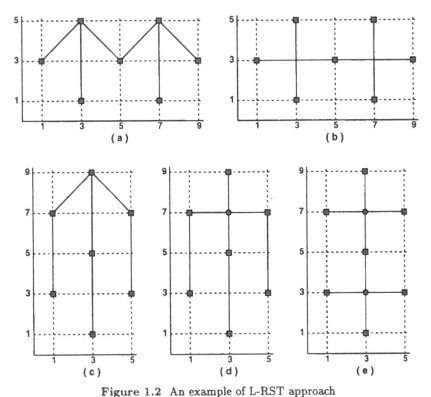

Figure 1.2 An example of L-RST approach
(a)A separable RMST; (b)Result of L-RST;
(c) Separable RMST if we swap the x,y coordinates of the points;
(d) Result of L-RST; (e) Optimal RST

The refinement approach starts from a rectilinear minimum spanning tree (RMST), and uses the refinement method to reduce the cost of the tree. The L-RST algorithm proposed by J-M. Ho, G. Vijayan and C. K. Wong is a typical refinement approach [10]. The main idea of L-RST is the separable RMST. An RMST is separable if the enclosing boxes of any two nonadjacent edges do not intersect or overlap. Starting from a separable RMST, L-RST selects an L-shaped layout of each nondegenerate edge, so as to maximize the total amount of overlap, thus minimizing the total cost of the resulting RST.

There are two weaknesses in the L-RST algorithm. First, the goodness of the resulting RST is limited by the separable RMST. For example, from the set of points in Figure 1.2a, the L-RST algorithm generates the separable RMST(Figure 1.2a) and the resulting RST(Figure 1.2b), which is optimal. However, if we exchange the x and y coordinates of each point in Figure 1.2a, the separable RMST and the resulting RST generated by the L-RST algorithm are shown in Figure 1.2c and Figure 1.2d, respectively. This RST is not optimal. The optimal one is shown in Figure 1.2e. A second weakness is that the connection cost of an RMST can be further reduced even if there is no nondegenerate edges in the RMST. Consider the separable RMST in Figure 1.3a. It is clear that the L-RST algorithm cannot improve the cost. However, by introducing the Steiner point p_s into the point set as shown in Figure 1.3b, and then regenerating an RMST, the connection cost can be improved, as shown in Figure 1.3c.

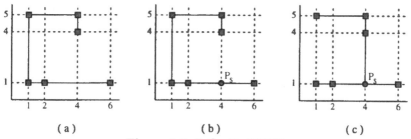

Figure 1.3 Refinable RMST
(a)RMST with no nondegenerate edge;
(b) Modified tree due to p_s;
(c) Refined Steiner tree

The iterated 1-Steiner approach [16] proposed by A. Kahng and G. Robins is to iteratively calculate 1-Steiner points which is the point p such that the cost of RMST of $P \cup \{p\}$ is minimized, and add them to the set of Steiner points. The time complexity of this method is $O(n^3)$ and empirical results indicate that the average cost improvement over the RMST is almost 11%.

1.2 LOCAL AND GLOBAL REFINEMENT

The motivation of our tree refinement is to alleviate the weaknesses of L-RST as described in Section 1.1. We propose a new approach which takes a more global view over that in the L-RST model. An RMST is first generated according to a weighted rectilinear metric and then refined iteratively by adding Steiner points and regenerating RMST's. Steiner points are introduced into the point set in two stages. In the first stage, Steiner points satisfying the "essential criterion" are generated from the local structure (adjacent edges) of the RMST. A refined RMST is constructed by including these points into the original point set. In the second stage, Steiner points are generated from a global view of the refined RMST so that they may be introduced between nonadjacent edges.

1.2.1 WEIGHT OF EDGES

Given a set of points, P, and $P = \{p_1, p_2, ..., p_n\}$. The *node priority* $n_p(p_i, p_j)$ between p_i and p_j is $\min(i, j)$. In our algorithm, we use an ordered tuple $(d_r(p_i, p_j), d_s(p_i, p_j), n_p(p_i, p_j))$ as the *weight*, $w(e(p_i, p_j))$, of edge $e(p_i, p_j)$. The weights of edges are compared in a lexicographic order. For example, in Figure 1.4, $w(e(p_0, p_1)) = (10, -5, 0)$, and $w(e(p_0, p_2)) = (10, 0, 0)$, and therefore $w(e(p_0, p_1)) < w(e(p_0, p_2))$. Similarly, $w(e(p_0, p_3)) = (12, -5, 1)$ $w(e(p_2, p_3)) = (12, -5, 3)$ and therefore $w(e(p_0, p_3)) < w(e(p_2, p_3))$. We use this metric in Prim's minimum spanning tree algorithm to generate our RMST [11]. For example, Figure 1.4 is the RMST generated by this metric using p_0 as the root node. The slant distance, the second key of the weight, is used to select the nondegenerate edges, and hence the cost of RMST has greater opportunity to be improved. The node priority, the third key of the weight, is used to select those edges which are connected to the original points in the enlarged points set.

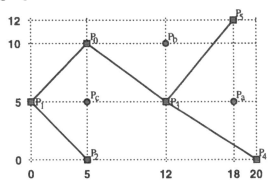

Figure 1.4 The RMST of TR

1.2.2 REFINED STEINER TREE AND MODIFIED TREE

Definition 1.1 *Refined Steiner Tree:* $T'(N', E')$ is a *refined Steiner tree* of $T(N, E)$ if $cost(T'(N', E')) < cost(T(N, E))$ and $N' \supseteq N$.

The underlying grid of N is the grid obtained by drawing a horizontal and a vertical line through each point of N. According to M. Hanan [2], only the grid points will be considered as Steiner points. A Steiner point is *valuable* if a refined Steiner tree can be constructed by adding it to the vertex set. The grid points on the boundary of an edge are called *candidate Steiner points* (CSP's) of the edge. The *slant distance* $d_s(p_i, p_j)$ between p_i and p_j is $- \min(|\ p_i.x - p_j.x\ |, |\ p_i.y - p_j.y\ |)$. An edge $e(p_i, p_j)$ is *degenerate* if $d_s(p_i, p_j) = 0$; otherwise, it is *non-degenerate*. A *layout* of an edge $e(p_i, p_j)$ is a shortest path between the points p_i and p_j on the underlying grid. For every nondegenerate edge $e(p_i, p_j)$ there are only two shortest paths with only one turning point. These two shortest paths are called *L-shaped layouts* of the nondegenerate edge and the two turning grid points are called the *corner points* of the nondegenerate edge. For example, in Figure 1.5a, p_a and p_c are the corner points of edge $e(p_1, p_2)$, and $p_1 - p_a - p_2$ is a L-shaped layout of $e(p_1, p_2)$.

Definition 1.2 *Modified Tree:* The *modified tree* of $T(P, E)$ due to a CSP, say p_s of $e(p_i, p_j)$, is a spanning tree of $P \cup p_s$, where edge $e(p_i, p_j)$ is replaced by the two edges $e(p_i, p_s)$ and $e(p_s, p_j)$.

For example, a modified tree due to the CSP, p_b, of $e(p_4, p_5)$ in Figure 1.5a is a spanning tree with $e(p_4, p_5)$ deleted, $e(p_5, p_b)$ and $e(p_b, p_4)$ inserted(Figure 1.5b). Since a CSP is on the boundary of an edge, the cost of the modified tree is the same as that of the original tree. Hence, a modified tree is not a refined Steiner tree.

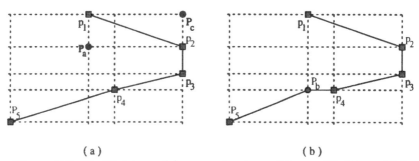

(a)   (b)

Figure 1.5 Modified tree: (a) a spanning tree; (b) a modified tree of (a)

1.2.3 LOCAL REFINEMENT

Let the *median point* p_m of three points p_i, p_j, and p_k, be the point whose x and y coordinates are the median of the x coordinates and y coordinates of the three points, respectively.

Lemma 1.1 *The median point, p_m, of points p_i, p_j and p_k is located* *within or on the boundaries of edges* $e(p_i, p_j)$, $e(p_i, p_k)$, *and* $e(p_j, p_k)$. *If* p_m *is a new point, then it is a CSP of edges* $e(p_i, p_j)$, $e(p_i, p_k)$ *and* $e(p_j, p_k)$.

Proof: We prove this lemma for edge $e(p_i, p_j)$ only. The proof for edges $e(p_i, p_k)$ and $e(p_j p_k)$ is similar.

Since the median m of three values i, j and k always satisfies $\min(i, j) \leq m \leq \max(i, j)$, we have

$$\min(p_i.x, \ p_j.x) \leq p_m.x \leq \max(p_i.x, \ p_j.x) \tag{1.1}$$

and

$$\min(p_i.y, \ p_j.y) \leq p_m.y \leq \max(p_i.y, \ p_j.y) \tag{1.2}$$

Equations 1.1 and 1.2 imply that p_m is located within or on the boundary of edge $e(p_i, p_j)$. If p_m is a new point, then at least one of the four less than or equal relations of Equations 1.1 and 1.2 must be an equal relation. Otherwise p_m and p_k must be the same point, it contradicts that p_m is a new point. So, p_m is on the boundary of $e(p_i, p_j)$ and is a CSP of edge $e(p_i, p_j)$. $\qquad\square$

Lemma 1.2 *For any two adjacent edges* $e(p_i, \ p_j)$ *and* $e(p_i, \ p_k)$ *in an RMST, if the median point* p_m *of* p_i, p_j *and* p_k *is a new point, then* p_m *is a valuable Steiner point.*

Proof: From Lemma 1.1, p_m is on the boundaries of both edges. We have

$$d_r(p_i, p_j) = d_r(p_i, p_m) + d_r(p_j, p_m)$$

and

$$d_r(p_i, p_k) = d_r(p_i, p_m) + d_r(p_k, p_m)$$

Since p_m is a new point, so $d_r(p_i, p_m) > 0$, These equations imply that

$$d_r(p_i, p_j) + d_r(p_i, p_k) > d_r(p_i, p_m) + d_r(p_j, p_m) + d_r(p_k, p_m)$$

So, by introducing p_m into the point set, a refined Steiner tree can be formed by deleting edges $e(p_i, p_j)$ and $e(p_i, p_k)$ and adding edges $e(p_i, p_m)$, $e(p_j, p_m)$ and $e(p_k, p_m)$. Hence, p_m is a valuable Steiner point. $\qquad\square$

The median point found by two adjacent edges is called a *local Steiner point* (LSP). The *gain* of an LSP is the length of the edge connected the LSP and the common terminal of two adjacent edges. These two adjacent edges are called the *dependent edges* of the LSP.

In an RMST, two LSP's may have a common dependent edge. Only one of them shall be selected to refine the tree. We need some strategies to select the LSP's to achieve a maximal improvement.

Definition 1.3 *Essential Leaf LSP:* If a leaf node, its parent node and other neighbors of the parent node form a unique LSP, then the LSP is called the *essential leaf LSP* of the leaf node.

Definition 1.4 *Essential Internal LSP:* For an internal node, if there is only one LSP generated among all pairs of adjacent edges of the node, then the LSP is called the *essential internal LSP* of the internal node.

In Figure 1.4, a leaf node p_4, its parent node p_3, and p_5, the only other neighbor of p_3, form a unique LSP, p_a. So p_a is the essential leaf LSP of p_4. The node p_5 does not have an essential leaf LSP because there are two LSP's, p_a and p_b, generated by p_5, p_5's parent node p_3 and other neighbors of p_3. For internal node p_1, p_c is its essential internal LSP. There is no essential internal LSP of p_3 since there are two LSP's, p_a and p_b, generated by p_3.

The essential LSP's are the necessary valuable Steiner points for deciding the layouts of their dependent edges. For example, in Figure 1.4, p_a is an essential leaf LSP of p_4, so if we want to reduce the cost of the RMST by deciding the layout of $e(p_4, p_3)$, then p_a should be selected as a Steiner point. Since these LSP's refer only to the local structure of an RMST, we call this kind of improvement a *local refinement*.

1.2.4 GLOBAL REFINEMENT

Given a spanning tree $T(V,E)$, a unique loop is formed after adding a *new edge* to the tree. If the length of the longest edge in the loop is greater than that of the new edge, the cost of the spanning tree can be reduced by adding the new edge and deleting the longest one. The difference between their lengths is the *gain* of the *new edge*. This tree refinement approach is called the *loop detection method*.

Definition 1.5 *Refinable tree:* A tree, T, is said to be *refinable* if a refined Steiner tree with the same vertex set can be obtained by using the loop detection method.

Obviously, a minimum spanning tree is not refinable. However, its modified tree may be refinable by a suitable selection of the CSP's. If a modified tree of an RMST is refinable, then the new edge must connect either two CSP's or a CSP and a vertex of two distinct edges in the original tree. It means that the new edge uses at least one CSP as its terminal and connects the boundaries of two edges. To achieve the maximum gain, only those shortest edges between the boundaries of two edges shall be considered as a new edge.

Lemma 1.3 *Under a rectilinear metric, if a point p_i is not within the boundary of an edge $e(p_j, p_k)$, then the shortest edge from p_i to the boundary of $e(p_j, p_k)$ is $e(p_i, p_m)$, where p_m is the median point of p_i, p_j and p_k.*

Proof: Let p_m be the median point of p_i, p_j and p_k, and p_q be any CSP of edge $e(p_j, p_k)$. We shall show that $d_r(p_i, p_q) \geq d_r(p_i, p_m)$.

First we claim that p_m is within or on the boundary of edge $e(p_i, p_q)$; i.e.,

$$\min(p_i.x, p_q.x) \leq p_m.x \leq \max(p_i.x, p_q.x)$$

and

$$\min(p_i.y, p_q.y) \leq p_m.y \leq \max(p_i.y, p_q.y)$$

Suppose it is not true, we may assume, without loss generality,

$$p_m.x < \min(p_i.x, p_q.x)$$

which implies $p_m.x < p_i.x$ and $p_m.x < p_q.x$. Since p_q is on the boundary of $e(p_j, p_k)$, $p_q.x \leq \max(p_j.x, p_k.x)$. We have $p_m.x < \max(p_j.x, p_k.x)$. Hence, $p_m.x$ is less then $p_i.x$ and one of $p_j.x$ and $p_k.x$. Thus $p_m.x$ is not the median of $p_i.x$, $p_j.x$ and $p_k.x$, and this is a contradiction.

Since p_m is within or on the boundary of edge $e(p_i, p_q)$, we have

$$d_r(p_i, p_q) = d_r(p_i, p_m) + d_r(p_m, p_q) \leq d_r(p_i, p_m)$$

Thus, $e(p_i, p_m)$ is the shortest edge from p_i to the boundary of $e(p_j, p_k)$. □

If the median point p_m of three points p_i, p_j and p_k is a new point, then p_m is a CSP of edge $e(p_j, p_k)$ from Lemma 1.1, and edge $e(p_i, p_m)$ is the shortest edge from p_i to the boundary of $e(p_j, p_k)$ from Lemma 1.3. So p_m is called the *closest CSP* of $e(p_j, p_k)$ with respect to p_i.

Lemma 1.4 *In an RMST, if the boundaries of two edges do not intersect, then among all shortest edges between the boundaries of two edges, there is an edge with the maximum gain with one terminal a corner point or a vertex and the other terminal the closest CSP of the edge it belongs to with respect to the corner point or the vertex.*

Proof: Assume $e(p_a, p_b)$ is an edge of the maximum gain among all shortest edges between the boundaries of two edges. If both p_a and p_b are not corner points or vertices, then they are at the middle of the boundary segment. There are two direction that one can move $e(p_a, p_b)$ parallel along the boundary segment. Since at least one direction will not shorten the length of the longest edge in the loop, we can move $e(p_a, p_b)$ along this direction until one of p_a and p_b hits a corner point or a vertex. This procedure leaves the maximum gain unchanged. The other terminal of the shortest edge is, from Lemma 1.3, the closest CSP of one edge with respect to this corner point or vertex.   □

Lemma 1.5 *In an RMST, if the boundaries of two edges intersect, then among all shortest edges between the boundaries of two edges, there is an edge $e(p_c, p_c)$ with the maximum gain and p_c is a closest CSP of one edge with respect to a vertex of the other edge.*

Proof: Since the boundaries of two edges intersect, all shortest edges are of zero length and their coincident terminals are the common CSP's of the two edges. We shall show that all these common CSP's are within the boundary of an edge $e(p_a, p_b)$, where p_a and p_b are closest CSP's of an edge with respect to a vertex of the other edge.

Suppose the boundaries of two edges $e(p_i, p_j)$, $e(p_k, p_l)$ intersect, then the x and y coordinates of their terminals must be interleaved. Assume

$$p_i.x \leq P_k.x \leq p_j.x \leq p_l.x \qquad (1.3)$$

Case 1. Suppose $p_i.y \leq p_j.y$. Since, in an RMST, p_k is not within the boundary of $e(p_i, p_j)$ and p_j is not within the boundary of $e(p_k, p_l)$, so the interleaved relation of y coordinates must be

$$p_k.y \leq p_i.y \leq p_l.y \leq p_j.y \qquad (1.4)$$

From Equations 1.3 and 1.4, each common CSP, p_d, of two edges satisfies

$$p_k.x \leq p_d.x \leq p_j.x$$

and

$$p_i.y \leq p_d.y \leq p_l.y$$

Let $p_a = (p_k.x, p_i.y)$ and $p_b = (p_j.x, p_l.y)$. Then p_d is within or on the boundary of $e(p_a, p_b)$. Since p_a is the median point of p_k, p_i, and p_j, by Lemma 1.3, p_a is the closest CSP of $e(p_i, p_j)$ with respect to p_k. Similarly, p_b is the closest CSP of $e(p_k, p_l)$ with respect to p_j.
Case 2. Suppose $p_i.y \geq p_j.y$. Then we have

$$p_k.y \geq p_i.y \geq p_l.y \geq p_j.y \qquad (1.5)$$

From Equations 1.3 and 1.5, define p_a and p_b as above. The same result can be obtained.

From the cases 1 and 2, we have shown that all common CSP's are within or on the boundary of an edge $e(p_a, p_b)$. Since both p_a and p_b are common CSP's of two edges, by the same argument used in Lemma 1.4, one of $e(p_a, p_a)$ and $e(p_b, p_b)$ has the maximum gain.   □

From Lemma 1.4 and Lemma 1.5, one of the terminal of the new edge with maximum gain is a closest CSP with respect to a corner point or a vertex. So we can start from a vertex or a corner point of an edge and then find the closest CSP's of all the other edges of the tree. These points (vertices and corner points) are called *generating points*. Using the closest CSP as one terminal of the new edge, and the generating point or the closest CSP, itself, as the other, we obtain the gain of the new edge. The gain of the new edge is the *gain* of the closest CSP. The longest edge, the edge of the closest CSP and the edge of the generating point (if it is a corner point) are the *dependent edges* of the closest CSP.

For example, in Figure 1.6a, p_a is a corner point of $e(p_1, p_2)$. The closest CSP of $e(p_4, p_5)$ with respect to p_a is the median point p_b of p_a, p_4 and p_5. After adding the new edge $e(p_a, p_b)$, a loop p_a-p_b-p_4-p_3-p_2-p_a is formed as shown by dashed lines in Figure 1.6b. Since $d_r(p_2, p_a)$-$d_r(p_a, p_b)$ is positive, a valuable CSP, p_b, is found. In this case, p_a is a generating point and p_b is a closest CSP generated by p_a. The gain of p_b is $d_r(p_2, p_a)$ - $d_r(p_a, p_b)$ and the dependent edges of p_b are $e(p_1, p_2)$ and $e(p_4, p_5)$. By including p_b in the vertex set and regenerating an RMST, a refined Steiner tree is formed, as shown in Figure 1.6c.

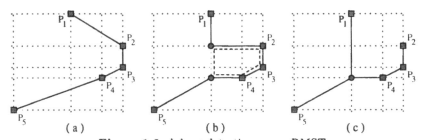

Figure 1.6 A loop detection on an RMST
(a) An example of RMST; (b) Modified tree and loop detection;
(c) Refined Steiner tree

It is seen from the above that a valuable Steiner point, produced by nonadjacent edges of an RMST, can be found by the loop detection method. Also, every LSP generated by two adjacent edges is a median point of the three terminals. By Lemma 1.3, it is also a closest CSP of a vertex, and so can be found by the loop detection method. Hence the loop detection method gives a global view of RST generation and detects all valuable Steiner points within the structure of an RMST. We thus call this kind of improvement a *global refinement*.

1.3 TREE REFINEMENT ALGORITHMS

The basic idea of our tree refinement algorithm is to deal with an RMST, choose a set of valuable Steiner points, add them to the vertex set, and generate an RMST of the enlarged point set. This procedure is repeated until no more valuable Steiner points can be found. Figure 1.7 shows a pseudo-code description of the algorithm.

We generate an RMST using Prim's algorithm [11] with the weighted metric mentioned in Section 1.3.2. In step (1), all the essential leaf LSP's are collected in a set S. A greedy method is used to select a maximal independent subset of S. It begins with an empty set S_c. The LSP, p_c, which has the largest gain in S, is moved to S_c, and the LSP's which have a common dependent edge with p_c are deleted from S. This procedure is repeated until S is empty.

Algorithm Tree_Refinement_algorithm_for_RST(P);
    step (1) generate an RMST $T(P, E)$ of P;
           collect all essential leaf LSP's into a set S;
           select a maximal independent set S_c of S;
           $V = P \cup S_c$.
    step (2) generate an RMST T(V,E) of V;
           for each internal node
               if there exists an essential internal LSP p_s
               then add p_s to V and generate the refined Steiner tree.
    step (3) generate an RMST $T(V, E)$ of V;
           collect closest CSP's of positive gain into a set S;
           if $S \neq \emptyset$
           then select a maximal independent set S_c of S;
               $V = V \cup S_c$;
               go to *step (3)*;
           else for each nondegenerate edge
               select one corner point and add it to V;
    step (4) add all closest CSP's of nonnegative gain to V
           generate the RMST $T(V, E)$;
           remove useless Steiner points and their incident edges.

Figure 1.7 Pseudo-code description of the algorithm

In step (2), the tree is refined immediately after an essential internal LSP is found. Since an internal node may have more than two incident edges, immediately refining the tree may provide opportunities to generate essential internal LSP's for later nodes and speed up the RST generation.

Consideration of the remaining LSP's is postponed after step (2) for the following reasons: (1) there is no good method to judge whether their inclusion will result in a better solution; (2) early binding of the local valuable Steiner points may miss the possibility for global improvement;

and (3) the loop detection method can also find all LSP's in a global manner.

In step (3), the loop detection method is used to find valuable Steiner points. The vertices of the tree and all corner points are the generating points. For each generating point p_g, we find a closest CSP, p_{s_e}, of each edge e in T with respect to p_g. Using $e(p_g, p_{s_e})$ or $e(p_{s_e}, p_{s_e})$ as a new edge in the loop detection method, we calculate the gain of p_{s_e}. The p_{s_e} which has positive gain is added to set S. For implementation, we do not have to consider all edges in the tree for each generating point; only those edges which intersect with the CSP search region have to be considered. After the valuable CSP's are collected in set S, a maximal independent set of S, S_c, is formed. The points in S_c are added to the vertex set and a refined RMST is generated. This procedure is repeated until no valuable CSP exists.

During step (3), only those CSP's with positive gain are generated, while in the last step, the CSP's with zero gain will also be considered. Since CSP's are on the boundaries of edges, and all the edges in the RST are degenerate at this time, the new CSP's will not destroy the rectilinear property of the tree and may help to find the useless Steiner points.

Example 1: Figure 1.8 shows the first example of an RST construction by TR algorithm. An RMST is constructed in step (1) as shown in Figure 1.8a. Two essential leaf LSP's, p_a and p_b, are generated by leaf nodes p_5 and p_9, respectively. Figure 1.8b is the refined RMST construct in step (2) after p_a and p_b are added to the vertex set. Three essential internal LSP's, p_c, p_d, and p_e, are generated sequentially by internal nodes p_3, p_2 and p_1, as shown in Figure 1.8b. For each internal LSP, we construct its corresponding refined tree before we generate the next LSP. Figure 1.8c is the refined tree for Figure 1.8b after p_e is included. By the loop detection method, two CSP's, p_f, and p_h, are found by generating points p_1 and p_g, respectively. The loops are shown as dashed lines in Figure 1.8c. After including p_f and p_h in the vertex set and generating the refined RMST, there are no more valuable Steiner points. We choose a corner point p_i to rectilinearize the nondegenerate edge $e(p_f, p_c)$. In last step, a CSP, p_j, with zero gain is found by generating point p_7 and is added to the vertex set. Since there are no useless Steiner points, Figure 1.8d is the final result.   ◇

Example 2: Figure 1.9 shows the second example of an RST construction by TR algorithm. We use this example to explain why the step (4) is included in our TR algorithm. As shown in Figure 1.9a, p_a was an essential leaf LSP generated in step (1). The refined RMST after step (3) is shown in Figure 1.9b. In step (4), using p_3 as a generating point, a zero gain CSP, p_j, was detected by the loop p_3-p_j-p_b-p_a-p_3 and included in the vertex set. In the last RMST construction, since the node priority of $e(p_3, p_j)$ is less than the node priority of $e(p_a, p_b)$, so $e(p_3, p_j)$ is selected as its edge, and p_a became a leaf node as shown in Figure 1.9c, therefore p_a and edge $e(p_a, p_3)$ were removed. The resulting RST is shown in Figure 1.9d.   ◇

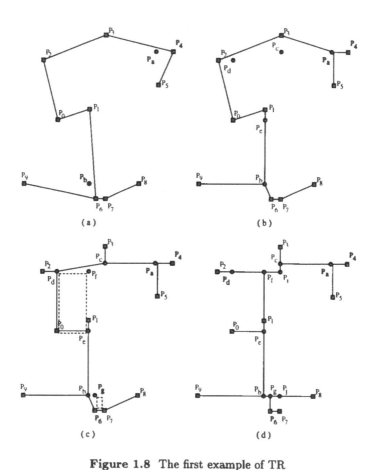

Figure 1.8 The first example of TR
(a) RMST and essential leaf LSP's;
(b) Refined RMST after step (1) and essential internal LSP's;
(c) Refined RMST and CSP's; (d) Final RST

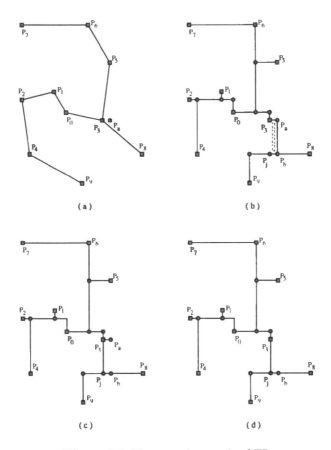

Figure 1.9 The second example of TR
(a) RMST and essential leaf LSP's p_a;
b) Refined RMST after step; (b) Refined RMST after step;
(c) Last RMST in step; (d) Final RST

1.4 IMPLEMENTATION AND EXPERIMENTS

We have implemented the TR algorithm in PASCAL on a VAX
8650. The following are some issues during the implementation of the
algorithm.

1.4.1 ROOT SELECTION AND NODE ORDERING

During loop detection, we have to trace back from the closest CSP
and the generating point to the root node separately. A loop is formed
with the first vertex they meet. To speed up the search, we select the
middle node in the longest path of the RMST as the root node. This

strategy is based on the intuitive observation that a tree may have a minimum height by selecting the middle node as the root. Also, during step (2) of the algorithm, internal nodes are processed by the inverse order from that of the generation of the nodes of the RMST.

1.4.2 THE EFFICIENCY OF LOOP DETECTION

In the loop detection method, the median of the generating point p_g and the terminals of each edge in T have to be calculated in order to find the valuable CSP's for p_g. To avoid unnecessary calculation, we focus on a subregion in the plane called the CSP search region, and only those CSP's in the region will be considered. The definition of the CSP search region of a generating point shall guarantee that the valuable CSP's will not be missed after all the generating points are processed.

Lemma 1.6 *If p_g is connected to p_i by a horizontal edge $e(p_g, p_i)$, and p_i is on the left of p_g, the CSP search region for p_g limited by $e(p_g, p_i)$ is the half plane bounded by $x \geq \frac{1}{2}(p_g.x + p_i.x)$.*

Proof: Suppose the shortest CSP, p_m, of edge $e(p_j, p_k)$ is outside the CSP search region; i.e., $p_m.x < \frac{1}{2}(p_g.x + p_i.x) < p_g.x$. Since $p_m.x$ is the median of $p_g.x$, $p_j.x$ and $p_k.x$, it implies that $p_j.x \leq p_m.x$, $p_k.x \leq p_m.x$, and $e(p_j, p_k)$ are not in the CSP search region.

(1) Suppose $p_m.y \neq p_i.y$ as shown in Figure 1.10a. This implies that p_m must be a vertex or a corner point, so, p_m will be treated as a generating point. Let p_l be the shortest CSP from p_m to edge $e(p_i, p_g)$ as shown in Figure 1.10a. The longest edge in the loop after inserting edge $e(p_g, p_m)$ is the same as that after inserting edge $e(p_m, p_l)$ except when the longest edge is $e(p_i, p_g)$ itself. In this special case, the gain for p_m is $d_r(p_i, p_g) - d_r(p_g, p_m)$, which is equal to $d_r(p_i, p_l) - d_r(p_m, p_l)$. This means the gain for p_l is no less than the gain for p_m. For the other cases, the length of the longest edge are equal and $d_r(p_m, p_l)$ is less than $d_r(p_g, p_m)$. Since p_l is in the CSP search region of p_m, it follows that if p_m is a valuable Steiner point generated by p_g, then the gain of p_l generated by p_m will be greater than the gain of p_m generated by p_g. Hence, p_m can be discarded from the CSP search region of p_g to avoid redundant evaluation.

(2) Suppose $p_m.y = p_i.y$ as shown in Figure 1.10b. This implies that $p_m.x \leq p_i.x$. We use p_i as a generating point, and p_m will be the shortest CSP from p_i to $e(p_j, p_k)$. Since $d_r(p_i, p_m) < d_r(p_g, p_m)$, the gain of p_m generated by p_i will be greater than that of p_m generated by p_g, so p_m can be discarded from the CSP search region of p_g.

Since in every case, the shortest CSP's outside the search region can be discarded, the search region is defined well enough to find the valuable CSP's for p_g. □

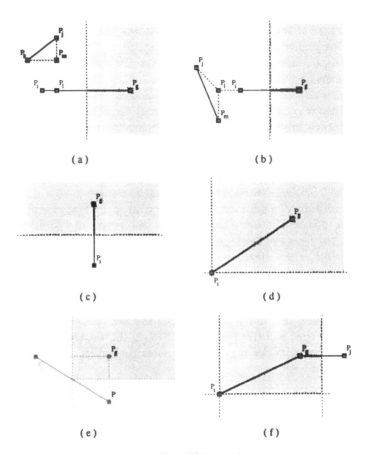

Figure 1.10 The CSP search region:
(a), (b) p_g is a terminal of a horizontal edge;
(c) p_g is a terminal of a vertical edge;
(d) p_g is a terminal of a nondegenerate edge;
(e) p_g is a corner point of $e(p_i,\ p_j)$;
(f) p_g is incident to two edges;

The CSP search region limited by a vertical edge is defined similarly to that limited by a horizontal edge. The shaded areas in Figures 1.10a to 1.10c are examples of the CSP search region of p_g limited by $e(p_g,\ p_i)$.

The CSP search region for p_g limited by a nondegenerate edge can be defined similarly to that limited by a degenerate edge. The CSP search region of a terminal of a nondegenerate edge is bounded by the other terminal, as shown in Figure 1.10d. The CSP search region for a corner point of a nondegenerate edge is shown in Figure 1.10e. For a generating point with more than one incident edge, the CSP search region is the intersection of all CSP search regions limited by its incident edges. Figure 1.10f is an example of the CSP search region of a vertex with a degree of two. As in the example shown, the larger the degree of a node, the smaller the CSP search region.

We use a horizontal and a vertical edge list as our data structure. While generating an MST, edges are recorded separately in a horizontal and vertical edge lists according the coordinates of the terminals of the edge. When a generating point, p_g, is used, the shortest incident edge of p_g is chosen as the limiting edge, and either the vertical or horizontal direction is decided as the search direction. Using edge lists and the coordinate of the boundary line of CSP search region, the edges intersecting with the search region can be easily retrieved sequentially and the shortest CSP's with respect to the edges calculated.

Two additional conditions are used to trim redundant closest CSP's. First, *the distance between the CSP and the generating point must be less than the length of the longest edge in the RMST.* Second, *if the generating point is a vertex, as opposed to a corner point, then the CSP cannot be a vertex.* The CSP's satisfying the above conditions are sorted according to the distance from p_g and only the four nearest CSP's with respect to the generating point are tested by loop detection to decide whether these CSP's are valuable. Since the average loop length in an RMST is $O(logn)$ and the number of CSP's in the search region is $O(n)$, the time complexity for a generating point to find its valuable closest CSP's is $O(nlogn)$.

1.4.3 USELESS STEINER POINTS

A Steiner point is useless if it is a leaf node in the final RMST. This may occur in early Steiner point inclusion as shown in the second example in Section 1.4.2. For removing the useless Steiner points and their incident edges, in the step (4) of TR, all zero gain CSP's are included in the vertex set, scan from the leaf nodes of the last RMST, the useless Steiner points can be detected by their node order greater than $| P |$. This procedure can be done in the linear time.

1.4.4 EXPERIMENTAL RESULTS

As mentioned in Section 1.4.3, the loop detection method takes a global view of the structure of an RMST, but does not give local information to guide the selection of CSP's. Methods which view the structure of RMST locally are able to find valuable Steiner points, but the LSP may be useless from the global view and affect the following refinement. To complement each other, our program has been implemented in such a way that the user can optionally skip step (1) or (2).

Let TR denote the tree refinement algorithm, TR1 denote the option of skipping step (2), TR2 denote the option of skipping step (1), and LD (loop detection) denote the option of skipping both step (1) and step (2). From our experiments, TR produces better cost results, on average, than the other options. This suggests that the selection criteria for LSP's are very good. However, each of the other options produces better results in certain cases, especially when the number of points increases.

The tree refinement algorithm was tested extensively on a number of random point sets with cardinalities ranging from 5 to 80. We generated 20 examples for each cardinality value. Steiner trees were constructed for each example by options TR, TR1, TR2, LD and L-RST [10]. For all the cases, our results are better or equal to those generated by the L-RST algorithm. For the TR option, 14% of the cases had generated useless Steiner points, and the average extra cost for these cases was 0.2%. Table 1.1 shows the average cost improvement over the RMST using the various options. The column named LBH refers to the results of [6]. The average cost improvement is 10.4% for TR, which is better than those reported in [6,10,12,13]. An average of 15% improvement was achieved over L-RST. The average number of iterations in step (3) is usually very small and can be treated as a constant. The time complexity of the algorithm is dominated by the complexity of the loop detection method, which is $O(n^2 logn)$. The average runing times for TR on a VAX 8650 are shown in column 8. Our experiments show that the time complexity of the algorithm (the last column) is better than $O(n^2 logn)$.

Table 1.1 Mean Ratios of Cost Improvement and Run Time

Pts	TR	TR1	TR2	LD	L-RST	LBH	Sec.
5	8.92%	8.92%	8.92%	8.92%	8.41%	8-10%	0.013
10	10.45%	10.26%	10.37%	10.11%	8.81%	7-8%	0.072
20	10.10%	10.10%	10.10%	10.12%	8.34%	8-10%	0.329
30	10.86%	10.91%	10.98%	10.97%	9.47%	8-10%	0.697
40	10.79%	10.78%	10.78%	10.70%	9.29%	—	1.323
60	10.58%	10.47%	10.51%	10.42%	8.96%	—	3.187
80	11.41%	11.40%	11.35%	11.33%	10.10%	—	5.480
Ave.	10.44%	10.40%	10.43%	10.37%	9.05%		

1.5 REMARKS

In this chapter, we presented some routing tree construction algorithms for the global routing problem in VLSI design. We presented a tree refinement algorithm, TR, for constructing rectilinear Steiner trees of a signal net. The idea of TR is to introduce Steiner points into an RMST incrementally according to their costs and essentiality, and to generate RMST's from the enlarged point set. The TR algorithm outperforms most other algorithms of its class by the fact that the average cost improvement over the rectilinear minimum spanning tree is 10.4%, and its time complexity is $O(n^2 log n)$.

REFERENCES

1. M. R. Garey, and D. S. Johnson, "The Rectilinear Steiner Tree Problem is NP-Complete," *SIAM Journal on Applied Mathematics*, Vol. 32, No. 4, 1977, pp. 826-834.

2. M. Hanan, "On Steiner's Problem with Rectilinear Distance," *SIAM Journal on Applied Mathematics*, Vol. 14, No. 2, 1966, pp. 255-265.

3. J. P. Cohoon, D. S. Richards, and J. S. Salowe, "An Optimal Steiner Tree Algorithm for a Net Whose Terminals Lie on the Perimeter of a Rectangle," *IEEE Transactions on Computer-Aided Design*, Vol. 9, No. 4, April 1990, pp. 398-407.

4. A. V. Aho, M. R. Garey, and F. K. Hwang, "Rectilinear Steiner Trees: Efficient Special-Case Algorithms," *Networks*, Vol. 7, 1977, pp. 37-58.

5. Y. Y. Yang, and O. Wing, "Suboptimal Algorithm for a Wire Routing Problem," *IEEE Transactions on Circuit Theory*, September 1972, pp. 508-511

6. J. H. Lee, N. K. Bose, and F. K. Hwang, "Use of Steiner's Problem in Suboptimal Routing in Rectilinear Metric," *IEEE Transactions on Circuits and Systems*, Vol. CAS-23, No. 7, July 1976, pp. 470-476.

7. F. K. Hwang, "An O(n log n) Algorithm for Suboptimal Rectilinear Steiner Trees," *IEEE Transactions on Circuits and Systems*, Vol. CAS-26, No. 1, January 1979, pp. 75-77.

8. K-W. Lee, and C. Sechen, "A New Global Router for Row-Based Layout," *Proceedings of International Conference on Computer Aided Design*, November 1988, pp. 180-183.

9. F. K. Hwang, "On Steiner Minimal Trees with Rectilinear Distance," *SIAM Journal on Applied Mathematics*, Vol. 330, No. 1, January 1976, pp. 104-114.

10. J-M. Ho, G. Vijayan, C. K. Wong, "New Algorithms for the Rectilinear Steiner Tree Problem," *IEEE Transactions on Computer-Aided Design*, Vol. 9, No. 2, February 1990, pp. 185-193.

11. R. C. Prim, "Shortest Connection Networks and Some Generalizations," *Bell System Technical Journal*, Vol. 36, November 1957, pp. 1389-1401.

12. Dana Richards, "Fast Heuristic Algorithms for Rectilinear Steiner Trees," *Algorithmica*, Vol. 4, 1989, pp. 191-207.

13. Marshall W. Bern, "Two Probabilistic Results on Rectilinear Steiner Trees," *Algorithmica*, Vol. 3, 1988, pp. 191-204.

14. J. M. Smith, D. T. Lee and J. S. Liebman, "An $O(n \log n)$ Heuristic Algorithm for the Rectilinear Steiner Minimal Tree Problem," *Engineering Optimization*, Vol. 4, 1980, pp. 179-192.

15. T.H. Chao and Y.C. Hsu, "Rectilinear Steiner Tree Construction By Local and Global Refinement," *IEEE Transactions on Computer-Aided Design of Integrated Circuits and Systems*, March 1994, pp. 303-309.

16. A. Kahng and G. Robins, "A New Family of Steiner Tree Heuristics With Good Performance: The Iterated 1-Steiner Approach," *Proceedings of International Conference on Computer-Aided Design*, November 1990, pp. 428-431.

Chapter 2

RECTILINEAR INTERCONNECTIONS IN THE PRESENCE OF OBSTACLES

James P. Cohoon and Joseph L. Ganley

Often, a VLSI routing instance includes obstacles that the routing must not intersect, such as logic cells and wires in previously routed nets. This chapter examines techniques for optimal and near-optimal routing in the presence of obstacles. Much of the literature on two-terminal interconnections in the presence of obstacles is surveyed, and both new and previously published results on multi-terminal interconnections in the presence of obstacles are presented. Also, the problem of routing in the presence of obstacles is examined in the context of several interesting special cases. Both theoretical and practical considerations are addressed.

2.1 INTRODUCTION

A VLSI circuit consists of a number of *logic cells*, each of which contains a number of *terminals*. There may also be terminals that are not part of any logic cell that act as, for example, contact points for other connections. A circuit description also contains a number of *nets*. A net is a set of terminals that must be interconnected. Wires in different nets must not intersect one another.

In VLSI physical design automation, a fundamental task is routing a net. Typically this routing is performed in the presence of obstacles that the wires of the net must not intersect, such as the logic cells and wires in previously routed nets. Furthermore, VLSI fabrication technology typically requires that wires consist only of horizontal and vertical segments, indicating the use of the *rectilinear* distance metric in which the distance between two points is the sum of the distances along each axis.

A special case that has received significant attention is the case in which only two terminals need to be connected. This problem is equivalent to finding a rectilinear shortest path in the presence of obstacles. A more general formulation, in which more than two terminals must be interconnected, forms a generalization of the rectilinear Steiner tree problem. The *rectilinear Steiner tree (RST)* problem is stated as follows: given a set of terminals in the plane, find a set of horizontal and vertical line segments of minimum total length that interconnects all the terminals. In constructing this interconnection, it is permissible to introduce new points called *Steiner points*. For example, Figure 2.1 depicts an optimal RST for a set of 27 terminals [19] (terminals are depicted as filled circles and Steiner points as open circles).

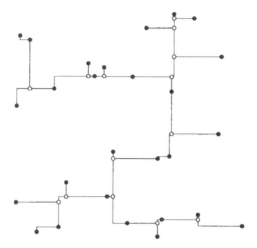

Figure 2.1 An optimal RST for a set of 27 terminals.

The *obstacle-avoiding rectilinear Steiner tree (OARST)* problem is identical to the RST problem except for the presence of rectilinear obstacles that the segments in the Steiner tree must not intersect. The RST problem is NP-complete [21], and therefore the OARST problem is as well. However, the two-terminal problem is efficiently solvable.

The remainder of this chapter consists of four sections. Section 2.2 surveys the literature on the special case of the OARST problem in which there are only two terminals. Section 2.3 presents results from the literature as well as new results on the general OARST problem. Section 2.4 discusses several special cases of the OARST problem, such as when all terminals lie on obstacle perimeters or when all terminals lie on the perimeter of the routing region. Finally, Section 2.5 concludes with a number of open problems.

An instance of the OARST problem consists of a set T of terminals and a set S of obstacle perimeter segments. We denote an instance as a pair (T, S). We let $n = |T|$ and $s = |S|$, and denote the instance size $c = n + s$.

2.2 TWO-TERMINAL INTERCONNECTIONS

The problem of two-terminal rectilinear interconnection in the presence of obstacles has been well studied. We refer to this special case of the OARST problem as the *obstacle-avoiding shortest path (OASP)* problem. The literature on the OASP and similar problems is quite vast; since other surveys have been written [26,33,52], and since our focus is on Steiner-tree routing of multi-terminal nets, we only overview some of the principal results for the two-terminal case.

2.2.1 GRID-BASED ALGORITHMS

The earliest techniques for solving the OASP problem are so-called *grid-based* algorithms. Grid-based routing techniques find their genesis in the *maze routing* algorithms of Lee [32] and Moore [41]. In a grid-based algorithm, the routing surface is typically divided into a grid in which each square is the size of the smallest feature that can be fabricated in the given technology (this distance is denoted λ). This grid divides the routing area into a number of *grid cells*. The two terminals, a and b, are each associated with a grid cell. An integer value v_{xy} is associated with each grid cell (x, y). Initially $v_{xy} = \infty$ for all x and y. Then v_a is assigned the value 0, and adjacent grid cells are visited in breadth-first search order. Each time a grid cell (x, y) is visited, for each of its neighbors (x', y'), $v_{x'y'}$ is set to the minimum of its current value and $v_{xy} + |x - x'| + |y - y'|$. When the process terminates, each grid cell has been labeled with its distance d from terminal a, as shown in Figure 2.2.

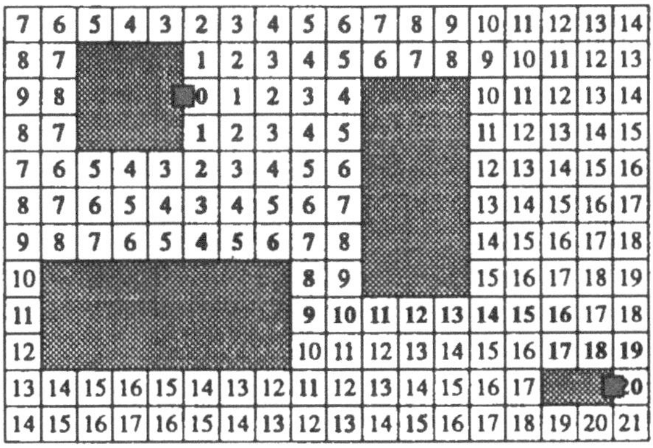

7	6	5	4	3	2	3	4	5	6	7	8	9	10	11	12	13	14
8	7			1	2	3	4	5	6	7	8	9	10	11	12	13	
9	8		0	1	2	3	4				10	11	12	13	14		
8	7			1	2	3	4	5			11	12	13	14	15		
7	6	5	4	3	2	3	4	5	6		12	13	14	15	16		
8	7	6	5	4	3	4	5	6	7		13	14	15	16	17		
9	8	7	6	5	4	5	6	7	8		14	15	16	17	18		
10								8	9		15	16	17	18	19		
11								9	10	11	12	13	14	15	16	17	18
12								10	11	12	13	14	15	16	17	18	19
13	14	15	16	15	14	13	12	11	12	13	14	15	16	17			20
14	15	16	17	16	15	14	13	12	13	14	15	16	17	18	19	20	21

Figure 2.2 The result of running a maze routing algorithm.

A shortest path from a to b can then be traced by moving from grid point b to the adjacent grid points with value $d - 1$, $d - 2$, etc., until the point a with value 0 is reached. One such path is shown in bold in Figure 2.2.

The major drawback of grid-based techniques is that they require time and space corresponding the the routing area, which can be arbitrarily large relative to the instance size c. When grid-based techniques were first devised, these demands were reasonable, but currently the routing area of state-of-the-art integrated circuits is becoming prohibitively large to use such methods.

2.2.2 LINE-BASED ALGORITHMS

To overcome the potentially large time and space requirements of grid-based algorithms, researchers turned to *gridless* or *line-based* algorithms. Research into line-based algorithms for rectilinear interconnection in the presence of obstacles began with the independent works of Hightower [25] and Mikami and Tabuchi [38]. Their algorithms were innovative in that they were the first line-based routing algorithms, but they suffer several disadvantages. Primary among them is that they do not necessarily find a shortest path between the two terminals, and furthermore, they sometimes do not find any solution even when one exists.

Soon after these works, other researchers devised algorithms that solve these problems. Many of these approaches, rather than describing an algorithm per se, describe the construction of a graph guaranteed to contain shortest paths between all pairs of terminals, to which shortest-path algorithms can then be applied. One such graph is the *escape graph* of Cohoon and Richards [10].

2.2.2.1 THE ESCAPE GRAPH

Figure 2.3(a) depicts a set of obstacles and terminals. The escape graph is constructed as follows. Draw a 'beltway' around each obstacle, at a distance of λ from the obstacle, and similarly inscribe a beltway around the interior of the routing region, as shown in Figure 2.3(b). Then extend each segment maximally, i.e., until it hits another beltway segment, as shown in Figure 2.3(c). (Some researchers assume that $\lambda = 0$, in which case step (b) is skipped, and step (c) amounts to simply extending each obstacle perimeter segment maximally.) Finally, extend segments from the terminals in all unobstructed directions, again maximally, as shown in Figure 2.3(d). The segments described by this construction are called *escape segments*. From the escape segments, a graph called the *escape graph* can be computed in a straightforward manner: the vertices correspond to the intersections of escape segments, and there is an edge between any pair of vertices that are adjacent along an escape segment. The weight of an edge is the rectilinear distance between its endpoints. We let $G_e(T, S)$ denote the escape graph for an instance (T, S), and let m denote the number of vertices in $G_e(T, S)$. Note that $m = O(c^2)$ in the worst case, and also that $G_e(T, S)$ is planar and thus contains $O(m)$

edges. The escape graph can be generated in $O(\max\{m, c \log c\})$ time using a plane-sweep algorithm [10].

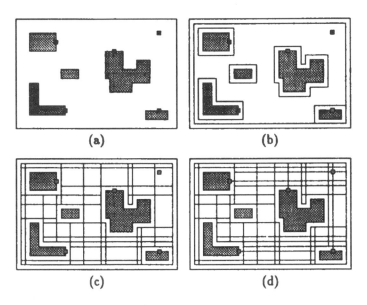

Figure 2.3 Constructing the escape graph.

It has been proven [10] that the escape graph contains a shortest path between every pair of terminals. Kanchanasut [30] describes an algorithm that computes a set of single-source shortest paths in a rectilinear graph in linear time, which can be applied to the escape graph to solve the OASP problem in $O(m) = O(c^2)$ time.

2.2.2.2 OTHER ALGORITHMS

Wu, Widmayer, Schlag, and Wong [56] describe a graph they call a *track graph*, which is identical to the escape graph defined above except that the segments adjacent to terminals extend only until they hit *some* escape segment (in the escape graph they are extended maximally). Thus, the track graph contains $O(n + s^2)$ vertices. Note that the track graph itself does not necessarily contain a shortest path between any pair of terminals. However, they describe an algorithm that, using the track graph, finds a set of single-source shortest paths in $O(n \log n + s^2 \log s)$ time. Depending on the values of n and s, this may or may not improve on the $O(c^2)$ time required by the technique of applying Kanchanasut's algorithm to the escape graph.

Several researchers [1,10,37] have devised techniques to find a shortest path between two vertices in a rectilinear graph such as the escape graph in $O(c \log c)$ time. (Note that this can be *sublinear* in the size of the escape graph!) This would seem to improve upon both the straightforward algorithm and the algorithm of Wu, Widmayer, Schlag, and Wong, but one must keep in mind that it can take as much as $O(c^2)$ time to *generate* the escape graph, so these algorithms can require $O(c^2)$ total time.

Mitchell [40] describes a more direct plane-sweep algorithm that does not explicitly use any graph structures. His algorithm requires $O(e \log s)$ time, where e is the number of "events" encountered during the plane-sweep algorithm. He proves that $e \leq O(s \log s)$, making the overall time complexity of the algorithm $O(s \log^2 s)$, which improves upon the times required by the algorithms described above. Furthermore, Mitchell conjectures that in fact $e = O(s)$, in which case his algorithm would have the optimal running time of $O(s \log s)$.

An elegant approach to the OASP problem is due to Widmayer [53], who describes the construction of a graph similar to the escape graph that contains $O(s \log s)$ vertices while still containing a shortest path between every pair of terminals. Widmayer calls this graph the *shortest-paths preserving graph*, or *spp graph*.

The *spp* graph is defined recursively in the following way. Let P denote the set of terminals and obstacle corner points. Choose a vertical line located at $x = x_d$ that divides P into two roughly equal-sized subsets P_1 and P_2. Let $G_1 = (V_1, E_1)$ be the *spp* graph for the set of points P_1, and let $G_2 = (V_2, E_2)$ be the *spp* graph for the set of points P_2. Initialize V' and E' to $\emptyset$. For each $p \in P_1 \cup P_2$, let v_p be a new vertex located at (x_d, p_y). For each v_p such that the straight line from p to v_p does not intersect any obstacle, add v_p to V' and add the edge (p, v_p) to E'. Also, add to E' the edge (v_{p_1}, v_{p_2}) for each v_{p_1} and v_{p_2} such that no obstacle and no other v_{p_i} lies between v_{p_1} and v_{p_2}. Now, the *spp* graph for P is $G = (V, E)$, where $V = V_1 \cup V_2 \cup V'$ and $E = E_1 \cup E_2 \cup E'$.

It can be shown that the *spp* graph contains a shortest path between every pair of terminals, and that it contains $O(s \log s)$ vertices and edges. Furthermore, the *spp* graph can be generated in $O(s \log s)$ time. The reader is referred to Widmayer [53] for details.

Widmayer also describes how to use the *spp* graph to compute a shortest path between two terminals in $O(s \log s \log \log s)$ time. However, we can apply Kanchanasut's algorithm [30] to the *spp* graph to compute a set of single-source shortest paths in $O(s \log s)$ total time, which is optimal [12].

Recent work by Lee, Yang, and Wong [34] gives an optimal-time plane-sweep algorithm that also solves the OASP problem in $O(s \log s)$ time.

2.2.3 MULTIPLE NET ROUTING

The problem of routing multiple nets has received substantially less attention than the single-net problem. The technique often used in practice is *sequential* routing, where the nets are ordered in some fashion, and then they are each routed in sequence, with each net becoming a set of obstacles for subsequently routed nets. If the instance is not entirely routed using this technique, then often *rip-up and reroute* techniques are applied, where one or more of the previously routed nets are removed and rerouted differently to make room for other nets [13,14,50]. Clearly these techniques are heuristic in nature, but the more elegant solution—simultaneously routing all the nets—is NP-complete even in a planar graph if all nets contain only two terminals [45,46].

Jájá and Wu [29] describe a technique for doing just that: simultaneously routing a number r of two-terminal nets. Their approach is to modify the escape graph described in Section 2.2.2.1 by replacing each escape segment with $O(r)$ parallel segments at distance λ from each other. A r-net escape graph G_e^r is constructed from these segments in the same manner as for the single-net escape graph; it contains $O(r^2 s^2)$ vertices and edges. The main result proven by Jájá and Wu is that the G_e^r correctly generalizes the single-net escape graph to the r-net case. Let $R = \{T_1, T_2, \ldots, T_r\}$ denote the set of nets; each net T_i consists of a pair of terminals. Thus, a problem instance is denoted (R, S), where S is the set of obstacle perimeter segments.

Theorem 2.1 (Jájá and Wu [29]) *A r-net routing instance (R, S) has a solution if and only if there are r vertex-disjoint paths in $G_e^r(R, S)$.*

They use algorithms by Robertson and Seymour [48] to find these vertex-disjoint paths in polynomial time for any fixed r. This approach is of largely theoretical interest, as the running time is exponential in r. However, Theorem 2.1 might instead be used in a heuristic fashion for routing multiple nets; to our knowledge, this has not yet been tried.

2.3 MULTI-TERMINAL INTERCONNECTIONS

The general OARST problem has received substantially less attention than the special case in which there are only two terminals. Several authors [36,56] have pointed out that if we can solve the OASP problem, then we can construct a minimum spanning tree of a multi-terminal instance. The minimum spanning tree has length not exceeding twice the length of an optimal OARST; thus, an obstacle-avoiding minimum spanning tree algorithm is a 2-approximation algorithm for the OARST problem [36].

For the standard RST problem, Hanan [23] proved that an optimal RST always exists that is a subgraph of the grid graph formed by passing a horizontal and vertical line through each terminal. Ganley and Cohoon [20] generalize this result to the OARST problem by showing that the escape graph (see Section 2.2.2.1) is guaranteed to contain an *optimal* OARST (note that if there are no obstacles, then the Hanan grid graph and the escape graph are identical).

Theorem 2.2 (Ganley and Cohoon [20]) *If an OARST exists for an instance (T, S), then there exists an optimal OARST that is a subgraph of the escape graph $G_e(T, S)$.*

Proof: Suppose there exists a problem instance (T, S) with $|T| > 2$, such that all optimal Steiner trees for (T, S) contain at least one segment that is not an escape segment. We show that this supposition leads to a contradiction.

Let τ be an optimal Steiner tree for (T, S) that contains a minimal number of non-escape segments among optimal Steiner trees for (T, S). Let segment ℓ be a routing segment in τ that is not an escape segment. Without loss of generality, assume that ℓ is horizontal.

Obviously, ℓ has two endpoints a and b beyond which no further collinear segment is incident. There may be segments incident and orthogonal to ℓ. In fact, there must be orthogonal segments incident to a and b. If either a or b did not have an orthogonal segment incident to it, then it would either be a terminal, contradicting the assumption that ℓ is not an escape segment, or else a portion of s could be removed, contradicting the assumption that τ is optimal.

Let u be the number of orthogonal segments incident to ℓ from above, and let d be the number of orthogonal segments incident from below. Colinear segments that are incident to ℓ both from above and below are considered two distinct segments separated by ℓ.

If u is equal to d, then slide ℓ up until it is collinear with some escape segment. We know there is room to slide, as ℓ is not an escape segment. An escape segment above ℓ must exist, since the routing region perimeter is itself inscribed by escape segments. Since the length of the segments above ℓ decreases by exactly the amount that the length of the segments below ℓ increases, the tree resulting from this maneuver has the same length as τ; hence, it is optimal. In addition, any vertical segment incident to ℓ that was an escape segment remains an escape segment. In fact, all segments that were escape segments before the slide remain so. Thus, the tree resulting from this sliding maneuver contradicts our assumption that τ contains a minimal number of non-escape segments.

If instead, u is greater than (less than) d, then we may slide ℓ up (down), decreasing the length of the tree and contradicting its optimality. We again know there is room to slide, since ℓ is not an escape segment.

This completes the case analysis. We have shown that every solvable instance of the OARST problem has an optimal solution composed only of escape segments, and thus an optimal solution to the graph Steiner tree problem in $G_e(T, S)$ is an optimal solution to the OARST instance (T, S). □

Theorem 2.2 is the first to allow computation of optimal OARSTs in time corresponding to the instance size rather than the size of the routing area. In addition, since the escape graph is guaranteed to contain an optimal OARST, applying approximation algorithms for the graph Steiner tree problem produces equivalent approximations for the OARST problem.

2.3.1 ESCAPE GRAPH REDUCTION

Often, many of the vertices in the escape graph can be deleted, along with their adjacent edges, while still guaranteeing that an optimal solution exists that is constrained to the escape graph. We now describe a few straightforward tests whose application typically eliminates many vertices from the escape graph. We call the resulting graph the *reduced escape graph*.

The first test is the *dimension reduction test* of Yang and Wing [57] for the standard RST problem. Yang and Wing prove that if a vertex v is a corner vertex, i.e., it is incident to exactly two orthogonal edges e_1 and e_2, and v is not a terminal, and edges exist that form the other two sides of a rectangle with e_1 and e_2, then v, e_1, and e_2 can be deleted. This configuration is illustrated below.

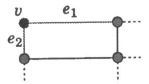

The proof of this theorem holds for escape graphs as well: any path that includes v can be replaced by a path of equal length that instead passes through the sides of the rectangle opposite e_1 and e_2.

The dimension reduction test can be implemented to run in $O(m)$ time. Start by storing all vertices of degree 2 in a queue. Then repeat the following process until the queue is empty: remove a vertex v from the front of the queue, and if it can be deleted according to the conditions above, then delete v and its adjacent edges from the graph, and add its neighbors to the back of the queue.

Any nonterminal vertex v of degree 2 that remains after the dimension reduction test can be eliminated, and its neighbors connected directly by an edge whose weight is the sum of the weights of the edges adjacent to v. In addition, any nonterminal of degree 1 can be deleted along with its adjacent edge. Finally, if any terminal has degree 1, then it can be deleted and its neighbor designated to be a terminal, and the appropriate edge added back into the completed solution.

Figure 2.4 shows the escape graph, the reduced escape graph, and an optimal OARST for a randomly generated instance with 10 terminals and 10 rectangular obstacles.

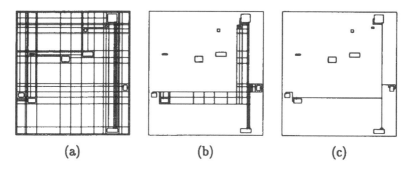

(a)	(b)	(c)

Figure 2.4 The escape graph (a), reduced escape graph (b), and optimal OARST (c) for an instance with 10 terminals and 10 rectangular obstacles.

2.3.2 EXACT ALGORITHMS

Typically, computing exact solutions to NP-complete problems is infeasible in practice. However, it is often the case that small instances can be solved practically. For the OARST problem, the escape graph model enables us to compute optimal Steiner trees for three- or four-terminal nets as efficiently as a typical heuristic solution.

A well-known folk theorem of VLSI routing is that most nets contain four or fewer terminals. In an effort to verify this claim, we examine the SIGDA Benchmark Suite [42] for standard-cell placement and routing. Figure 2.5 illustrates the distribution of terminals per net in the SIGDA benchmarks. As can be seen in the figure, in these benchmarks three- and four-terminal nets comprise the vast majority of the nets with more than two terminals.

For a three-terminal net, an optimal Steiner tree can have only one of two topologies. It can be a simple path between the terminals, or all three terminals can be connected to a single Steiner point. Thus, an optimal OARST for a three-terminal net may be computed as follows. The length of each of the three possible simple paths is checked, as well as the length of every tree formed by connecting the three terminals to each candidate Steiner point, and the tree with minimum length is optimal. The latter topology—where the tree contains a Steiner point—dominates the computation time, and is examined in $O(m)$ time, where m is the number of candidate Steiner points, assuming all-pairs shortest paths information is available. All-pairs shortest paths can be computed in $O(m^2)$ time by applying the $O(m)$ single-source shortest paths algorithm of Kanchana-sut [30] to each vertex in the escape graph.

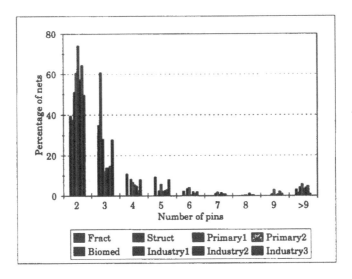

Figure 2.5 Distribution of terminals per net in the SIGDA benchmarks.

A similar observation can be made for four-terminal nets. For four terminals, the following topologies are possible:

- A simple path through the four terminals.
- A *star* in which three terminals are all connected to the fourth.
- A *cross* in which all four terminals are connected to a single Steiner point.
- A T in which three terminals are connected to a single Steiner point, and the fourth terminal is connected to one of those three terminals.
- An H in which two terminals are connected to each of two Steiner points, which are connected to each other.

These topologies are illustrated in Figure 2.6.

There are twelve possible orderings of the simple path topology, four for the star, one for the cross, twelve for the T, and six for the H, totaling 35 distinct topological instances—a sufficiently small number to examine explicitly. An optimal Steiner tree for a four-terminal net can be efficiently computed by enumerating these topologies; the path and star topologies are examined, the cross and T topologies are examined with respect to every single candidate Steiner point, and the H topology is examined with respect to every pair of candidate Steiner points. The shortest tree seen is returned as the optimal tree. This computation incurs a time complexity of $O(m^2)$, dominated by checking the H topology.

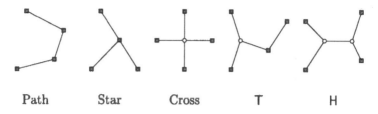

| Path | Star | Cross | T | H |

Figure 2.6 The possible topologies for four terminals.

These algorithms are similar to Hakimi's spanning tree enumeration algorithm [22]; however, identifying and examining the various topologies explicitly has significant dividends. If T is the set of terminals and S is the set of nonterminals (candidate Steiner points), then the spanning tree enumeration algorithm proceeds as follows: For every subset $S' \subseteq S$ such that $|S'| \leq |T| - 2$, compute a minimum spanning tree of $T \cup S'$. The shortest tree seen is an optimal Steiner tree. The spanning tree enumeration algorithm is many times slower than our explicit enumeration algorithms. For example, for randomly generated instances with 10 rectangular obstacles, Table 2.1 shows the average running time of the explicit enumeration (EE) algorithm versus the average running time of the spanning tree enumeration (STE) algorithm. As can be seen from the table, spanning tree enumeration is far slower than explicit enumeration.

Table 2.1 Average running times (seconds).

$n = 3$		$n = 4$	
EE	STE	EE	STE
0.04	1.45	0.08	9.06

It is possible to perform the case analyses and construct similar explicit enumeration algorithms for exact solution of problem instances with more than four terminals. However, the number of possible topologies increases superexponentially, and examining them all rapidly becomes too expensive.

A *full topology* is one in which every terminal has degree 1. Let $f(n, k)$ denote the number of full topologies with n terminals and k Steiner points. Since no vertex in the escape graph has degree exceeding 4, a Steiner point can have degree 3 or 4. A full topology with n terminals and k Steiner points can be formed by any pair of terminals connected to a vertex in a full topology with $n - 1$ terminals and $k - 1$ Steiner points, or by any three terminals connected to a vertex in a full topology with $n - 2$ terminals and $k - 1$ Steiner points. These facts lead to the following recurrence for $f(n, k)$:

$$f(n, 1) = 1$$
$$f(n, k) = \binom{n}{2} f(n - 1, k - 1) + \binom{n}{3} f(n - 2, k - 1)$$

A full topology can have anywhere from 1 to $n - 2$ Steiner points, so the total number $f(n)$ of full topologies on n terminals is $\sum_{i=1}^{n-2} f(n, i)$.

A general topology on n terminals can take one of the following forms:

- A full topology on n terminals,
- A topology on $n - 1$ terminals, with the nth terminal connected by an edge to one of the other $n - 1$ terminals, or
- A full topology on i terminals, $3 \leq i \leq n - 1$, connected at any of its terminals to a topology on the remaining $n - i + 1$ terminals.

Thus, the total number $F(n)$ of topologies on n terminals is given by the following recurrence:

$$F(2) = 1$$
$$F(n) = f(n) + nF(n - 1) + \sum_{i=3}^{n-1} f(i) \binom{n}{i} iF(n - i + 1)$$

The values of $f(n)$ and $F(n)$ for $3 \leq n \leq 8$ are given in Table 2.2. As the table suggests, the explicit enumeration approach might be practically applicable for five terminals, but almost certainly not for six. The time complexity of explicit enumeration can be improved by examining only full topologies and allowing degenerate full topologies (i.e., allowing more than one Steiner point to map to the same vertex, and allowing Steiner points to map to terminal vertices). However, this still requires the examination of $f(n)$ different topologies, which rapidly becomes prohibitively large.

Table 2.2 The number of full topologies and total topologies.

n	3	4	5	6	7	8
$f(n)$	1	7	81	1356	31312	952673
$F(n)$	4	35	516	10662	285398	9496937

2.3.3 GRAPH-BASED HEURISTICS

Since explicit enumeration is probably impractical for nets with more than four terminals, heuristics for the graph Steiner tree problem can instead be used to quickly find good solutions for such nets. Given the exact three- and four-terminal algorithms in Section 2.3.2, a natural approach is *Steinerization*. In a Steinerization heuristic, portions of a minimum spanning tree that contain a few adjacent terminals are replaced with an optimal Steiner subtree for those terminals. Typically, such heuristics examine subsets of a fixed size, i.e., subsets of size N for some small N. In light of the results in Section 2.3.2, we examine heuristics for $N = 3$ and $N = 4$.

The first heuristic, *greedy Steinerization*, starts with a *minimum spanning tree (MST)* of the terminals. It then repeatedly examines vertex subsets of size N that are adjacent in the MST, Steinerizing the one that improves the minimum spanning tree the most. The Steiner points introduced by the Steinerization are candidates for further Steinerization in later iterations. For $N = 3$, greedy Steinerization is an oft-repeated idea whose genesis is unclear. For the standard RST problem, Richards (see Hwang, Richards, and Winter [28]) first investigated 3-Steinerization in this greedy form, and more complex variants appear in Chao and Hsu [5], Lee, Bose, and Hwang [35] and Smith, Lee, and Liebman [51]; these and others are summarized in Hwang, Richards, and Winter [28]. For the OARST problem, greedy 3-Steinerization (henceforth called *G3S*) has time complexity $O(n^2 m)$, since $O(nm)$ time is required to find each locally optimal 3-Steinerization and at most $O(n)$ of these operations are performed. For $N = 4$, greedy Steinerization is similar to the algorithm of Beasley [2], though Beasley's algorithm computes a new MST at each iteration rather than locally modifying the current MST. For the OARST problem, greedy 4-Steinerization (henceforth, *G4S*) has time complexity $O(n^3 m^2)$, since $O(n^2 m^2)$ time is required to find each locally optimal 4-Steinerization and at most $O(n)$ of these operations are performed.

We can speed up Steinerization heuristics by a batching technique similar to the heuristic of Hasan, Vijayan, and Wong [24] for the standard RST problem. In their *neighborhood Steinerization* heuristic, each vertex v is assigned a weight that is the amount of improvement over the MST that is gained by Steinerizing v and its neighbors. Since any vertex in a rectilinear minimum spanning tree can have at most 8 neighbors, each Steinerization can be performed in constant time. Since we cannot efficiently Steinerize large neighborhoods for the OARST problem, our heuristic instead sets the weight of a vertex v to the best improvement gained by 3-Steinerizing v and any two of its neighbors.

The heuristic then finds a *maximum-weight independent set (MWIS)* of the tree, which can be computed in $O(n)$ time by dynamic programming. The best 3-neighborhood of each vertex in the MWIS is then Steinerized, and the process is repeated for the new tree produced by replacing each neighborhood with its Steiner subtree. The time complexity of this algorithm, which we call *batched 3-Steinerization (B3S)* is $O(bnm)$, where b is the number of iterations required, which is a function of n. In the worst case, b is $O(n)$, so the worst-case time complexity is the same as for G3S; however, this bound is quite pessimistic. Table 2.3 shows the average value of b for various values of n, for randomly generated instances containing 10 rectangular obstacles.

Table 2.3 Average iteration count for B3S.

n	3	5	7	9	11	13	15
b	1.0	1.5	1.8	2.0	2.1	2.2	2.3

Empirically it appears that b is $O(\log n)$, giving B3S a time complexity of $O(mn \log n)$ in practice.

We can also perform another optimization to reduce the running time of Steinerization heuristics. Before computing the Steiner subtree for each subset T' of the terminals, perform the reductions described in Section 2.3.1 on the escape graph, considering only members of T' to be terminals. Since the size N of the subsets considered is small, the reduction in the size of escape graph is typically substantial. Since the number N of terminals is a fixed constant, the reductions have time complexity $O(m)$ for each subset. For 3-Steinerization, this is asymptotically the same as the time complexity of actually Steinerizing the subset, so performing the reductions is not productive. However, for 4-Steinerization, the cost of Steinerizing each subset is $O(m^2)$, so linear-time preprocessing can be effective if it substantially reduces m. For $N = 4$, we expect many of the vertices to be eliminated, and experimentally this approach does indeed dramatically improve the running time of G4S—for the 20-terminal instances tested, this optimization improves the average running time by a factor of almost 2.5.

Table 2.4 shows the result quality (percent improvement over the minimum spanning tree) and runtime for G3S and B3S, and for G4S with the reduction optimization described above, for randomly generated instances containing 10 rectangular obstacles and the indicated numbers of terminals.

For the standard RST problem, the average improvement of optimal RSTs over the minimum spanning tree is roughly 12% (see Hwang, Richards, and Winter [28]). For the instances of the OARST problem tested here, this value is somewhat lower. We can compute an optimal Steiner tree in the escape graph using the algorithm of Dreyfus and Wagner [15]. Table 2.5 gives the average improvement of the optimal OARST over the minimum spanning tree for the instances in Table 2.4 with 10 or fewer terminals. Thus, the improvement values in Table 2.4 should not be compared with those reported in the literature for standard RST heuristics.

The reader should note that B3S produces trees roughly as good as, and sometimes better than, those produced by G3S. In addition, note that G4S consistently produces better trees than G3S, but that the difference in running times is not nearly as pronounced as one would expect from their relative time complexities—in particular, the running time of G4S *decreases* relative to the running time of G3S as n increases. We attribute this phenomenon to the reduction technique described above.

Table 2.4 Average result quality (percent improvement over MST) and running time (in seconds) for the heuristics.

	G3S		B3S		G4S	
n	Qual.	Time	Qual.	Time	Qual.	Time
4	7.83	0.25	7.83	0.25	8.19	0.40
5	8.58	0.48	8.59	0.46	9.21	0.97
6	8.55	0.81	8.61	0.75	9.12	1.81
7	8.81	1.26	8.82	1.10	9.35	2.94
8	8.44	1.92	8.40	1.56	9.02	4.55
9	8.74	2.75	8.72	2.10	9.33	6.61
10	9.02	4.17	9.03	2.96	9.53	9.82
12	8.72	7.59	8.69	4.78	9.14	17.64
14	8.93	13.63	8.90	7.48	9.40	31.58
16	8.99	22.66	8.95	11.51	9.48	51.33
18	9.03	34.63	9.04	16.15	9.46	78.79
20	9.02	50.21	8.99	21.53	9.43	112.9

Table 2.5 Improvement of optimal OARST over minimum spanning tree for test instances.

n	4	5	6	7	8	9	10
Qual.	8.19	9.48	9.46	9.95	9.87	10.10	10.32

Note that the worst-case ratio of the length of a minimum spanning tree to the length of an optimal Steiner tree (called the *Steiner ratio*) for the OARST problem is 2 [36]. All three of these heuristics always produce trees at least as short as the MST, and thus produce trees that are no more than twice the length of an optimal tree.

Aside from allowing computation of optimal OARSTs, Theorem 2.2 has important implications with respect to approximate OARSTs. Since the escape graph is guaranteed to contain an optimal OARST, any approximation algorithm for the graph Steiner tree problem with an approximation bound of $\bar{\rho}$ is an approximation algorithm with the same bound for the OARST problem. As was mentioned previously, a minimum spanning tree is an approximate solution to the graph Steiner tree problem, and thus the OARST problem, with $\bar{\rho} = 2$. Recently some researchers have devised approximations with better bounds for the graph Steiner tree problem: Zelikovsky [58] describes an algorithm with $\bar{\rho} = 11/6 \approx 1.83$, and Berman and Ramaiyer [3] improve Zelikovsky's bound to $\bar{\rho} = 16/9 \approx 1.78$, with better approximations possible at the expense of increased running time. Theorem 2.2 implies that these results provide equivalent approximations for the OARST problem.

2.3.4 OTHER HEURISTICS

It is possible to compute heuristic OARSTs without using a graph model. For example, the standard routing techniques of global followed by detailed routing do exactly that. However, such algorithms are designed to find a feasible routing of a large number of nets, rather than a good routing of a single net. A detailed discussion of these techniques is beyond the scope if this chapter; interested readers are referred to Preas and Lorenzetti [43].

Also, as mentioned previously, one can use any of the two-terminal interconnection techniques described in Section 2.2 to compute heuristic OARSTs.

Chen [6] describes a somewhat different technique. He describes an algorithm that, given any heuristic OARST, performs iterative improvements that possibly reduce the length of the OARST without changing its topological structure. If certain conditions are met, then the algorithm computes an *optimal* OARST with the given topology.

2.4 SPECIAL CASES

One common solution technique when examining NP-complete problems is to find special cases that might be solvable in polynomial time. In this section we examine two special cases. In the case where all terminals lie on the perimeters of obstacles, as is often true in practice, we observe that the problem remains NP-complete. In the case where all terminals lie on the perimeter of the routing region, forming a switchbox with obstacles, we show that the OARST problem is solvable in polynomial time.

2.4.1 TERMINALS ON OBSTACLE PERIMETERS

One can easily show that the OARST problem remains NP-complete even if all terminals lie on obstacle perimeters (as is often the case in practice). The proof is by reduction from the standard RST problem. Given an RST instance, simply attach to each terminal a square of side length ϵ, where ϵ is very small. The escape graph for such an OARST instance is equivalent to the Hanan grid graph (see Section 2.3) with each segment replaced by two or three parallel segments at distance ϵ from one another. For sufficiently small ϵ (or by scaling the original terminals up), an optimal solution to this OARST problem provides an optimal solution to the original RST instance. Thus, the OARST problem remains NP-complete even if all terminals lie on obstacle perimeters.

However, if the number of obstacles is small relative to the number of terminals, then we can use results due to Bern [4] to compute an optimal OARST more efficiently than by applying the Dreyfus-Wagner algorithm to the escape graph. Among the results in Bern's paper is an algorithm for computing optimal Steiner trees in a graph when all the terminals lie on the boundaries of a small number of faces. Each obstacle forms a face in the escape graph, so if all terminals lie on obstacle perimeters, then they all lie on the boundaries of the corresponding faces. If there are f obstacles (i.e., the terminals lie on the boundaries of f faces), then Bern's algorithm can be used to find an optimal OARST in $O(mn^{2f+1} + (m \log m)n^{2f})$ time.

2.4.2 SWITCHBOXES WITH OBSTACLES

Another special case of the OARST problem is when the terminals lie on the perimeter of the routing region. This special case is roughly analogous to the standard RST problem when the terminals lie on the perimeter of a convex polygon. In the standard RST problem, this restriction renders the problem solvable in polynomial time. For example, Cohoon, Richards, and Salowe [11] describe an algorithm to compute an optimal RST in linear time for a set of terminals that lie on the perimeter of a rectangle. More generally, Richards and Salowe [47] present an algorithm that computes, in $O(k^4n)$ time, an optimal RST for terminals on the perimeter of a k-sided convex polygon. Cheng, Lim, and Wu [7] describe an algorithm that computes an optimal RST of such an instance in $O(n^3)$ time regardless of k (note that this is an improvement if $k = \Omega(\sqrt{n})$). Further discussion of such special cases of the RST problem can be found in Chapter 3.

Mirayala, Hashmi, and Sherwani [39] present a linear-time algorithm that computes an optimal OARST when the terminals lie on the perimeter of a rectangle and there is only one rectangular obstacle. They also present an approximation algorithm for the more general case where there can be any number of rectangular obstacles. If ℓ is the length of an OARST computed by their approximation algorithm, and ℓ^* is the length of an optimal OARST, then they prove that $\ell \leq \ell^* + w$, where w is the maximum length of any obstacle perimeter segment.

Chiang, Sarrafzadeh, and Wong present algorithms that compute an optimal OARST if the terminals lie on two parallel lines (i.e., on opposite sides of a channel) [8] and if the terminals lie on the perimeter of a rectangle [9]. These algorithms run in time linear in the number of terminals but exponential in the number of obstacles.

Ganley and Cohoon [17] observe that the OARST problem is solvable in polynomial time if the terminals lie on the perimeter of the routing region, regardless of the shape of the routing region. Recall from Section

2.2.2.1 that the escape graph contains an optimal OARST. The escape graph for any routing instance is clearly planar. Furthermore, if the terminals lie on the perimeter of the routing region, then the terminals in the escape graph all lie on the boundary of the infinite face of the escape graph. A planar graph in which all terminals lie on the boundary of the infinite face is called *1-outerplanar*. Figure 2.7 shows an instance with the terminals on the perimeter of the routing region, and its escape graph.

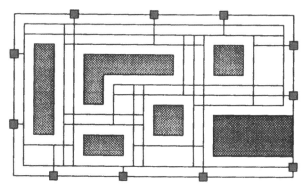

Figure 2.7 The escape graph for an instance with terminals on the perimeter of the routing region.

Erickson, Monma, and Veinott [16] (and independently Provan [44]) show that the graph Steiner tree problem is solvable in polynomial time for a 1-outerplanar graph. One way to achieve polynomial-time solution is with a modification of the Dreyfus-Wagner dynamic programming algorithm [15]. The key to the Dreyfus-Wagner algorithm, and the source of its exponential time complexity, is that it examines all possible subsets of the set of terminals. For the case where the graph is 1-outerplanar, the Dreyfus-Wagner algorithm need only consider those subsets of the set of terminals that are adjacent along the boundary of the infinite face. The resulting algorithm has time complexity $O(n^3 m^2 + n^2 m^2 \log m + m^2)$.

Recently, Kaufmann, Gao, and Thulasiraman [31] have modified this dynamic programming approach, using geometric properties of grid graphs to speed up the algorithm. Their algorithm finds an optimal OARST in a 1-outerplanar grid graph in $O(n^2 \cdot \min\{n^2 \log n, m\})$ time. However, their algorithm is not applicable to a grid graphs with "holes," such as most escape graphs.

2.5 OPEN PROBLEMS

The OARST problem has received little attention compared to, for example, the OASP problem. In this section we describe several possible directions for future work on the OARST problem.

2.5.1 GRAPH REDUCTIONS

One interesting open problem is whether one can devise a graph with fewer than $O(c^2)$ vertices that is guaranteed to contain an optimal OARST (or equivalently, a graph reduction that can be proven to remove $\Omega(f(c))$ vertices). Such a result would have significant impact on the standard rectilinear Steiner tree problem as well, since to our knowledge it has not been proven that a graph with fewer than $O(n^2)$ vertices exists that is guaranteed to contain an optimal RST.

Winter [55] has devised a number of graph reductions intended for the standard RST problem. These reductions appear empirically to be much more effective than the standard reductions such as the dimension reduction test described in Section 2.3.1. Furthermore, the reductions are applicable to other rectilinear graphs such as the escape graph, and Winter [54] conjectures that they will be at least as effective in this domain as for the standard RST problem. The application of Winter's reductions to escape graphs should be empirically examined.

2.5.2 COMPUTING OPTIMAL OARSTS

A *full Steiner tree* is one in which every terminal has degree 1. A *full set* of terminals is one for which every optimal Steiner tree is a full Steiner tree. For the standard RST problem, Hwang [27] proved that an RST of a full set can have only one of two simple topologies, enabling computation of an optimal full Steiner tree in linear time. This characterization enables the construction of algorithms that compute optimal rectilinear Steiner trees more efficiently than by using an algorithm for the graph Steiner problem [18,19,49].

Unfortunately, no such characterization is known for the OARST problem. In particular, Hwang's theorem does not necessarily hold in the presence of obstacles. For example, Hwang's theorem indicates that the Steiner points in a full Steiner tree induce a chain. This is not the case for the OARST problem; Figure 2.8 illustrates an instance for which the optimal OARST is a full Steiner tree, but in which the Steiner points do not induce a chain.

In Section 2.4.1 we observed that the OARST problem remains NP-complete even if all terminals lie on obstacle perimeters, in which case every optimal Steiner tree is a full Steiner tree. This would seem to indicate that the problem of computing a full OARST is NP-complete.

Even in the absence of characterizations as strong as Hwang's theorem, it might be possible to use the geometric structure of the escape graph to compute optimal OARSTs more efficiently than by simply applying the Dreyfus-Wagner algorithm for the graph Steiner tree problem to the escape graph. For example, Hwang's theorem does not apply in a 1-outerplanar grid graph with a non-convex boundary. However, Kaufman, Gao, and Thulasiraman [31] use the geometric structure of the grid graph to implement tie-breaking rules that eliminate from consideration many Steiner trees of equal length. The resulting algorithm is faster than the 1-outerplanar version of the Dreyfus-Wagner algorithm. We believe that applying such a strategy to escape graphs will result in faster algorithms for computing optimal OARSTs.

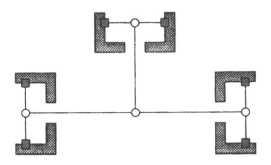

Figure 2.8 An OARST instance for which Hwang's theorem does not hold.

2.5.3 MULTIPLE-NET STEINER TREE ROUTING

An important problem is how to compute (heuristic) OARSTs simultaneously for many nets, such that the wires in each net do not cross one another. It seems likely that the results of Jájá and Wu [29] (see Theorem 2.1) can be generalized to Steiner tree routing. However, the *multiple Steiner tree* problem is much more difficult than the disjoint-paths problem.

ACKNOWLEDGMENTS

The authors' work has been supported in part through National Science Foundation grants MIP-9107717 and CDA-8922545. In addition, JLG is partially supported by a Virginia Space Grant Fellowship. Their support is greatly appreciated. Also, thanks are due to Hans Bodlaender, Andrew Kahng, and Dana Richards for helpful discussions on various aspects of this work, and to Shaodi Gao for some helpful comments on an earlier draft of this chapter, particularly for pointing out the applicability of Bern's results in Section 2.4.1.

REFERENCES

1. T. Asano, "Generalized Manhattan Path Algorithm with Applications," *IEEE Transactions on Computer-Aided Design*, Vol. 7, 1988, pp. 797-804.

2. J. E. Beasley, "A heuristic for Euclidean and Rectilinear Steiner Problems," *European Journal of Operational Research*, Vol. 58, 1992, pp. 284-292.

3. P. Berman and V. Ramaiyer, "Improved Approximations for the Steiner Tree Problem," *Journal of Algorithms*, Vol. 17, 1994, pp. 381-408.

4. M. Bern, "Faster Exact Algorithms for Steiner Trees in Planar Networks," *Networks*, Vol. 20, 1990, pp. 109-120.

5. T. Chao and Y. Hsu, "Rectilinear Steiner Tree Construction by Local and Global Refinement," *Proceedings of the International Conference on Computer-Aided Design*, 1990, pp. 432-435.

6. D. S. Chen, "Constrained Wirelength Minimization of a Steiner Tree," Technical report, Department of Electrical Engineering and Computer Science, Northwestern University, Evanston, Illinois, 1994.

7. S. Cheng, A. Lim, and C. Wu, "Optimal Rectilinear Steiner Tree for Extremal Point Sets," *Proceedings of the International Symposium on Algorithms and Computation*, Volume 762 of *Lecture Notes in Computer Science*, Springer-Verlag, Berlin, Germany, 1993, pp. 523-532.

8. C. Chiang, M. Sarrafzadeh, and C. K. Wong. "An Optimal Algorithm for Rectilinear Steiner Trees for Channels with obstacles," *International Journal of Circuit Theory and Applications*, Vol. 19, 1991, pp. 551-563.

9. C. Chiang, M. Sarrafzadeh, and C. K. Wong, "An Algorithm for Exact Rectilinear Steiner Trees for Switchbox with Obstacles," *IEEE Transactions on Circuits and Systems*, Vol. 39, 1992, pp. 446-455.

10. J. P. Cohoon and D. S. Richards, "Optimal Two-Terminal α-β Wire Routing," *Integration: the VLSI Journal*, Vol. 6, 1988, pp. 35-57.

11. J. P. Cohoon, D. S. Richards, and J. S. Salowe, "An Optimal Steiner Tree Algorithm for a Net Whose Terminals Lie on the Perimeter of a Rectangle," *IEEE Transactions on Computer-Aided Design*, Vol. 9, 1990, pp. 398-407.

12. P. J. de Rezende, D. T. Lee, and Y. F. Wu, "Rectilinear Shortest Paths with Rectangular Barriers," *Proceedings of the Seventeenth Symposium on Computational Geometry*, 1985, pp. 204-213.

13. W. A. Dees and P. G. Karger, "Automated Rip-Up and Reroute Techniques," *Proceedings of the Nineteenth Design Automation Conference*, 1982, pp. 432-439.

14. W. A. Dees and R. J. Smith II, "Performance of Interconnection Rip-Up and Reroute Strategies," *Proceedings of the Eighteenth Design Automation Conference*, 1981, pp. 382-390.

15. S. E. Dreyfus and R. A. Wagner, "The Steiner Problem in Graphs," *Networks*, Vol. 1, 1972, pp. 195-207.

16. R. E. Erickson, C. L. Monma, and A. F. Veinott Jr., "Send-and-Split Method for Minimum-Concave-Cost Network Flows," *Mathematics of Operations Research*, Vol. 12, 1987, pp. 634-664.

17. J. L. Ganley and J. P. Cohoon, "Complexity Results for Special Cases of the Rectilinear Steiner Tree Problem," Technical Report CS-93-15, Department of Computer Science, University of Virginia, Charlottesville, Virginia, March 1993.

18. J. L. Ganley and J. P. Cohoon. "A Faster Dynamic Programming Algorithm for Exact Rectilinear Steiner Minimal Trees," *Proceedings of the Fourth Great Lakes Symposium on VLSI*, 1994, pp. 238-241.

19. J. L. Ganley and J. P. Cohoon, "Optimal Rectilinear Steiner Minimal Trees in $O(n^2 2.62^n)$ Time," *Proceedings of the Sixth Canadian Conference on Computational Geometry*, 1994, pp. 308-313.

20. J. L. Ganley and J. P. Cohoon, "Routing a Multi-Terminal Critical Net: Steiner Tree Construction in the Presence of Obstacles," *Proceedings of the International Symposium on Circuits and Systems*, 1994, pp. 113-116.

21. M. R. Garey and D. S. Johnson, "The Rectilinear Steiner Tree Problem Is NP-complete," *SIAM Journal of Applied Mathematics*, Vol. 32, 1977, pp. 826-834.

22. S. L. Hakimi, "Steiner's Problem in Graphs and its Implications," *Networks*, Vol. 1, 1971, pp. 113-133.

23. M. Hanan, "On Steiner's Problem with Rectilinear Distance," *SIAM Journal of Applied Mathematics*, Vol. 14, 1966, pp. 255-265.

24. N. Hasan, G. Vijayan, and C. K. Wong, "A Neighborhood Improvement Algorithm for Rectilinear Steiner Trees," *Proceedings of the International Symposium on Circuits and Systems*, 1990, pp. 2869-2872.

25. D. W. Hightower, "A Solution to the Line-Routing Problem on the Continuous Plane," *Proceedings of the Sixth Design Automation Workshop*, 1969, pp. 1-24.

26. D. W. Hightower, "The Interconnection Problem – A Tutorial," *Proceedings of the Tenth Design Automation Workshop*, 1973, pp. 1-21.

27. F. K. Hwang, "On Steiner Minimal Trees with Rectilinear Distance," *SIAM Journal of Applied Mathematics*, Vol. 30, 1976, pp. 104-114.

28. F. K. Hwang, D. S. Richards, and P. Winter, *The Steiner Tree Problem*, Volume 53 of *Annals of Discrete Mathematics*, North-Holland, Amsterdam, Netherlands, 1992.

29. J. JáJá and S. A. Wu, "On Routing Two-Terminal Nets in the Presence of Obstacles," *IEEE Transactions on Computers*, Vol. 8, 1989, pp. 563-570.

30. K. Kanchanasut, "A Shortest-Path Algorithm for Manhattan Graphs," *Information Processing Letters*, Vol. 49, 1994, pp. 21-25.

31. M. Kaufmann, S. Gao, and K. Thulasiraman, "On Steiner Minimal Trees in Grid Graphs and Its Application to VLSI Layout," *Proceedings of the International Symposium on Algorithms and Computation*, Volume 834 of *Lecture Notes in Computer Science*, Springer-Verlag, Berlin, Germany, 1994, pp. 351-359.

32. C. Y. Lee, "An Algorithm for Path Connections and Its Applications," *IRE Transactions on Electronic Computers*, Vol. 10, 1961, pp. 346-365.

33. D. T. Lee, C. D. Yang, and C. K. Wong, "Rectilinear Paths Among Rectilinear Obstacles," Technical Report 92-AC-123, Department of Electrical Engineering and Computer Science, Northwestern University, Evanston, Illinois, September 1992.

34. D. T. Lee, C. D. Yang, and C. K. Wong, "On Bends and Distances of Paths Among Obstacles in 2-Layer Interconnection Model," *IEEE Transactions on Computers*, Vol. 43, 1994, pp. 711-724.

35. J. H. Lee, N. K. Bose, and F. K. Hwang, "Use of Steiner's Problem in Suboptimal Routing in Rectilinear Metric," *IEEE Transactions on Circuits and Systems*, 1976, pp. 470-476.

36. T. Lengauer, *Combinatorial Algorithms for Integrated Circuit Layout*, John Wiley and Sons, Chichester, England, 1990.

37. W. Lipski Jr., "An $O(n \log n)$ Manhattan Path Algorithm," *Information Processing Letters*, Vol. 19, 1984, pp. 99-102.

38. K. Mikami and K. Tabuchi, "A Computer Program for Optimal Routing of Printed Circuit Connectors," *IFIPS Proceedings*, Vol. H47, 1968, pp. 1475-1478.

39. S. Mirayala, J. Hashmi, and N. Sherwani, "Switchbox Steiner Tree Problem in Presence of Obstacles," *Proceedings of the International Conference on Computer-Aided Design*, 1991, pp. 536-539.

40. J. S. B. Mitchell, "L_1 Shortest Paths Among Polygonal Obstacles in the Plane," *Algorithmica*, Vol. 8, 1992, pp. 55-88.

41. E. F. Moore, "Shortest Path Through a Maze," *Annals of the Computational Laboratory of Harvard University*, Vol. 30, 1959, pp. 285-292.

42. B. T. Preas, "Benchmarks for Cell-Based Layout Systems," *Proceedings of the Twenty-fourth Design Automation Conference*, 1987, pp. 319-320.

43. B. T. Preas and M. J. Lorenzetti, editors, *Physical Design Automation of VLSI Systems*, Benjamin/Cumming Publishing Company, Menlo Park, California, 1988.

44. J. S. Provan, "A polynomial Algorithm for the Steiner Tree Problem on Terminal-Planar Graphs," Technical Report 83/10, Department of Operations Research, University of North Carolina, Chapel Hill, North Carolina, 1983.

45. R. Raghavan, J. P. Cohoon, and S. Sahni, "Single Bend Wiring," *Journal of Algorithms*, Vol. 7, 1986, pp. 232-257.

46. D. S. Richards, "Complexity of Single-Layer Routing," *IEEE Transactions on Computers*, Vol. 33, 1984, pp. 286-288.

47. D. S. Richards and J. S. Salowe, "A Linear-Time Algorithm to Construct a Rectilinear Steiner Minimal Tree for k-extremal Point Sets," *Algorithmica*, Vol. 7, 1992, pp. 247-276.

48. N. Robertson and P. D. Seymour, "Graph Minors XIII: The Disjoint Paths Problem," *Journal of Combinatorial Theory Series B*, Vol. 63, 1995, pp. 65-110.

49. J. S. Salowe and D. M. Warme, "35-point Rectilinear Steiner Minimal Trees In a Day," *Networks*, Vol. 25, 1995, pp. 69-87.

50. H. Shin and A. Sangiovanni-Vincentelli, "A Detailed Router Based on Incremental Routing Modifications: Mighty," *IEEE Transactions on Computer-Aided Design*, Vol. 6, 1987, pp. 942-955.

51. J. M. Smith, D. T. Lee, and J. S. Liebman, "An $O(n \log n)$ Heuristic Algorithm for the Rectilinear Steiner Minimal Tree Problem," *Engineering Optimization*, Vol. 4, 1980, pp. 179-192.

52. J. Soukup, "Circuit Layout," *Proceedings of the IEEE*, Vol. 69, 1981, pp. 1281-1304.

53. P. Widmayer, "On Graphs Preserving Rectilinear Shortest Paths in the Presence of Obstacles," *Annals of Operations Research*, Vol. 33, 1991, pp. 557-575.

54. P. Winter, Personal Communication, 1994.

55. P. Winter, "Reductions for the Rectilinear Steiner Tree Problem," Manuscript, 1994.

56. Y. Wu, P. Widmayer, M. D. F. Schlag, and C. K. Wong, "Rectilinear Shortest Paths and Minimum Spanning Trees in the Presence of Rectilinear Obstacles," *IEEE Transactions on Computers*, Vol. 36, 1987, pp. 321-331.

57. Y. Y. Yang and O. Wing, "Optimal and Suboptimal Solution Algorithms for the Wiring Problem," *Proceedings of the International Symposium on Circuit Theory*, 1972, pp. 154-158.

58. A. Z. Zelikovsky, "An 11/6-Approximation Algorithm for the Network Steiner Problem," *Algorithmica*, Vol. 9, 1993, pp. 463-470.

Chapter 3

THE GENERAL SOLUTION OF
THE RECTILINEAR STEINER'S PROBLEM
AND ITS APPLICATIONS

Yeun Tsun Wong and Michael G. Pecht

With the rapid increase in device functions and improvements in electron-beam drawing techniques, both the wire density and the number of routing layers in electronic devices have skyrocketed. Providing new routing techniques for minimizing the number of layers (or the routing area in each layer) and the number of vias is increasingly critical for reducing manufacturing costs, easing testing and maintenance, and improving reliability [18, 19, 44, 47, 48, 53, 70, 74, 77, 95, 106, 107, 111-114, 117, 123].

The number of layers is determined by the area occupied by wires and wasted routable area. Improperly selected and poorly connected wires are the major reasons for the expanding number of layers and vias. Thus, reducing layers and vias requires

- reducing the total length of electronic wires; and

- reducing the wasted routable area (generally up to 50% of the total area).

A tree whose length can be reduced by adding extra points is called a *Steiner tree*, as proposed by Jacob Steiner in the early nineteenth century. A Steiner tree is called a *rectilinear (Manhattan) Steiner tree* if the distance between points (x_p, y_p) and (x_q, y_q) is measured by $|x_p - x_q| + |y_p - y_q|$. A straightforward way to reach the goals of routing is to find and use all minimum rectilinear Steiner trees (MRSTs), because

- the minimum rectilinear Steiner tree has the minimum length of all tree types, and precisely simulates the connection requirements in a multilayer printed circuit; and

- different MRSTs that connect the same set of terminals (pins) can be selected to reduce wire congestion pertaining to the number of layers, vias, special requirements, and wasted area.

Similarly, it is important to supply MRSTs for the optimal setting designs of communication networks, cable TV networks, pipelines, and sewage systems in cities or buildings [6, 7, 71, 80, 97-98]. MRSTs are also useful for simulating phylogenetic trees in biological systematics and for other applications [12-14, 20, 25-28, 32-35, 40, 49, 91, 104].

Finding the MRST connecting a set of nodes is called a *rectilinear Steiner's problem* [24, 45, 59, 63]. Although many authors have proposed theoretical approaches to solve this problem, no successful result was

49

achieved [15, 30, 37, 38, 43, 54, 55, 59, 61, 62, 76, 77, 81, 82, 85, 86, 89, 92, 93, 100, 110, 118, 120-122]. Thus, the rectilinear Steiner's problem has been simplified into a special problem, in that the number of nodes is limited or the locations of nodes are restricted [1, 23, 56, 88, 90, 105, 118, 120-122]; alternatively, it has been heuristically approximated by connecting nodes based on three-node and four-node subgraphs [3, 5, 50-52, 73, 87, 99], by improving the connections in the minimum spanning tree [8, 16, 50-52, 57, 69, 123], by rectangle trees, arborescence trees or other graphs [2, 4, 15, 29, 31, 32, 33, 35, 72, 75, 83, 86, 109], by probability methods[9, 21, 29, 60, 64, 68, 103, 107, 108], or by reducing length with various greedy methods [4, 5, 11, 22, 41, 43, 49, 78, 87, 94, 94, 99, 101, 102].

The major flaws in these approaches can be summarized as follows:

- A rectilinear Steiner tree was directly selected as the researched object. Because there are many bends in an edge, many edges can be selected, and an edge may be dynamically moved in a rectilinear Steiner tree, it is too difficult to express mathematically the dynamic and multichoice properties of an edge in a rectilinear Steiner tree.

- Conservation subsystems were not considered. A rectilinear Steiner tree is also a complex dynamic system. To simplify the research, a directly researched object should be a conservation subsystem, rather than the dynamic system itself. Some approaches have focused on a full tree, in that one-degree points are nodes; however, this is a special tree, rather than a conservation subsystem.

- Many researchers are too constricted by the concept of non-deterministic polynomial (NP) completeness [38]. Thus, they cannot consider the proposition that having no efficient algorithm for the worst case in computing the exact solution does not mean there is no efficient theoretical algorithm for an accuracy-estimable solution.

To eliminate these theoretical problems, we extended a rectilinear edge into an edge set that contains all the shortest rectilinear edges connecting the same point pair; extended every edge in a rectilinear tree to an edge set to generate a graph called an edge-set tree; and extended each Steiner point into a dynamic Steiner point that can be moved within specific locations without changing the length of each tree contained in the edge-set tree [115, 116]. Thus, selecting bends for an edge can be avoided, and the problem of finding an MRST is transferred to finding an edge-set tree containing a set of MRSTs.

For both engineering and academic purposes, we further defined a set of all global MRSTs connecting the same set of given nodes to be the general solution of the rectilinear Steiner's problem. Thus, finding the general solution becomes a matter of finding a set of edge-set trees containing all MRSTs.

Each edge set has a rectangular boundary. Thus, Steiner points can be determined by the intersection of adjacent rectangular boundaries of edge sets in an edge-set tree. If the intersecting area of the boundaries of two edge sets contains more than one point, a redundant length is considered to exist in each tree contained in this edge-set tree, and these trees must not be MRSTs. A redundant length can be detected from the connections of length-conservation units called *cliques* in an edge-set tree; and it can be removed by replacing cliques (or edge sets) in an edge-set tree with new edge sets (or cliques). Proposing a theory to support an algorithm for generating the general solution by removing every redundant length is one of the purposes of this research.

According to the complexity of each clique, the algorithm for the general solution can be expanded into a series of subalgorithms, whose two subalgorithms can efficiently reduce 99.9% of the total redundant length. Thus, another purpose is to approach an efficient and accuracy-estimable algorithm expansion for an inefficient polynomial (P) or NP problem.

Our third purpose has been to prove and propose efficient algorithms for special cases and for algorithm expansions in the rectilinear Steiner's problem. We also introduce an MRST routing based on these efficient algorithms.

3.1 BASIC DEFINITIONS

Let points $p = (x_p, y_p)$ and $q = (x_q, y_q)$ be moved within locations occupied by point sets M_p and M_q, called *moving domains* of p and q, respectively. A set of all rectilinear edges from p to q is contained in an *edge set*, E_{pq} (Figure 3.1a), whose boundary is a rectangle, R_{pq}, connecting p and q (Figure 3.1b). The total number of edges in E_{pq} is $|E_{pq}| = C(|x_p - x_q| + |y_p - y_q|, |x_p - x_q|)$. The length, $l_{e_{pq}}$, of edge-set $e_{pq} \in E_{pq}$ equals half of the perimeter of the boundary of E_{pq}. The length, $l_{E_{pq}}$, of E_{pq} is defined as the length of $e_{pq} \in E_{pq}$.

(a) E_{pq} (b) R_{pq}

Figure 3.1 A bent-edge set and a rectangle

When $x_p \equiv x_q$ or $y_p \equiv y_q$, E_{pq} is called a *straight-edge set* (Figure 3.2). E_{pq} is a straight-edge set if and only if every edge in E_{pq} is straight.

A straight-edge set that can be moved parallel to itself without changing its length is called a *parallel moving edge set*.

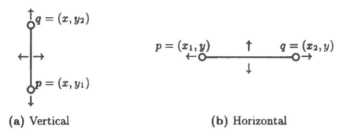

(a) Vertical   (b) Horizontal

Figure 3.2 Straight-edge sets

When $x_p \neq x_q$ and $y_p \neq y_q$, E_{pq} is called a *bent-edge set*. E_{pq} is a bent-edge set if and only if there is one bent edge in E_{pq} (Figure 3.1a). A straight edge may be contained in bent-edge set E_{pq} if $|M_p| > 1$ or $|M_q| > 1$; all edges in bent-edge set E_{pq} are bent edges if $|M_p| = |M_q| = 1$. The bent-edge set E_{pq} in Figure 3.1a contains a set of six bent edges when $|M_p| = |M_q| = 1$.

Finding R_{pq} is equivalent to finding E_{pq}, without regard to a specific edge, because R_{pq} is exactly filled by all edges from p to q. For convenience, E_{pq} is simply expressed as a rectangle in all figures in this chapter. If $p = q$, E_{pq} is reduced to a point.

If $|M_s| = 1$, s is either a given node or a fixed Steiner point (Figure 3.3a). When $|M_s| > 1$, s is a *dynamic Steiner point*. When M_s is a set of points lying on a straight edge, such that s can be moved either horizontally or vertically, s is a *one-dimensional (1-D) Steiner point*. Steiner points s_1 and s_2 in Figure 3.3b are 1-D. The moving domain of s_1 is the projection of $E_{p_2 q_2}$ on $E_{p_1 q_1}$, and the moving domain of s_2 is the projection of $E_{p_1 q_1}$ on $E_{p_2 q_2}$. When s can be moved along both directions, s is a *two-dimensional (2-D) Steiner point* (Figure 3.3c).

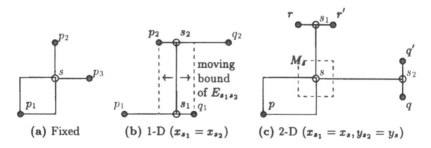

(a) Fixed   (b) 1-D $(x_{s_1} = x_{s_2})$ (c) 2-D $(x_{s_1} = x_s, y_{s_2} = y_s)$

Figure 3.3 Steiner points

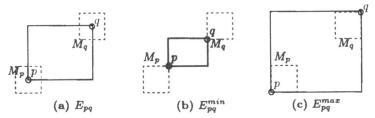

Figure 3.4 Minimized and maximized edge sets

Based on the moving domain of each vertex, an edge set can be classified as either fixed or moving. E_{pq} is a fixed edge set only if $|M_p| = |M_q| = 1$. If $|M_p| \geq 1$ and $|M_q| = 1$, E_{pq} contains $|M_p|$ different fixed edge sets. If $|M_p| \geq 1$, $|M_q| \geq 1$, and p and q are moved independently, E_{pq} contains a total of $|M_p||M_q| - C(|M_p \cap M_q|, 2)$ fixed edge sets. In all fixed edge sets of a moving edge set, E_{pq}, the one with the minimum length is called the *minimized edge set* of E_{pq}, denoted as E_{pq}^{min}; the one with the maximum length is called the *maximized edge set* of E_{pq}, denoted as E_{pq}^{max} (Figure 3.4).

Extending edges into edge sets in a tree, an *edge-set tree* with n_e edge sets can be expressed as a connected graph:

$$
\begin{aligned}
T &= (E_T, V_T) \\
&= (E_T, V_n \bigcup V_s(T)) \\
&= \{t_j = (E(t_j), V_T) \mid j = 1, ..., n_t\}
\end{aligned}
\tag{3.1}
$$

where $E_T = \{E_{p_i q_i} \mid i = 1, ..., n_e\}$ is a set of all edge sets in T, $V_T = \cup_{i=1}^{n_e} \{p_i, q_i\}$, V_n is a set of given nodes, $V_s(T)$ is a set of Steiner points in T, $E(t_j) = \{e_{p_i q_i} \in E_{p_i q_i} \mid i = 1, ..., n_e\}$ is a set of all edges in tree t_j, and n_t is the total number of different trees contained in T. Denoting l_T as the length of T and l_t as the length of t, $l_t = l_T = \sum_{i=1}^{n_e} l_{e_{p_i q_i}}$. Because T is a connected graph, $\{p_i, q_i\} \cap (\cup_{j=1, \neq i}^{n_e} \{p_j, q_j\}) \neq \emptyset$. Implicitly expressing V_T by the subscripts of each edge set, an edge-set tree can also be expressed as a union of a set of edge sets:

$$
T = \bigcup_{i=1}^{n_e} E_{p_i q_i}
\tag{3.2}
$$

According to Equation 3.1, an edge-set in T is still considered a member and is expressed as $E_{p_i q_i} \in T$. In an edge-set tree, an edge set *adjacent to* another edge set is referred to as an edge set intersecting another edge set at an area including a point or a set of points. Similarly, a point adjacent to another point is referred to as a point directly connecting another point by an edge set.

An *edge-set subtree* is a connected subgraph of T. A vertex in an edge-set subtree is either an end vertex or an inner vertex. A vertex adjacent to only one other vertex in an edge-set subtree is an end vertex. If an end vertex is a dynamic Steiner point, then it is called a *dynamic end vertex*, otherwise, it is called a *fixed end vertex* (or an *end node*, if it is a node).

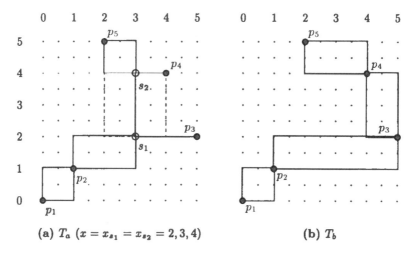

(a) T_a $(x = x_{s_1} = x_{s_2} = 2, 3, 4)$ (b) T_b

Figure 3.5 Edge-set trees

Example: In Figures 3.5, $T_a = (E_{T_a}, V_{T_a}) = E_{p_1 p_2} \cup \cup_{i=2}^{3} E_{p_i s_1} \cup E_{s_1 s_2} \cup \cup_{i=4}^{5} E_{s_2 p_i}$ and $T_b = (E_{T_b}, V_{T_b}) = \cup_{i=1}^{4} E_{p_i p_{i+1}}$, where $E_{T_a} = \{E_{p_1 p_2}, E_{p_2 s_1}, E_{s_1 p_3}, E_{s_1 s_2}, E_{s_2 p_4}, E_{s_2 p_5}\}$, $V_{T_a} = \{p_i, s_j \mid i = 1, ..., 5; j = 1, 2\}$, $E_{T_b} = \{E_{p_i p_{i+1}} \mid i = 1, ..., 4\}$, and $V_{T_b} = \{p_i \mid i = 1, ..., 5\}$. In T_a, $E_{p_2 s_1}$, $E_{p_3 s_1}$, $E_{p_4 s_2}$, and $E_{p_5 s_2}$ are moving edge sets, $E_{s_1 s_2}$ is a parallel moving edge set, and s_1 and s_2 are inner vertices of the edge-set subtree with p_2, p_3, p_4 and p_5 as end vertices, where $2 \leq x_{s_1} = x_{s_2} \leq 4$.   ◇

Similar to a cycle in graph theory, an *edge-set cycle* is a connected graph in that every edge set is exactly adjacent to two edge sets. It can be expressed as $C_y = \cup_{i=1}^{n_c - 1} E_{p_i p_{i+1}} \cup E_{p_1 p_{n_c}}$, and can be generated by connecting an edge set to an edge-set tree. Removing an edge set from an edge-set cycle generates an edge-set tree. No edge-set cycle can be embedded in another edge-set cycle.

If there are two or more points in the intersecting area of R_{pq} and $R_{p'q'}$, a redundant length must exist between adjacent edge sets whose boundaries are R_{pq} and $R_{p'q'}$. Figure 3.6 shows the intersection combinations of two adjacent edge sets. In Figure 3.6a, there is no redundancy between E_{pq} and $E_{p'q'}$.

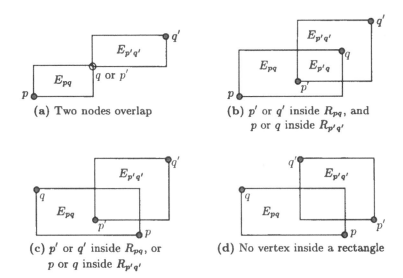

(a) Two nodes overlap

(b) p' or q' inside R_{pq}, and p or q inside $R_{p'q'}$

(c) p' or q' inside R_{pq}, or p or q inside $R_{p'q'}$

(d) No vertex inside a rectangle

Figure 3.6 Possible intersections of two adjacent edge sets

If an edge set, E_{pq}, in an edge-set tree, T, can be replaced by another edge set, $E_{p'q'}$, with $l_{E_{p'q'}^{min}} < l_{E_{pq}^{max}}$, to form a new edge-set tree, then $l_{E_{pq}^{max}} - l_{E_{p'q'}^{min}}$ is called a *redundant length* of E_{pq}. Replacement rules are further described in Section 3.5.

An edge-set tree in which no redundant length can be found by replacing a single edge set with a new edge set is called a *no-redundancy edge-set tree (NR edge-set tree)*, and is denoted as T_{NR}. A set of MRSTs must be contained in an NR edge-set tree. Similarly, a no-redundancy edge-set subtree is an *NR edge-set subtree*. An NR edge-set subtree is called a *node-ended NR edge-set subtree (NN edge-set subtree)* if each end vertex is a node or a dynamic Steiner point whose moving domain includes a node, and if each inner vertex is a Steiner point whose moving domain does not include a node.

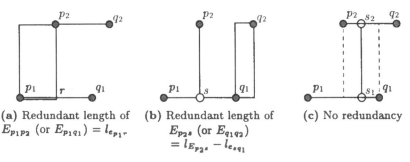

(a) Redundant length of $E_{p_1p_2}$ (or $E_{p_1q_1}$) $= l_{e_{p_1r}}$

(b) Redundant length of E_{p_2s} (or $E_{q_1q_2}$) $= l_{E_{p_2s}} - l_{e_{sq_1}}$

(c) No redundancy

Figure 3.7 Redundant length

Example: Figures 3.7a and 3.7b show redundant lengths of two cases. In Figure 3.7b, the redundant length of $E_{p_2 s}$ can be found from an edge-set cycle generated by connecting p_2 and $E_{q_1 q_2}$. Figure 3.7c shows an NR edge-set tree, in that $E_{p_1 s_1} \cup E_{s_1 s_2} \cup E_{q_1 s_1}$ and $E_{p_2 s_2} \cup E_{s_1 s_2} \cup E_{q_2 s_2}$ are NN edge-set subtrees because $p_2 \in M_{s_2}$ and $q_1 \in M_{s_1}$.   ◇

3.2 BASIC CONNECTIVITIES
IN AN NR EDGE-SET TREE

This section proves properties describing the basic connectivities of edge sets adjacent to either one Steiner point or one edge set in an NR edge-set tree. Proofs in this section are more concise than those given previously [115]. A Steiner point is determined by the mid-values of the coordinates of three points. To express the location of a Steiner point, *mid(a, b, c)* is defined to be a *selection function* selecting the mid-value from a, b and c. If $a \leq c \leq b$ or $b \leq c \leq a$, *mid(a, b, c)* = c.

Property 1 *A Steiner point is an intersection of three or four edge sets in an NR edge-set tree.*

Property 2 *At most, four edge sets intersect at a node in an NR edge-set tree.*

Property 3 *Four edge sets intersecting at a Steiner point or at a node in an NR edge-set tree must be straight.*

Property 4 *A Steiner point must be an intersection of exactly three edge sets in an NR edge-set tree if one of the edge sets is a bent-edge set.*

Property 5 *If one of three edge sets intersecting at a Steiner point is a bent-edge set in an NR edge-set tree, the other two edge sets must be straight.*

Property 6 *A parallel moving edge set must not be shorter than an edge occupied by the moving domain of its vertex in an NR edge-set tree.*

Property 7 *Point (s_x, s_y) must be a Steiner point or a node in an NR edge-set tree for p, q and r, where $s_x = mid(x_p, x_q, x_r)$, and $s_y = mid(y_p, y_q, y_r)$.*

Property 8 *A dynamic Steiner point is an intersection of exactly three edge sets in an NR edge-set tree.*

Property 9 *There are, at most, four edge sets that must be adjacent to an edge set in an NR edge-set tree.*

Property 10 *For n nodes, there are, at most, n − 2 Steiner points in an NR edge-set tree.*

Proof of Properties 1 - 5: Let s be the origin of the x-axis and the y-axis. If E_{ps} is a straight-edge set in an NR edge-set tree, then p must lie on one of four half-axes: the x^+-axis, x^--axis, y^+-axis, or y^--axis. If E_{ps} is a bent-edge set in an NR edge-set tree, then E_{pq} must occupy two perpendicular half-axes. If E_{ps} and E_{qs} occupy the same half-axis, redundant length exists in E_{ps} or E_{qs}. Thus, it is impossible for five edge sets to intersect at a Steiner point in an NR edge-set tree. Furthermore, the existence of a Steiner point that is an intersection of only two edge sets is equivalent to the existence of a Steiner point inside the boundary of an edge set. Thus, a Steiner point that is an intersection of only two edge sets is exhibited in an edge-set tree. Therefore, Properties 1 - 5 are proven. □

According to Properties 1 and 2, an NR edge-set tree can be expressed as a binary tree if a node or a Steiner point that is an intersection of four edge sets is considered to be a special edge set with overlapping vertices.

In the case of four arbitrary points, there must be at least a pair of separate rectangles connecting two pairs of points each, and at most two different pairs of such rectangles. Figure 3.8 shows combinations of a pair of separate rectangles connecting two pairs of points. If the rectangles in Figure 3.8a and 3.8b are considered to be boundaries of edge sets, it can also be proven that $l_{E_{aa'}} + l_{E_{bb'}} = l_{E_{ab'}} + l_{E_{a'b}}$, if the projection of $E_{aa'}$ on $E_{bb'}$ equals the projection of $E_{a'b}$ on $E_{ab'}$ (Figure 3.9).

In Figure 3.9, a pair of rectangles is connected by the *minimum rectangle*, a rectangle with a minimum perimeter. If the rectangles are considered to be the boundaries of edge sets, edges disconnected from the newly connected edge set must be discarded. For example, edges contained in $E_{bb'}$ that are not adjacent to q must be discarded (Figure 3.10a). Discarding the disconnected edges in each edge set shown in Figure 3.9 yields the edge-set trees connecting four nodes shown in Figure 3.10.

Proof of Property 6: If the projection of an edge set (or rectangle) on another edge set (or rectangle) is not empty (Figure 3.9a and 3.9b), the projection is exactly occupied by the moving domain of a vertex of the parallel moving edge set (Figure 3.10a and 3.10b). If $l_{E_{pq}} = l_{E_{p'q'}}$, there exist two different NR edge-set trees connecting the four nodes a, b, c, and d (Figure 3.10a and 3.10b), and the general solution for these nodes is contained in two NR edge-set trees. If $l_{E_{pq}} > l_{E_{p'q'}}$, the general solution is contained only in the NR edge-set tree shown in Figure 3.10a. In this case, the redundant length of $E_{b'p'}$ in Figure 3.10b can be found because $E_{b'p'}^{max}$ can be replaced by $E_{b'b}$, where $l_{E_{b'b}} = l_{E_{b'b''}} + l_{E_{p'q'}} <$ $l_{E_{b'p'}}^{max} = l_{E_{b'b''}} + l_{E_{pq}}$. This proves Property 6. $\qquad\qquad\square$

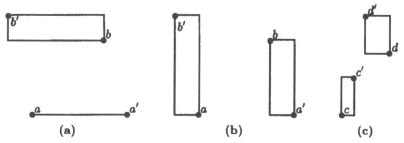

Figure 3.8 Separate rectangles connecting two pairs of points

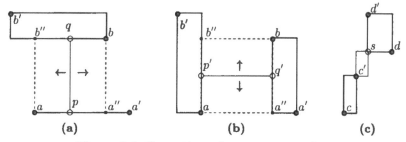

Figure 3.9 Connections of separate rectangles

- Dashed box is the maximum moving area of E_{pq} or $E_{p'q'}$.
- $e_{b''b}$ ($e_{aa''}$) is the projection of $E_{aa'}$ ($E_{bb'}$) on $E_{bb'}$ ($E_{aa'}$).
- $e_{a''b}$ ($e_{ab''}$) is the projection of $E_{ab'}$ ($E_{a'b}$) on $E_{a'b}$ ($E_{ab'}$).

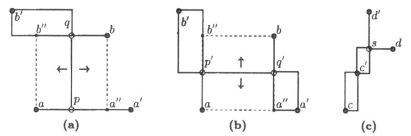

Figure 3.10 Connecting two separate edge sets into an NR edge-set tree

Proof of Property 7: If the projection of an edge set (or rectangle) on the other edge set (or rectangle) is empty (Figure 3.9c), two separate edge sets must be connected by a bent-edge set to form an NR edge-set tree (Figure 3.10c). In Figure 3.10c, if $c' = s$ and c' is not a given node, s is the only Steiner point that connects c, d, and d' into an NR edge-set tree. Let $c' = s$ be an origin of the x-axis and y-axis. According to Properties 1 - 5, at least two straight-edge sets intersect at s. Letting these straight-edge sets be E_{ds} and $E_{d's}$, d and d' must lie on two perpendicular half-axes, and E_{cs} occupies one or two other half-axes. Because s is the origin, $s_x = mid\{c_x, d_x, d'_x\}$ and $s_y = mid\{c_y, d_y, d'_y\}$, proving Property 7. □

Edge set $E_{cc'}$ and new edge set $E_{dd'}$ in Figure 3.8c, also NR edge-set trees for two subsets of nodes, are generated by removing $E_{c's}$ from the NR edge-set tree in Figure 3.10c. New edge sets generated in edge-set trees for subsets of nodes by removing edge sets from an NR edge-set tree are called *descendant edge sets* of the removal edge sets. $E_{dd'}$ is a descendant edge set of $E_{c's}$. However, $E_{bb'}$ in Figure 3.8a is not a descendant edge set of E_{pq} in Figure 3.10a because $E_{bb'} \subset E_{b'q}$ is not new. Similarly, $E_{ab'}$ and $E_{a'b}$ in Figure 3.8b are not descendant edge sets of $E_{p'q'}$ in Figure 3.10b. Due to descendant edge sets and Steiner points, an edge-set subtree is distinct from an edge-set tree for a subset of nodes.

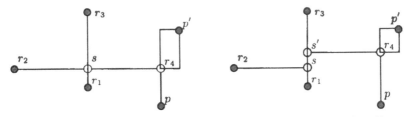

(a) E_{sr_4} adjacent to five edge sets   (b) E_{r_4s} can be moved to $E_{r_4s'}$

Figure 3.11 A dynamic Steiner point is a common vertex
of exactly three edge sets in an NR edge-set tree

Proofs of Properties 8 and 9: According to Property 1, a Steiner point may be an intersection of four edge sets. In Figure 3.11a, dynamic Steiner point s is a common vertex of edge sets E_{r_2s} and E_{r_4s}. If an edge set is adjacent to more than four edge sets (Figure 3.11a), it must be a moving edge set. When it is moved, s becomes the intersection of exactly three edge sets (Figure 3.11b). This proves Property 8. Because no end vertex of a moving-edge set is an intersection of four edge sets, Property 9 is proven. □

Proof of Property 10: Let n be the number of nodes in an NR edge-set tree. If $n = 2$, no Steiner point exists. When $n = 3$, there exists, at most, one Steiner point, according to Property 1. Increasing n by 1 generates at most one additional Steiner point (Property 7). Thus, Property 10 is proven. □

3.3 CLIQUES

To simplify the optimization process and find an optimal solution in a complex dynamic system, it is important to find and understand basic conservation units (subsystems) in the system. Considering an NR edge-set tree as a dynamic length-conservation system, a length-conservation unit is defined as a *clique*. More specifically, a clique is an NR edge-set subtree whose inner vertices are Steiner points only, whose length is minimal and constant independent of movements of Steiner points, and whose total number of edge sets is the minimum for maintaining the maximum number of points in each moving domain.

A clique with i edge sets is called an *i-edge-set clique*, denoted as C_i. The length of clique C_i equals the total length of all edge sets in C_i, and is denoted as l_{C_i}. An NN i-edge-set subtree consisting of either a clique or cliques and their interfaces is called an *(i-edge-set) NN clique*, denoted as C_{in}. Although a parallel moving edge set in an NR edge-set tree has the minimum length and the maximized moving domains of its vertices, it is merely an edge set in C_i ($i > 1$) because there is no inner vertex in it. For convenience, it is still considered to be a special clique, denoted as C_1.

A moving edge set must be included in a clique, or in an interface among cliques and nodes. Otherwise, there must exist a redundant length in the moving edge set, that equals the difference of the maximum and minimum lengths of this moving edge set.

Clique Theorem 1 *The moving domain of a dynamic Steiner point in an NR edge-set tree must be a set of points enclosed by a rectangle; a dynamic end vertex of a clique must be a 1-D Steiner point whose moving domain occupies an edge perpendicular to its adjacent straight-edge set in the clique.*

Proof: According to Property 8, a dynamic Steiner point must be the intersection of exactly three edge sets. Let s be a dynamic Steiner point and $E_{ps} \cup E_{rs} \cup E_{r's}$ be an NR edge-set subtree in a clique (Figure 3.12a). When E_{ps} is a bent-edge set, E_{rs} and $E_{r's}$ must be straight-edge sets (Property 5).

When both e_{sa} and e_{sb} are projections of other edge sets on E_{ps} (Figure 3.12a), E_{rs} and $E_{r's}$ can be moved parallel to two perpendicular directions, respectively, and s can be moved between s and a and between s and b. When E_{rs} is moved to $E_{r_1 s_1}$, projection e_{sb} is parallel moved to $e_{s_1 b_1}$. Moving both E_{rs} and $E_{r's}$, s becomes s_2, and the total edge-set length for connecting p, r, and r' is not changed. In moving E_{rs} to $E_{r_2 a}$, e_{sb} is moved to $e_{a s_3}$. Because s can be an arbitrary point in e_{sb}, s can be an arbitrary point in rectangle R_{ab}. If the moving domains of r and r' occupy edges perpendicular to E_{sr} and $E_{sr'}$, respectively, the total edge-set length for connecting p, r, and r' is constant. According to Property 6, the length of an edge occupied by M_r and the length of an edge occupied by $M_{r'}$ are equal to or less than $l_{E_{rs}^{min}}$ and $l_{E_{r's}^{min}}$, respectively. Thus, $E_{ps} \cup E_{rs} \cup E_{r's}$ is a clique (Figure 3.12b).

If r' is 2-D, the length of edge-set subtree $E_{ps} \cup E_{rs} \cup E_{r's}$ is not constant, and there must exist a dynamic Steiner point, r'', adjacent to r'. If r'' is also 2-D, there must exist a dynamic Steiner point, r''', adjacent to r'', and so forth. Because the number of Steiner points in an NR edge-set tree is finite, the end-vertex of a clique must not be 2-D. Similarly, an edge occupied by its moving domain must be perpendicular to its adjacent straight-edge set in the clique. $\square$

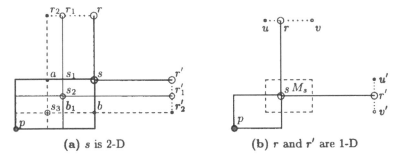

(a) s is 2-D   **(b)** r and r' are 1-D

Figure 3.12 Moving domains of dynamic Steiner points

NR edge-set subtree $E_{ps} \cup E_{sr} \cup E_{sr'}$ in Figure 3.12b is a *three-edge-set clique*, in that r and r' are 1-D, and inner vertex s can be moved in the rectangular moving domain M_s. Two edges, e_{uv} and $e_{u'v'}$, occupied by the moving domains of r and r', are perpendicular to straight-edge sets E_{sr} and $E_{sr'}$, respectively. Furthermore, $l_{e_{uv}} \le l_{E_{sr}^{min}}$ and $l_{e_{u'v'}} \le l_{E_{s'r'}^{min}}$.

If the fixed end vertex of a three-edge-set clique is also 2-D, a *five-edge-set clique* (C_5) is generated (Figure 3.13). A C_5 includes two inner Steiner points and their adjacent edge sets, and four end vertices lying on four sides of a rectangle whose perimeter equals $2l_{C_5}$.

Like the moving domain of an inner Steiner point of a three-edge-set clique, moving domains of two inner Steiner points of a five-edge-set clique depend on moving domains of four end vertices. However, movements of two inner Steiner points may or may not depend upon each other. If $C_5 = \cup_{i=1}^2 E_{sr_i} \cup E_{ss'} \cup \cup_{i=1}^2 E_{s'r_i'}$, the projection of an edge set in $\{E_{sr_1}, E_{sr_2}\}$ on an edge set in $\{E_{s'r_1'}, E_{s'r_2'}\}$ cannot occupy more than one point; otherwise, $\cup_{i=1}^2 E_{sr_i} \cup E_{ss'} \cup \cup_{i=1}^2 E_{s'r_i'}$ is not a clique, or a redundant length exists in this edge-set subtree.

Let r_1, r_2, r_1' and r_2' be four end vertices of a five-edge-set clique, C_5, and let M_{r_1}, M_{r_2}, $M_{r_1'}$ and $M_{r_2'}$ lie on four sides of a rectangle (Figure 3.13). There exists another five-edge-set clique connecting the same set of edge-set trees for four subsets of nodes, if and only if the following three conditions are satisfied:

(a) the projection of an edge occupied by M_{r_1} on an edge occupied by $M_{r_1'}$ is not empty;

(b) the projection of an edge occupied by M_{r_2} on an edge occupied by $M_{r_2'}$ is not empty; and

(c) at least one of the above projections occupies more than one point.

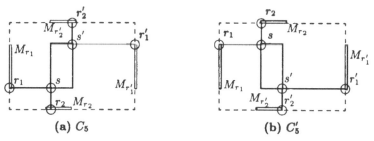

(a) C_5 (b) C_5'

Figure 3.13 Five-edge-set cliques
$$(C_5 = C_5' = \cup_{i=1}^2 E_{r_{i}s} \cup E_{ss'} \cup \cup_{i=1}^2 E_{s'r_i'})$$

Interfaces are needed to connect cliques and given nodes. The structures of interfaces are described by Clique Theorem 2.

Clique Theorem 2 *If $E_{rr'}$ is an edge set in an NR edge-set tree and if r and r' are end-vertices of two cliques in two edge-set subtrees, then*

(a) *r and r' must be adjacent to given nodes, if $E_{rr'}$ is a bent-edge set; otherwise,*

(b) *$E_{rr'}$ and other straight-edge sets lie on the same straight line terminated at two given nodes.*

Proof: Let r and r' be end vertices of cliques C_i and C_j in an NR edge-set tree, respectively; let E_{rs} and $E_{r's'}$ be edge sets in C_i and C_j, respectively; and let E_{pr}, $E_{rr'}$ and E_{rs} intersect at r (Figures 3.14a and 3.14c). Edge set $E_{rr'}$ is either a bent-edge set or a straight-edge set.

(a) When $E_{rr'}$ is a bent-edge set:

According to Property 5, E_{pr} and E_{rs} are straight-edge sets. Assuming p is not a given node, then p must connect at least three edge sets (Property 1). Thus, a projection of one of these edge sets on E_{rs} or $E_{rr'}$ occupies more than one point. Hence, E_{pr} becomes a parallel moving edge set and r can be moved along the direction of straight-edge set E_{rs}. According to Clique Theorem 1, an end vertex of a clique cannot be moved along the direction of its adjacent straight-edge set in the clique. Therefore, p must be a given node.

(b) When $E_{rr'}$ is a straight-edge set:

When $E_{rr'}$ is a straight-edge set, p can be a Steiner point. According to the proof of (a) and Property 1, p must be an intersection of three straight-edge sets; a projection of an arbitrary one of these straight-edge sets on an arbitrary edge set with r as a vertex occupies only one point. Thus, Steiner points $p^{(m)}, ..., p', p, r, r', ..., r^{(n)}$ in the NR edge-set tree lie on a straight line. Because the number of Steiner points in an NR edge-set tree is finite (Property 10), there must exist two given nodes that terminate the straight line. □

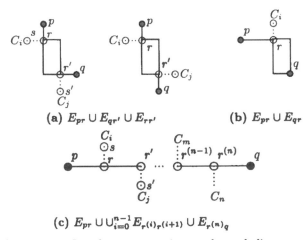

(c) $E_{pr} \cup \bigcup_{i=0}^{n-1} E_{r^{(i)} r^{(i+1)}} \cup E_{r^{(n)} q}$

Figure 3.14 Interfaces among given nodes and cliques

Example: The structures of the interfaces described by conclusion (a) of Clique Theorem 2 are shown in Figure 3.14a, where p and q are given nodes, and r and r' are end vertices of C_i and C_j, respectively. Figure 3.14b shows a special case of a structure in Figure 3.14a. Figure 3.14c shows the structure of an interface described by conclusion (b) of Clique Theorem 2, where p and q are given nodes, and Steiner points $r, r', ...,$ and $r^{(n)}$ are end vertices of n cliques adjacent to the interface. ◊

Note that a straight-edge set adjacent to a Steiner point in an interface is degenerated. In the same interface, the projection of a *degenerated edge set* is, at most, a point on its next degenerated edge set. Considering each Steiner point to be both an inner vertex and an edge set with zero length, each interface has a constant length and satisfies the conditions of a clique. Such an interface is called a *degenerated clique*, and a Steiner point in it is called a *degenerated Steiner point*. A degenerated clique is enclosed by a rectangle connecting two end nodes, and its length equals half the perimeter of the rectangle. A degenerated clique with i edge sets is denoted as C_{id}. According to Clique Theorem 2, there are three types of degenerated cliques:

(1) A degenerated clique including only one degenerated Steiner point is a two-edge-set degenerated clique, C_{2d}.

(2) A degenerated clique including a bent-edge set with two degenerated Steiner points as vertices is a three-edge-set degenerated clique, C_{3d}.

(3) A degenerated clique including i straight-edge sets on the same line terminated at two given nodes is an i-edge-set degenerated clique, C_{id} $(i \geq 2)$.

A degenerated clique lying on a line is called a *linear degenerated clique (LD clique)*. A C_{id} $(i > 3)$ must be an LD clique. To distinguish them from degenerated cliques, other cliques, such as C_3 and C_5, are also called *nondegenerated cliques*.

Clique Lemma 1 *If there are more than four edge sets in a nondegenerated clique, then every inner Steiner point of the clique is an intersection of three edge sets.*

Proof: According to Property 1, there are at most four edge sets intersecting at a Steiner point in an NR edge-set tree. Let $\cup_{i=1}^{4} E_{p_i s} \cup E_{p_s p_j}$ $(1 \leq j \leq 4)$ be an edge-set subtree in a nondegenerated clique (Figure 3.15). Thus, $E_{p_j s}$ must be a parallel moving edge set and s must be a dynamic Steiner point. According to Property 8, s is an intersection of three edge sets. □

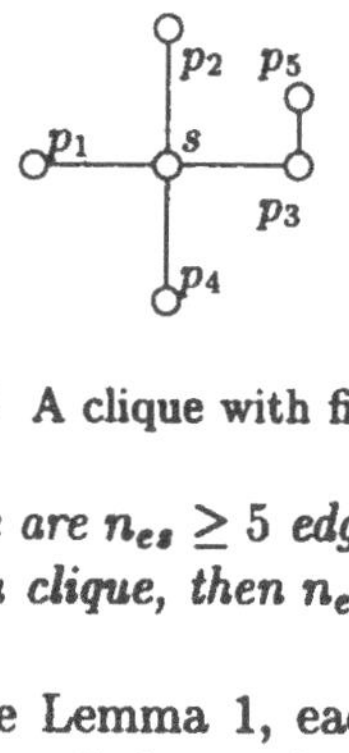

Figure 3.15 A clique with five edge sets

Clique Theorem 3 *If there are $n_{es} \geq 5$ edge sets, n_{ev} end vertices and n_{is} inner Steiner points in a clique, then $n_{es} = 2n_{is} + 1$, $n_{ev} = n_{is} + 2$ and $n_{ev} = (n_{es} + 3)/2$.*

Proof: According to Clique Lemma 1, each inner Steiner point of a clique, C_i, is an intersection of three edge sets if $n_{es} \geq 5$. In $3n_{is}$ edge sets connected by n_{is} inner Steiner points, $(n_{is} - 1)$ edge sets, whose vertices are inner Steiner points only, are counted twice. Thus, $n_{es} = 3n_{is} - (n_{is} - 1) = 2n_{is} + 1$. Because $n_{es} = n_{is} + n_{ev} - 1$ and $n_{es} = 2n_{is} + 1$, $n_{ev} = n_{is} + 2$ and $n_{ev} = (n_{es} + 3)/2$. □

If a nondegenerated clique with four fixed edge sets intersecting at one fixed Steiner point is considered a special case of C_5 in that two inner Steiner points overlap, then the number of edge sets in a nondegenerated clique is always odd. More complex nondegenerated cliques are discussed in Section 3.4.

3.4   EVOLVED CLIQUES

Based on the discussion of cliques C_3, C_5, and degenerated cliques in Section 3.3, this section describes evolutions of a degenerated clique and its adjacent degenerated edge sets. Based on these evolutions, the structures of all NN cliques and all nondegenerated cliques with more than five edge sets can be determined.

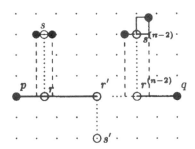

Figure 3.16   An LD clique, C_{id}, and its adjacent edge sets
form an NN clique, $C_{(2i-1)n}$

Clique Theorem 4 *If E_{sr} is a degenerated edge set adjacent to r in an LD clique, and if the moving domain of r does not include a node, then s is either a node or a 1-D Steiner point whose moving domain includes a node and occupies an edge parallel to the LD clique; the LD clique and its adjacent degenerated edge sets form an NN clique.*

Proof: Let p and q be end nodes of an LD clique in an NR edge-set tree, T_{NR}, (Figure 3.16). Then, $E_{sr} \in T_{NR}$. According to Properties 1, 3, and 5, if s is a Steiner point, there must exist a straight-edge set, $E_{us} \in T_{NR}$, parallel to the LD clique. Thus, E_{rs} is a parallel moving edge set. If u or s is 2-D, then E_{us} can be moved parallel. If no node is included in M_r, then the projection of the line from p to q on E_{us} is E_{us} itself, and s can be moved to M_u. Because $l_{E_{us}^{min}} = 0$ and E_{us} can be moved parallel, Property 6 is violated. Thus, neither s nor u can be 2-D. If u is 1-D, then its moving domain must occupy an edge parallel to the LD clique, and a 1-D Steiner point, u', must exist, where s, u, and u' must lie on the same straight line. Because no node exists in M_r, M_s must include a node on the straight line. Thus, the LD clique and its adjacent edge sets form an NN clique.   □

Example: In Figure 3.16, $C_{id} = E_{pr} \cup \bigcup_{i=0}^{n-3} E_{r^{(i)}r^{(i+1)}} \cup E_{r^{(n-2)}q}$ is an LD clique in an NN clique, where $r, r', ..., r^{(n-2)}$ are degenerated Steiner points; parallel moving edge sets E_{rs}, $E_{r's'}$, ..., $E_{r^{(n-2)}s^{(n-2)}}$ are degenerated edge sets adjacent to C_{id}; and each point in $\{s, s', ..., s^{(n-2)}\}$ is a node or a 1-D Steiner point whose moving domain includes a node. Thus, C_{id} and its adjacent degenerated edge sets form an NN clique. ◇

Clique Theorem 5 *An LD clique and its adjacent degenerated edge sets form a clique if the moving domain of each degenerated Steiner point includes more than one point, but not a node.*

Proof: According to Clique Theorem 4, an LD clique and its adjacent degenerated edge sets form an NN clique, C_{in} ($i > 3$). Let $S \subset C_{in}$ be an arbitrary edge-set subtree with two to $(i - 2)$ edge sets. At least, an end vertex of S is a degenerated Steiner point. Because no degenerated Steiner point is fixed, the length, l_S, of S is not constant; alternatively, the number of points in a moving domain of the degenerated Steiner point is decreased if l_S becomes constant. Therefore, S is not a clique. However, the length of C_{in} is constant, independent of movements of Steiner points. Thus, C_{in} is an i-edge-set clique.   □

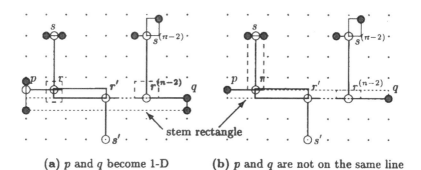

(a) p and q become 1-D   (b) p and q are not on the same line

Figure 3.17 Clique evolutions based on
the evolutions of two end nodes of each LD clique

When two end nodes of an LD clique of an edge-set subtree become (*evolve* into) 1-D end vertices that can be moved perpendicular to the LD clique (Figures 3.16 and 3.17a), each degenerated Steiner point in the LD clique evolves into a dynamic Steiner point whose dimension is increased by one. If an edge-set subtree formed by an LD clique, C_{id}, and its adjacent degenerated edge sets is a clique, C'_m ($m = 2i - 1$), then the edge-set subtree, consisting of all edge sets adjacent to the Steiner points evolved from the degenerated Steiner points of C_{id}, is still an m-edge-set clique, C_m. Thus, C_m is evolved from C'_m.

Two end nodes of the LD clique are called *main vertices* of C'_m, and two end vertices evolved from the end nodes of the LD clique are called main vertices of C_m. Vertices p and q in the cliques in Figures 3.16 and 3.17 are main vertices. Figure 3.17b shows a special case of the evolution, where two end nodes of an LD clique are relocated and are not on the same straight line.

Because C'_m does not exist in the NR edge-set tree including C_m, C'_m is called an *imagined clique* of C_m. The LD clique $C_{id} \subset C'_m$ and the edge-set subtree evolved from C_{id} are called *stems* of C'_m and C_m, respectively. There are, at most, two nodes in a stem of a clique. The stem of C'_m in Figure 3.16 and the stem of C_m in Figure 3.17 can be expressed by $S_T = E_{pr} \cup \cup_{i=0}^{n-3} E_{r(i)r(i+1)} \cup ... \cup E_{r(n-2)q}$. If the degenerated Steiner point, r', in the LD clique in Figure 3.16 is fixed, it can be considered an end vertex of a stem, $\cup_{i=1}^{n-3} E_{r(i)r(i+1)} \cup ... \cup E_{r(n-2)q} \subset S_T$. Thus, a clique can exist in another clique. But a clique cannot be formed by cliques and their interfacing degenerated cliques.

When p and q are the main vertices of a clique, the stem of the clique can be evolved from an LD clique lying on a side of a rectangle called a *stem rectangle* of the clique. This stem rectangle is the minimum rectangle connecting the projection of M_p on M_q, if the projection of M_p on M_q is empty (Figure 3.17b); otherwise, it is a rectangle whose two sides occupy the projections of M_p on M_q and M_q on M_p (Figure 3.17a). According to the moving domains of the two main vertices, the stem rectangle, and thus the LD clique, can be determined. Given two main vertices and the edge sets that will be adjacent to the clique being found, whether the main vertices and those edge sets can be connected by a clique can be determined.

The location of the LD clique of C'_m is determined by the number, $(m-1)/2$, of degenerated edge sets in C'_m, and by the projection of M_p on M_q, where C'_m is the imagined clique of C_m being found, and M_p and M_q are moving domains of main vertices p and q of C_m. If $(m-1)/2$ is even, then the LD clique of C'_m lies on a side of the stem rectangle of C_m, and it is perpendicular to an edge occupying M_p or M_q. Otherwise, those edge sets that will be adjacent to C_m are separated into two sets of $(m-3)/4$ and $(m+1)/4$ edge sets, respectively, by the stem rectangle. To avoid redundant length, the side of the stem rectangle that is close to the set of $(m+1)/4$ edge sets is selected to locate the LD clique in C'_m. In Figure 3.17, the LD clique lies on the upper side of the stem rectangle. Algorithms for finding cliques are given in Section 3.7.

A clique evolved from one LD clique and its adjacent degenerated edge sets is called a *single clique*. Similarly, a clique evolved from a set of adjacent LD cliques and their adjacent degenerated edge sets is a *compound clique*, which can be found by maximizing the moving domains of Steiner points of the single cliques evolved from adjacent LD cliques and their degenerated edge sets. Figure 3.18 shows two cliques evolved from adjacent LD cliques and their adjacent degenerated edge sets. Maximizing the moving domain of each Steiner point in the edge-set tree in Figure 3.18 yields the compound clique in Figure 3.19.

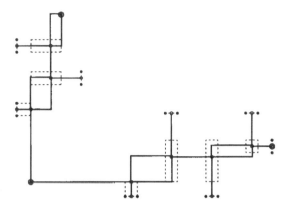

Figure 3.18 Two adjacent cliques evolved from two imagined cliques

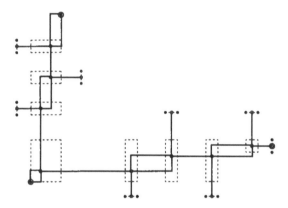

Figure 3.19 A seventeen-edge-set clique evolved from two adjacent cliques

Clique Lemma 2 *If a single clique, C_i, is evolved from an imagined clique and can be generated by removing an edge set from another single clique, C_m ($m = i + 2 > 3$), then C_m must be evolved from an imagined clique.*

Proof: Let $E_{r's'} \in C_i$ be a descendant edge set of $E_{rs} \in C_m$; let C_i be generated by removing E_{rs} from C_m; and let p and q be main vertices of C_i. C_m is determined by where $E_{r's'}$ is located in C_i.

(1) $E_{r's} \in C_m$ becomes a parallel moving edge set, when $E_{r's'}$ is

    (a) an edge set connected to an edge set not in the stem of C_i (Figure 3.20a); or

    (b) an edge set in the stem of C_i, where $\{r', s'\} \cap \{p, q\} = \emptyset$ (Figure 3.20b); or

    (c) an edge set in the stem of C_i, where $s' \in \{p, q\}$ and there is more than one point in the projection of E_{rs} on an edge set with $r' \notin \{p, q\}$ as the vertex (Figure 3.20c).

Thus, E_{rs} is an edge set of a degenerated clique, rather than an edge set of C_m, and s is an end vertex of a clique with i or less than i edge sets. This conflicts with the given condition.

(2) If $E_{r's'}$ is an edge set in the stem of C_i, where $s' \in \{p, q\}$ and $\{r', s'\} \cap \{r, s\} = \emptyset$, and if the projection of E_{rs} on any edge set with $r' \notin \{p, q\}$ as vertex occupies only one point (Figure 3.20d), C_m is a single clique and is also evolved from an imagined clique whose main vertices are also p and q. According to (1) and (2), C_m must be a single clique evolved from an imagined clique, if both C_m and C_i are single cliques and $m = i + 2$. □

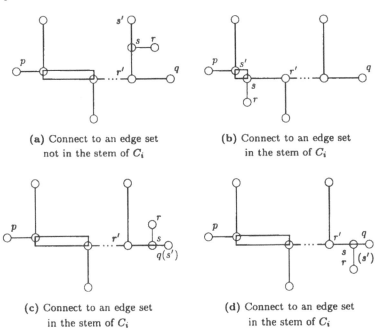

(a) Connect to an edge set
not in the stem of C_i

(b) Connect to an edge set
in the stem of C_i

(c) Connect to an edge set
in the stem of C_i

(d) Connect to an edge set
in the stem of C_i

Figure 3.20 C_m and C_i $(m = i + 2)$

Clique Theorem 6 *Every single clique can be evolved from an imagined clique.*

Proof: The induction method is applied here. According to Clique Theorem 3, only those cliques with an odd number of edge sets are considered.

(1) C_3 is evolved from an imagined clique including one degenerated Steiner point, and can be generated by connecting one edge set to another.

(2) A single clique, C_5, is evolved from an imagined clique including two degenerated Steiner points, and can be generated by connecting an edge set to a C_3 under condition (2) in the proof of Clique Lemma 2.

(3) Assume C_i is an arbitrary single clique evolved from an imagined clique including $(i-1)/2$ degenerated Steiner points, and single clique C_m is generated by connecting an edge set to C_i. According to Clique Lemma 2, C_m must be a single clique evolved from an imagined clique, because C_i is a single clique and can be generated by removing an edge set from C_m.   □

Like the analogous terms for a single clique, main vertices, imagined NN cliques, stems and stem rectangles for an NN clique can be defined. According to Clique Theorem 4, Clique Lemma 2 and the proof of Clique Theorem 6, Clique Corollary 1 can be proven because the evolution of an NN clique is the same as that of a single clique.

Clique Corollary 1 *An arbitrary NN clique can be evolved from an edge-set subtree formed by an LD clique and its adjacent degenerated edge sets.*   □

A clique is either imagined, single, or compound. Figure 3.21 expresses cliques and their evolutions. An inner vertex, r', in a clique may be a 1-D Steiner point that can be moved parallel to an edge occupying the moving domain of a main vertex of the clique. Thus, a clique may include other single cliques. Among NN cliques are two- or three-edge-set degenerated cliques in which each degenerated Steiner point can be moved to a node, because LD cliques are evolved to stems of NN cliques. Thus, an interface between NN cliques is also an NN clique, and an NR edge-set tree can be considered to be formed by NN cliques only.

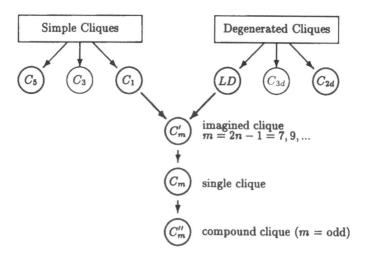

Figure 3.21 Cliques and their evolutions

3.5 EDGE SET AND CLIQUE REPLACEMENTS

To test a redundant length, an edge-set tree is broken into separate edge-set trees for subsets of nodes by removing edge sets. After testing, these separate edge-set trees must be connected into an edge-set tree by edge sets. The connecting is the negative operation of the breaking.

Example: The NR edge-set tree in Figure 3.22b can be formed by connecting p_1 and p_2 (Figure 3.22a), then generating Steiner point s and connecting p_3 to s, and finally discarding edges from $E_{p_1 p_2}$ that are disconnected from s. Removing $E_{p_1 s}$ from the edge-set tree in Figure 3.22b generates two edge-set trees for two subsets of nodes (Figure 3.22c). Thus, the NR edge-set tree in Figure 3.22b can also be generated by connecting p_1 to $E_{p_2 p_3}$.   ⋄

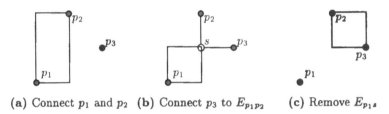

(a) Connect p_1 and p_2 (b) Connect p_3 to $E_{p_1 p_2}$ (c) Remove $E_{p_1 s}$

Figure 3.22 Generating an NR edge-set tree for three nodes

Descendant edge sets may exist in the edge-set trees generated by removing edge sets, and a Steiner point may be generated by connecting a new edge set to an edge-set tree. Let p_3 and $E_{p_1 p_2}$ be in two edge-set trees for two subsets of nodes. According to Property 7 and Clique Theorem 1, a *connecting rule* for forming an NR edge-set tree by connecting p_3 in an edge-set tree to $E_{p_1 p_2}$ in another edge-set tree includes:

(1) maximizing $E_{p_1 p_2}$ and computing $s = (x_s, y_s)$, where
    $x_s = mid(x_{p_1}, x_{p_2}, x_{p_3})$ and $y_s = mid(y_{p_1}, y_{p_2}, y_{p_3})$;
(2) breaking $E_{p_1 p_2}^{max}$ into $E_{p_1 s}$ and $E_{p_2 s}$;
(3) connecting p_3 and s as $E_{p_3 s}$, and maximizing their moving domains.

A *connecting operator*, "+", is defined as "is connected with." It maximizes $E_{p_1 p_2}$, determines Steiner point s, splits $E_{p_1 p_2}$ into $\{E_{p_1 s}, E_{p_2 s}\}$ by s, and maximizes the moving domains of s and its newly adjacent point, p_3. The operation for forming an NR edge-set tree by connecting p_3 and $E_{p_1 p_2}$ in two edge-set trees can be expressed as

$$
\begin{aligned}
E_{p_1 p_2} + p_3 &= \{E_{p_1 p_2}^{max}, p_3\} \bigcup E_{p_3 s} \\
&= \bigcup_{i=1}^{3} E_{p_i s}
\end{aligned}
\tag{3.3}
$$

If edge sets connecting edge-set trees including $E_{p_1 p_2}$ and p_3, respectively, are expressed implicitly, then "+" is used; otherwise, "$\cup$" is used. For convenience, we also use $\{E_{p_1 p_2}, p_3\} \cup E_{p_3 s}$, instead of $\{E_{p_1 p_2}^{max}, p_3\} \bigcup E_{p_3 s}$, to express that $E_{p_3 s}$ connects two edge-set trees including $E_{p_1 p_2}$ and p_3, respectively.

According to the proof of Clique Theorem 1, the moving domain of a dynamic Steiner point is determined by the movement of each edge set adjacent to this Steiner point. In Figure 3.23a, E_{sf}, E_{sg}, and E_{sh} are adjacent to Steiner point s. Because the projection of E_{sg} on $E_{ff'}$ is edge $e_{ff'}$, edge set E_{sf} can be moved upward until f overlaps f'. Similarly, E_{sg} can be moved rightward until g overlaps g'. Thus, M_s is a set of points enclosed by a rectangle, R_{ab}, where $a = (x_{g''}, y_{f''})$ and $b = (x_{g'}, y_{f'})$ (Figure 3.23b). Note that $x_g = x_s$ and $y_f = y_s$. If h is also a dynamic Steiner point, $x_s \leq x_h$ and $y_h \leq y_s$.

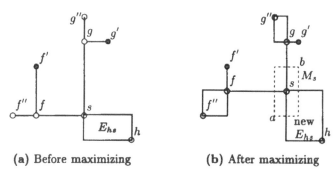

(a) Before maximizing   (b) After maximizing

Figure 3.23 Maximizing moving domains

To break an edge-set tree into edge-set trees for subsets of nodes, a *breaking rule* for removing one of three edge sets intersecting at a Steiner point includes:

(1) maximizing and removing the edge set;
(2) generating a descendant edge set;
(3) removing two edge sets overlapping the descendant edge sets.

Removing a redundant length of a descendant edge set is explained by Clique Theorem 7. A *breaking operator*, "$-$", is defined as "where to remove." Removing one of three edge sets intersecting at a Steiner point for generating two edge-set trees can be expressed as:

$$\bigcup_{i=1}^{3} E_{p_i s} - E_{p_3 s} = E_{p_1 s}^{min} \bigcup E_{p_2 s}^{min} \bigcup E_{p_3 s}^{max} - E_{p_3 s}^{max}$$

$$= \{E_{p_1 p_2}, p_3\} \qquad (3.4)$$

where $E_{p_3 s}$ is removed from an edge-set tree including $\cup_{i=1}^{3} E_{p_i s}$.

Example: Figure 3.24 shows a removal process. When $E_{p_2 s_2}$ is removed from an edge-set tree in Figure 3.24a, two edge-set trees are generated for two subsets of nodes (Figure 3.24b). Because the descendant edge set, $E_{s_1 s_3}$, of $E_{p_2 s_2}$ is a parallel moving edge set, possible redundant length can be found by moving it. Maximizing $E_{s_1 s_3}$ (Figure 3.24c) and removing the redundant length, $l_{E_{s_1 s_3}^{max}} - l_{E_{p_5 p_6}}$, of $E_{s_1 s_3}$, generates new edge-set trees for two subsets of nodes (Figure 3.24d).   ◇

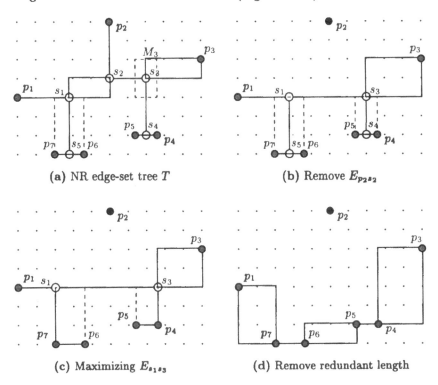

Figure 3.24 Removing an edge set from an edge-set tree

Applying the connecting and breaking rules on each edge set adjacent to an end vertex of an m-edge-set subtree (m is odd), a set of $(m+3)/2$ edge-set trees can be connected by this edge-set subtree, and an arbitrary m-edge-set subtree can be removed from an edge-set tree. If two of three edge sets intersecting at a Steiner point are removed, the third edge-set cannot exist. Removing two edge sets intersecting at a Steiner point is covered by removing a clique or an edge-set subtree with an odd number of edge sets. Removing a set of edge sets to generate edge-set trees for i subsets of nodes and connecting the edge-set trees for i subsets of nodes with another set of edge sets is also called *replacing* a set of edge sets with another set, where "a set of edge sets" refers to:

- *an edge-set subtree with $2i - 3$ edge sets; or*
- *$i - 1$ edge sets without common Steiner points; or*
- *i_0 edge sets without common Steiner points, and edge-set subtrees with $i_1, ..., i_k$ edge sets, where $i = (i_0 + 1) + \sum_{j=1}^{k}(i_j + 3)/2$ and i_j is odd.*

Therefore, an edge-set subtree in an NN clique can be considered as a single clique in the replacement. If a clique, C_i, in an NR edge-set tree can be replaced by a set of edge sets whose total length is less than l_{C_i}, then a redundant length exists in C_i. Similarly, a redundant length exists in a set of edge sets if it can be replaced by a clique whose length is less than the total length of the removal edge sets.

If no i-edge-set NN clique in edge-set tree T can be replaced by a set of edge sets whose total length is less than the replaced clique, and if no edge sets in T can be replaced by an i-edge-set NN clique whose length is less than the total length of the replaced edge sets, then T is defined as an *edge-set tree tested by i-edge-set NN cliques*, denoted as $T(C_{in})$. An edge-set tree tested in turn by $C_{1n}, C_{3n}, ...,$ and C_{mn} is denoted as $T(C_{1n}, ..., C_{mn})$. Substituting C_{in} by an i-edge-set subtree in an NN clique defines an *edge-set tree tested by cliques $C_1, C_3, ...,$ and C_m*, denoted as $T(C_1, ..., C_m)$. According to the definition of an NR edge-set tree, $T(C_1) \equiv T_{NR}$. A $T(C_1, ..., C_m)$ must be a $T(C_{1n}, ..., C_{mn})$ because $C_i \subset C_{in}$ $(i = 1, ..., m)$.

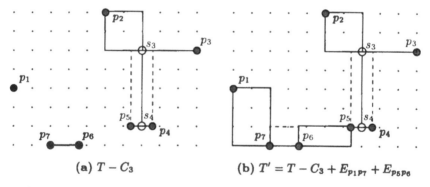

(a) $T - C_3$ (b) $T' = T - C_3 + E_{p_1 p_7} + E_{p_5 p_6}$

Figure 3.25 Replacing $C_3 = E_{p_1 s_1} \cup E_{s_1 s_2} \cup E_{s_1 s_5} \subset T$ by $\{E_{p_1 p_7}, E_{p_5 p_6}\}$

Example: Figure 3.24a shows a three-edge-set clique, $C_3 = E_{p_1 s_1} \cup E_{s_1 s_2} \cup E_{s_1 s_5}$, in edge-set tree T. Figure 3.25 shows that C_3 is replaced by a pair of edge sets, $\{E_{p_1 p_7}, E_{p_5 p_6}\}$, in which no common Steiner points exist between $E_{p_1 p_7}$ and $E_{p_5 p_6}$. After the replacement, a new edge-set tree, $T' = (T - C_3) \cup E_{p_1 p_7} \cup E_{p_5 p_6}$, is generated. Because $l_{E_{p_1 p_7}} + l_{E_{p_5 p_6}} < l_{C_3}$, redundant length exists in T. Further testing indicates T' is an edge-set tree tested by $C_1, C_3, ...,$ and C_m, where $m = 11$. However, no cliques with more then seven edge sets can be found in this example.   ◇

Clique Theorem 7 *If a redundant length exists in descendant edge sets generated by replacing a clique in an NN clique of edge-set tree T with a set of edge sets, then the redundant length can be eliminated in a new edge-set tree generated by replacing the NN clique with another set of edge sets in T.*

Proof: Let a set of descendant edge sets, $E_d' \subset T'$, be generated by replacing $C_i \subset C_{mn} \subset T$ with $E' \subset T'$. Because $C_i \subset C_{mn}$ and no descendant edge sets are generated by removing an NN clique, $C_{mn} \supset E_d' \cup C_i$. To remove a redundant length of an edge set in E_d', E_d' must be replaced by a set of edge sets, $E'' \subset T''$, in generating T'' from T'. Thus, T'' is generated by replacing an edge-set subtree, $E_d' \cup C_i \subset C_{mn} \subset T$, with a set of edge sets, $\{E', E''\}$. That is, $T'' = [T - (E_d' \cup C_i)] \cup E' \cup E''$. Therefore, T'' can be generated by replacing $C_{mn} \subset T$ with $\{E', E'', C_{mn} - (E_d' \cup C_i)\} \subset T''$, and a redundant length of an arbitrary edge set in $E_d' \subset T'$ can be eliminated. □

According to Clique Theorem 7, the redundant length of descendant edge sets generated by removing edge-set subtrees can be ignored in finding a $T(C_{1n}, ..., C_{mn})$, because the possible redundant length of descendant edge sets can be removed by replacing an NN clique with a set of edge sets.

Example: If the redundant length of the descendant edge set of $E_{p_2 s_2}$ in Figure 3.24c is ignored, this redundant length can be eliminated in generating edge-set tree T' by replacing $C_{7n} \subset T$ with $E' \subset T'$, where $C_{7n} = E_{p_1 s_1} \cup E_{s_1 s_5} \cup E_{s_1 s_2} \cup E_{p_2 s_2} \cup E_{s_2 s_3} \cup E_{s_3 s_4} \cup E_{p_3 s_3}$ (Figure 3.24a), and $E' = \{E_{p_1 p_7}, E_{p_5 p_6}, C_3 = E_{p_2 s_3} \cup E_{p_3 s_3} \cup E_{s_3 s_4}\}$ (Figure 3.25b). ◇

Clique Theorem 8 *If the length of an edge-set tree, T, cannot be reduced by replacing an edge set in T with another edge set, then the length of T cannot be reduced by replacing a pair of edge sets in T with another pair of edge sets.*

Proof: Let l_X be the length of X. Assume that $T' = (T - E) \cup E'$ is an edge-set tree, where $l_{T'} < l_T$, $E = \{E_{p_1 q_1}, E_{p_2 q_2}\} \subset T$ and $E' = \{E_{p_1' q_1'}, E_{p_2' q_2'}\} \subset T'$. Because $T - E$ and $T' - E'$ include the same set of edge-set trees for three subsets of nodes, $E_1 \in E$ must exist in an edge-set cycle in $T \cup E_1'$, and $E_2 = E - E_1$ must exist in an edge-set cycle in $T \cup E_2'$, where $E_1' \in E'$ and $E_2' = E' - E_1'$. According to the given condition, $l_{E_1} \le l_{E_1'}$ and $l_{E_2} \le l_{E_2'}$. Thus, $l_T - l_{T'} = \sum_{i=1}^2 l_{E_i} - \sum_{i=1}^2 l_{E_i'} \le 0$. Therefore, $l_{T'} < l_T$ is impossible. □

The replacement in the proof can be extended to two sets of i edge sets. Therefore, replacing two sets of i edge sets is no longer considered.

3.6 THE GENERAL SOLUTION

The *general solution* of the rectilinear Steiner's problem consists of a set of NR edge-set trees containing all MRSTs for a set of given nodes. Based on the clique theorems described in the previous sections, the relationship between the general solution and edge-set trees tested by a series of cliques can be explored. In this section, l_X indicates the length of graph X.

Clique Theorem 9 *An edge-set tree tested by $C_{1n}, C_{3n}, ...,$ and C_{mn} can be considered to be formed by NN cliques only.*
Proof: According to Clique Theorem 3, $T = T(C_{1n}, ..., C_{mn})$ is an edge-set tree tested by all possible NN cliques because there are $(m + 3)/2$ nodes. According to Clique Theorem 7, T is an NR edge-set tree. According to Clique Theorems 2 and 4, Clique Lemma 2 and Clique Corollary 1, an arbitrary i-edge-set degenerated clique for $i > 3$ is, or is evolved to, an edge-set subtree in an NN clique. Thus, among NN cliques are two- or three-edge-set degenerated cliques, in which each degenerated Steiner point is also a main vertex of an NN clique and can be moved to a node. Therefore, each two- or three-edge-set degenerated clique is also a one-edge-set NN clique, C_{1n}. That is, T is formed by NN cliques only. □

Clique Lemma 3 *The total length of a set, E, of edge sets in an NN clique of an edge-set tree tested by $C_{1n}, C_{3n}, ...,$ and C_{mn} for $(m + 3)/2$ nodes is equal to or less than the total length of edge sets that can replace all the edge-sets in E.*
Proof: According to Clique Theorem 9, $T = T(C_{1n}, ..., C_{mn})$ is considered to be formed by NN cliques only. Let $T' = (T - E) \cup E'$ be a connected graph, where $E \subset C_{jn} \subset T$ and $S' = (C_{jn} - E) \subset T'$. Thus, $T' = (T - C_{jn}) \cup E' \cup S'$ is also a connected graph. According to the definition of an edge-set tree tested by $C_{1n}, C_{3n}, ...,$ and C_{mn}, $l_{C_{jn}} \leq l_{E'} + l_{S'}$. Because $l_{C_{jn}} = l_{S'} + l_E$, $l_E \leq l_{E'}$. □

Clique Lemma 4 *If edge-set tree T' is generated by replacing a set, E, of edge sets in an edge-set tree, T, tested by $C_{1n}, C_{3n}, ...,$ and C_{mn} for $(m + 3)/2$ nodes with a set, E', of edge sets in an NN clique of T', then the total length of all edge sets in E is equal to or less than the total length of all edge sets in E'.*
Proof: According to Clique Theorem 9, $T = T(C_{1n}, ..., C_{mn})$ is considered to be formed by NN cliques only. Let $T' = (T - E) \cup E'$, where $E' \subset C'_{jn} \subset T'$ and $S = C'_{jn} - E' \subset T$. Thus, $T' = (T - S - E) \cup C'_{jn}$ is also a connected graph. According to the definition of an edge-set tree tested by $C_{1n}, C_{3n}, ...,$ and C_{mn}, $l_S + l_E \leq l_{C'_{jn}}$. Because $l_{C'_{jn}} = l_S + l_{E'}$, $l_E \leq l_{E'}$. □

Clique Theorem 10 *An edge-set tree tested by $C_{1n}, C_{3n}, ...,$ and C_{mn} contains a set of MRSTs for $(m+3)/2$ nodes; an edge-set tree containing a set of MRSTs for $(m + 3)/2$ nodes must be an edge-set tree tested by $C_{1n}, C_{3n}, ...,$ and C_{mn}.*

Proof: Let T be an edge-set tree tested by $C_{1n}, C_{3n}, ...,$ *and* C_{mn} and T' be an arbitrary NR edge-set tree for the same set of $(m+3)/2$ nodes. According to Clique Theorem 9, both T and T' are formed by NN cliques only.

Further let $E \subset T$ and $E' \subset T'$ be two sets of edge sets, where $E \cap E' = \emptyset$. Because each edge set in both T and T' must be included in an NN clique, E can be divided into two subsets: $E_1 = \{S_{1i} \mid i = 1, ..., |E_1|\}$ and $E_2 = \{S_{2i} \mid i = 1, ..., |E_2|\}$, and E' can be divided into two subsets: $E_1' = \{S_{1i}' \mid i = 1, ..., |E_1|\}$ and $E_2' = \{S_{2i}' \mid i = 1, ..., |E_2|\}$, where S_{1i} is a set of edge sets in an NN clique and can be replaced with a set of edge sets, S_{1i}'; and S_{2i} is a set of edge sets that can be replaced with a set of edge sets, S_{2i}', in an NN clique.

According to Clique Lemma 3, $l_{S_{1i}} \leq l_{S_{1i}'}$, where $i = 1, ..., |E_1|$. According to Clique Lemma 4, $l_{S_{2i}} \leq l_{S_{2i}'}$, where $i = 1, ..., |E_2|$. Therefore, $l_T - l_{T'} = (\sum_{i=1}^{|E_1|} l_{S_{1i}} + \sum_{i=1}^{|E_2|} l_{S_{2i}}) - (\sum_{i=1}^{|E_1|} l_{S_{1i}'} + \sum_{i=1}^{|E_2|} l_{S_{2i}'}) = \sum_{i=1}^{|E_1|} (l_{S_{1i}} - l_{S_{1i}'}) + \sum_{i=1}^{|E_2|} (l_{S_{2i}} - l_{S_{2i}'}) \leq 0$. This means that T has the minimum length of all edge-set trees for the same set of nodes. Thus, T contains a set of MRSTs.

If $l_T = l_{T'}$, then $l_{S_{1i}} = l_{S_{1i}'}$ and $l_{S_{2j}} = l_{S_{2j}'}$, where $i = 1, ..., |E_1|$, and $j = 1, ..., |E_2|$. Therefore, T' must also be an edge-set tree tested by $C_{1n}, C_{3n}, ...,$ and C_{mn} for $(m+3)/2$ nodes; otherwise, trees contained in T' are not MRSTs, proving Clique Theorem 10. □

To find an i-edge-set NN clique to replace a set of edge sets in an edge-set tree, T, for n nodes, T must be broken into edge-set trees for $(i+3)/2$ subsets of nodes, where $i = 1, 3, 5, ..., 2n - 3$. The total number of these sets is about 2^n. Taking other computations into account, we cannot find a polynomial algorithm for the solution (see Section 3.7). Although the algorithm for the exact solution is computationally inefficient, an efficient algorithm for an accuracy-estimable solution can be found from this inefficient algorithm (see Section 3.8).

Because an i-edge-set subtree used to test an edge-set tree is an i-edge-set NN clique or an edge-set subtree in an i'-edge-set NN clique $(i < i' \leq m)$, an edge-set tree tested by cliques $C_1, C_3, ...,$ *and* C_m is also an edge-set tree tested by cliques $C_{1n}, C_{3n}, ...,$ *and* C_{mn}. According to Clique Theorem 10 and its proof, Clique Corollaries 2 and 3 can be proven.

Clique Corollary 2 *An edge-set tree contains a set of MRSTs for n*

nodes if and only if it is tested by cliques C_1, C_3, *...,* *and* C_m, *where* $m = 2n - 3$.

Clique Corollary 3 *The general solution for* $(m + 3)/2$ *nodes consists of all edge-set trees tested by* $C_{1n}, C_{3n}, ...,$ *and* C_{mn} *that are generated from the same edge-set tree,* T, *tested by* $C_{1n}, C_{3n}, ...,$ *and* C_{mn} *by replacing NN cliques in* T *with edge sets, or by replacing edge sets in* T *with NN cliques.*

The general solution is independent of an initial edge-set tree, but the first edge-set tree containing a set of MRSTs depends on an initial edge-set tree. According to the proof of Clique Theorem 10, the general solution generated from an edge-set tree, T, containing a set of MRSTs consists of the following edge-set trees:

(1) a set of edge-set trees, $S_1 = \{S_{1i}|\ i = 1, ..., |S_1|\}$, where $S_{1i} = \{T_{1ij}|\ j = 1, ..., |S_{1i}|\}$, and T_{1ij} is an edge-set tree for n nodes generated by a replacement, R_{1ij}, replacing each set of edge-set subtrees in the ith NN clique of T with the jth set of edge sets whose total length equals the total length of the replaced edge sets in the ith NN clique;

(2) a set of edge-set trees, $S_2 = \{S_{2i}|\ i = 1, ..., |S_2|\}$, where $S_{2i} = \{T_{2ij}|\ j = 1, ..., |S_{2i}|\}$, and T_{2ij} is an edge-set tree for n nodes generated by a replacement, R_{2ij}, replacing the ith set of edge sets in T with each set of edge-set subtrees in the jth NN clique whose total length equals the total length of the replaced edge sets;

(3) a set of edge-set trees, S_3, including all edge-set trees generated by all combinations of $R_{1i1}, R_{1i2}, ..., R_{1i|S_{1i}|}$, where $i = 1, ..., |S_1|$;

(4) a set of edge-set trees, S_4, including all edge-set trees generated by all combinations of $R_{2i1}, R_{2i2}, ..., R_{2i|S_{2i}|}$, where $i = 1, ..., |S_2|$, and the replaced edge sets in T for different replacements are different;

(5) a set of edge-set trees, S_5, including all edge-set trees generated by all combinations taking a subset of replacements from $\{R_{1ij}|\ i = 1, 2, ..., |S_1|; j = 1, 2, ..., |S_{1i}|\} \cup \{R_{2ij}|\ i = 1, 2, ..., |S_2|; j = 1, 2, ..., |S_{2i}|\}$, where the replaced edge sets used in different replacements of the same combination are different.

Example: Figure 3.26 shows the general solution, consisting of T_1, T_2, and T_3 for a node set, $\{p_i|\ i = 1, ..., 7\}$. Any couple of edge-set trees can be generated by replacing NN cliques in the third edge-set tree with a set of new edge sets, or by replacing edge sets in the third edge-set tree with NN cliques.

The relations among T_1, T_2 and T_3 can be expressed as $T_2 = (T_1 - C_{3n}) \cup C'_{3n}$, and $T_3 = (T_1 - E) \cup C''_{7n}$, where $C_{3n} = E_{p_2 s_3} \cup E_{s_3 s_4} \cup E_{p_3 s_3} \subset T_1$, $C'_{3n} = E_{p_2 s'_3} \cup E_{s'_3 s'_4} \cup E_{p_3 s'_3} \subset T_2$, $E = \{E_{s_2 s_4}, E_{p_1 s_1}, C_{3n}\} \subset T_1$, and $C_{7n}'' = E_{p_1 s_4''} \cup E_{s_3'' s_4''} \cup \bigcup_{i=2}^{3} E_{p_i s_3''} \cup \cup E_{s_4'' s_5''} \cup \bigcup_{i=5}^{6} E_{p_i s_5''} \subset T_3$. These relations indicate that T_2 can be generated by replacing $C_{3n} \subset T_1$ with C_{3n}', and that T_3 can be generated by replacing $E \subset T_1$ with C_{7n}''. Because $C_{3n} \subset E$, replacing both C_{3n} and E with C_{3n}' and C_{7n}'' cannot generate an edge-set tree containing a set of MRSTs. ◇

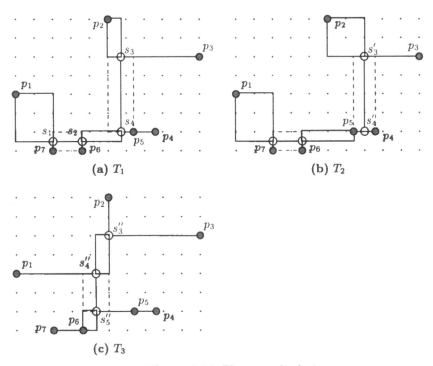

(a) T_1

(b) T_2

(c) T_3

Figure 3.26 The general solution

According to Clique Theorem 7, finding an edge-set tree tested by C_1, C_3, ..., and C_m consumes more time than finding an edge-set tree tested by C_{1n}, C_{3n}, ..., and C_{mn}. According to Clique Lemmas 3 and 4, using Clique Corollary 2 will increase the number of duplicate results in finding an edge-set tree containing a set of MRSTs. However, testing a complex NN clique may be substituted for by testing simpler single cliques. In the general solution shown in Figure 3.26, the relationship between T_1 and T_3 can be expressed as $T_3 = (T_1 - E_{p_1 s_1} - E_{s_3 s_4}) \cup C_3''$, where $C_3'' = E_{p_1 s_4''} \cup E_{s_3'' s_4''} \cup E_{s_4'' s_5''} \subset C_7''$, which indicates that T_2 can be generated by replacing $\{E_{p_1 s_1}, E_{s_3 s_4}\} \subset E$ with C_3''. Finding a seven-edge-set clique is much more time consuming than finding a three-edge-set clique (see Section 3.7). This feature can be used for finding efficient algorithm expansions for inefficient P and NP problems (see Section 3.8).

3.7 ALGORITHMS FOR THE GENERAL SOLUTION

According to Clique Theorem 7 or Clique Corollary 2, a set of MRSTs for a set, V_n, of n nodes is contained in an edge-set tree tested by $C_{1n}, C_{3n}, ...,$ and C_{mn}, or tested by $C_1, C_3, ...,$ and C_m, where $m = 2n-3$. According to Clique Corollary 3, replacing edge sets with NN cliques, or cliques from the same edge-set tree containing a set of MRSTs, generates the general solution for the rectilinear Steiner's problem. Therefore, finding the general solution involves two major steps:

(1) finding edge-set tree $T(C_{1n}, ..., C_{mn})$, or $T(C_1, ..., C_m)$; and
(2) finding the general solution from $T(C_{1n}, ..., C_{mn})$, or $T(C_1, ..., C_m)$.

To implement these steps, it is critical to find an i-edge-set clique, C_i $(i = 3, ..., m)$, connecting $(i+3)/2$ edge-set subtrees for replacing a clique with a set of edge sets or replacing a set of edge sets with a clique. Except for the moving domain of an end vertex, finding a C_i or an i-edge-set subtree in an NN clique is the same as finding a C_{in}. In the following algorithms, C_i is used primarily.

3.7.1 FINDING AN NR EDGE-SET TREE

Applying the connecting and breaking rules, an algorithm generating an NR edge-set tree, T_{NR}, can be described as follows:

PROCEDURE *FIND_T_{NR}(input: n nodes; output: T_{NR})*
{

    (1) Initiate a set of edge sets to form an edge-set tree,
        $T = \cup_{i=1}^{n_e} E_{p_i q_i}$;

    (2) For $i = 1$ to n_e, do (a) to (c):

        (a) Remove edge set $E_{p_i q_i}$ from T to generate edge-set trees T_1 and T_2;

        (b) Maximize the length of a moving bent-edge set and the area swept by a moving straight-edge set in each pair of compared edge sets in T_1 and T_2; find an edge set, $E_{p_a q_a} \in T_1$, nearest to an edge set, $E_{p_b q_b} \in T_2$;

        (c) If the distance between $E_{p_a q_a}$ (or $R_{p_a q_a}$) and $E_{p_b q_b}$ (or $R_{p_b q_b}$) is less than $l_{E_{p_i q_i}}$, then generate a bent-edge set or a straight-edge set, $E_{p_i' q_i'}$, in terms of the projection of $E_{p_a q_a}$ (or $R_{p_a q_a}$) on $E_{p_b q_b}$ (or $R_{p_b q_b}$); connect T_1 and T_2 by $E_{p_i' q_i'}$. Otherwise, re-connect $E_{p_i q_i}$;

    (3) If no removal edge set is replaced by a new edge set and no moving domain is expanded, then stop; else go to (2).

}

An initial edge-set tree for step (1) can be arbitrary. The time complexity of this algorithm is $O(n^3)$. A more efficient algorithm is presented in Section 3.9.

3.7.2 FINDING A THREE-EDGE-SET CLIQUE

According to Clique Theorem 1, three end vertices of a three-edge-set clique, C_3, lie on the perimeter of a rectangle. Thus, finding C_3 can be substituted for by finding a rectangle that connects edge-set trees for three subsets of nodes. Let edge-set trees for three subsets of nodes be included in $\{T_i = \cup_{j=1}^{n_{e_i}} E_{p_j^{(i)} q_j^{(i)}} \mid i = 1, 2, 3\}$, where there are n_{e_i} edge sets in T_i; and let S_R be a record holding a rectangle connecting three edge sets. An algorithm for finding a three-edge-set clique can be expressed as follows:

PROCEDURE *FIND_C_3(input: T_1, T_2, T_3; output: C_3)*
{

    (a) Clear S_R, and let clique_length = preset_max_number;

    (b) For $i = 1$ to 3 and $j_i = 1$ to n_{e_i}, do (b1) and (b2):

        (b1) Find the minimum rectangle, R, that connects three edge sets in $S_E = \{E_{p_{j_i}^{(i)} q_{j_i}^{(i)}} \mid i = 1, 2, 3\}$, where $\{p_{j_i}^{(i)}, q_{j_i}^{(i)}\} \cap V_n \neq \emptyset$ ($p_{j_i}^{(i)}$ or $q_{j_i}^{(i)}$ is a node);

        (b2) If R is found and $l_R <$clique_length, $S_R = \{R, S_E\}$ and clique_length $= l_R/2$, where l_R is the perimeter of R;

    (c) According to the overlaps of R and S_E, determine the moving domains of three end vertices and the inner Steiner point for C_3.

}

If $S_R = \emptyset$, T_1, T_2 and T_3 are connected by two edge sets in finding $T(C_1, ..., C_m)$. Similarly, a five-edge-set clique connecting four edge-set trees can be determined by finding the minimum rectangle connecting four edge-set trees. According to the moving domain of an end vertex of a clique, a vertex of an edge set directly connected by an NN clique must be a node. Therefore, the time complexity of the algorithm is $O((n/3)^3)$ for finding C_{3n}, or $O((n/4)^4)$ for finding C_{5n}. Finding a C_i is more time consuming than finding a C_{in}.

3.7.3 FINDING AN m-EDGE-SET CLIQUE OR A SET OF EDGE SETS CONNECTING $(m + 3)/2$ EDGE-SET TREES

According to clique evolution and Clique Theorems 4 and 5, finding an m-edge-set clique, C_m, connecting edge-set trees for $(m+3)/2$ subsets

of nodes requires finding a line connecting the main vertices (nodes) for
the stem of the imagined clique, and finding projections on the line from
edge sets that will be adjacent to C_m (Figure 3.27a). The stem and
its adjacent edge sets in C_m can be determined by the line and the
projections (Figure 3.27b). This line is a side of the stem rectangle,
which is either the minimum rectangle connecting two edge sets (Figure
3.27a), or a rectangle whose opposite sides are the projections of two
edge sets on each other.

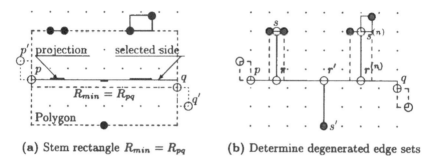

(a) Stem rectangle $R_{min} = R_{pq}$ (b) Determine degenerated edge sets

Figure 3.27 Finding a clique

Let S_P be a record for holding $(m + 3)/2$ edge sets connected by
a polygon, P (Figure 3.27a). The number of elements in a set, S_P,
is expressed as $|S_P| = (m + 3)/2$. According to the relations among
main vertices, stem rectangle, and edge sets whose projections on the
stem rectangle are not empty, an algorithm, FIND_CLIQUE(m, S_P, C_m),
finding an m-edge-set clique connecting edge sets in S_P, is expressed as
follows:

PROCEDURE *FIND_CLIQUE(input: m, S_P; output: C_m)*
{

    (a) *Find edge sets $E_{pp'}$ on $E_{qq'}$ from S_P that have the maximum
    horizontal (vertical) distance, and separate other edge sets in
    S_P into two subsets, S_{E1} and S_{E2}, where $||S_{E1}| - |S_{E2}|| = 0$
    or 1;*

    (b) *If the projection of $E_{pp'}$ on $E_{qq'}$ is empty, then generate the
    minimum rectangle, R_{min}, connecting $E_{pp'}$ and $E_{qq'}$; other-
    wise, generate a rectangle, R_{min}, whose opposite sides exactly
    occupy the projections of $E_{pp'}$ and $E_{qq'}$ on each other;*

    (c) *Select a side closest to S_{E2} from R_{min} if $|S_{E1}| \leq |S_{E2}|$; other-
    wise select a side closest to S_{E1};*

    (d) *Find projections of edge sets in S_{E1} and S_{E2} on the selected
    side of R_{min};*

(e) *If the two closest projections on the selected side are projected by edge sets in the same subset, S_{E1} or S_{E2}, of edge sets, or if the two closest projections on the selected side overlap, go to (g);*

(f) *Generate a clique, C_m, by connecting edge sets in S_{E1} and S_{E2} to the selected line by a set of straight-edge sets, and by maximizing moving domains of the end vertices of the straight-edge sets (Figure 3.27b), if the minimum length of any straight-edge set is not less than the length of its adjacent projection;*

(g) $C_m = \emptyset$.

}

Let $S_T = \{T_i = \cup_{j_i=1}^{n_{e_i}} E_{p_{j_i}^{(i)} q_{j_i}^{(i)}} \mid i = 1, ..., n_T\}$, where $n_T = (m+3)/2$ and there are n_{e_i} edge sets in T_i; let S be a set of edge sets whose total length is the minimum for connecting n_T edge set trees; and let $S'_{P_i} \in S_{P_i}$ be a record holding a set of i edge sets in i edge-set trees in S_T, where edge sets are adjacent to, but not enclosed by, a polygon. Based on *FIND_CLIQUE*, finding a set of edge sets connecting n_T edge-set trees is expressed as follows:

PROCEDURE *FIND_A_REPLACEMENT_SET(input: S_T, n_T;*

output: S)

{

(a) *For $i = 2$ to n_T, find a set, S_{p_i}, of all subsets of i edge sets in i edge-set trees in S_T;*

(b) *For $i = 2$ to n_T, $FIND\_CLIQUE(m, S'_{p_i}, C_m)$ from each $S'_{p_i} \in S_{p_i}$, where $m = 2i - 3$;*

(c) *Select a set, S, of edge sets and cliques connecting n_T edge-set trees with the minimum length.*

}

The time complexity of this algorithm is $O((n/n_T)^{n_T})$ when an m-edge-set subtree in S is a C_{mn} only, where $m = 2n_T - 3$.

3.7.4 FINDING SETS OF EDGE-SET TREES FOR n_T SUBSETS OF NODES

Finding a set of edge sets, E, that can replace a clique, C_m, where $l_E \leq l_{C_m}$, requires identifying all sets of edge-set trees for n_T subsets of nodes. Let t_T be a binary tree whose nodes store the nodes and Steiner points of edge-set tree T; let n_T be the number of edge-set trees generated by removing a set of edge sets from T; and let r_t be the root of a tree, t_r, where a node, f, stores a removal edge set from T. *Traversing* a tree means moving a pointer through each node exactly once starting from and terminating at the root for processing data in each node. Letting $f = r_t$ be a node of a binary tree and $t'_T = \emptyset$ be another binary tree for recursion, a procedure building a tree, t_r, that holds removal edge sets for breaking T into n_T edge-set trees, is expressed as follows:

PROCEDURE *BREAK_TREE(input: t_T, t'_T, n_T, f; output: t_r)*
{
    Traversing t_T and t'_T from their roots, do (a) to (d):
    (a) Get a current edge set, E, from t_T or t'_T;
    (b) Generate t_1 and t_2 for two subsets of nodes by removing E from t_T or t'_T;
    (c) Link E to its parent node, f;
    (d) If $n_T > 1$, then $t_T = t_1$; $t'_T = t_2$; $n_T = n_T - 1$; $f = E$; BREAK_TREE(t_T, t'_T, n_T, f).
}

The time complexity of this algorithm is $O(n^2)$. The algorithm generates an n_T-level tree, t_r, whose root, $r_t = \emptyset$, is level 0. A set of $n_T - 1$ removal edge sets consists of one edge set in each level. A set of edge-set trees for n_T subsets of nodes is generated by removing a set of removal edge sets starting from T (see Section 3.7.7).

Considering the maximum number of edge sets in T is $2n - 3$ and each edge-set tree is a node after running the $(n-1)$th recursion in *BREAK_TREE*, there are $(2n-3)/(n-1) \approx 2$ edge sets reduced in each breaking. Using $2n - 2$ as the maximum number of edge sets, the number of sets of removal edge sets stored in an n_T-level t_r is $\prod_{i=1}^{n_T-1}(2n - 2 - 2i)/(n_T - 1)! = 2^{n_T-1} \prod_{i=1}^{n_T-1}(n - 1 - i)/(n_T - 1)! = 2^{n_T-1}C(n - 1, n_T - 1)$.

3.7.5 FINDING AN EDGE-SET TREE TESTED BY C_m

A set of edge sets connecting each set of edge-set trees in t_r, generated by *BREAK_TREE*, can be found by *FIND_A_REPLACEMENT_SET*. An edge-set tree tested by C_m can be found by finding a set of edge sets whose length is the minimum for connecting a set of edge-set trees.

PROCEDURE *FIND_T(C_m) (input: $T(C_1, ..., C_{m-1})$;*
                            output: $T(C_1, ..., C_m)$)
{
    (1) Set $S_T = T = T(C_1, ..., C_{m-1})$;
    generate t_T from T; max_length=0; $n_T = (m + 3)/2$;
    $f = r_t = \emptyset$; BREAK_TREE($t_T, t'_T, n_T, f; t_r$);

(2) *Traversing tree t_r,*
retrieve a set, $S_r = \{E_i|\ i = 1, ..., n_r\}$, of removal edge sets;

(3) *For $i = 1$ to n_r,*
remove edge set E_i from S_T: $S_T = S_T - E_i$;

(4) *Using FIND_A_REPLACEMENT_SET(S_T, n_T, S_E),*
find a set of edge sets, S_E, connecting edge-set trees in S_T;
compute the total length, l_{S_E}, of edge sets in S_E;

(5) *If $max\_length < \Delta l = \sum_{i=1}^{n_r} l_{E_i} - l_{S_E}$, then*
$max\_length = \Delta l$, and an edge-set tree generated by connecting
S_T with S_E is substituted for T;

(6) *If not end of traversing on t_r, go to (2);*
else $T(C_1, ..., C_m) = T$.

}

According to the time complexity of *FIND_A_REPLACEMENT_SET* and the time complexity of *BREAK_TREE*, the time complexity of this algorithm is $O(2^{n_T-1}C(n-1, n_T-1)(n/n_T)^{n_T})$, where $n_T = (m+3)/2$.

3.7.6 FINDING THE GENERAL SOLUTION

Based on *FIND_T(C_m)* for finding an NR edge-set tree, and *FIND_T$_{NR}$* for finding an edge-set tree tested by C_i ($i = 1, 3, ..., 2n - 3$), finding an edge-set tree containing a set of MRSTs can be expressed as follows:

PROCEDURE *FIND_T($C_1, ..., C_m$)(input: n nodes;*
output: $T(C_1, ..., C_m)$)

{

(1) *Build an NR edge-set tree, $T(C_1) = T_{NR}$.*

(2) *For an odd index, $i = 3$, to $2n - 3$ do:*
Using FIND_T(C_i)($T(C_1, ..., C_j)$, $T(C_1, ..., C_i)$),
generate $T(C_1, ..., C_i)$ by testing C_i in $T(C_1, ..., C_j)$,
where $j = i - 2$.

}

According to the time complexity of *FIND_T(C_m)*, the time for running this algorithm is proportional to $N = \sum_{i=1}^{n-1} 2^i C(n-1, i)(n/(i+1))^{i+1}$. Because $n/(i+1) \geq 1$ and $\sum_{i=1}^{n-1} 2^i C(n-1, i) = 3^{n-1}$, $N > 3^{n-1}$. Provided that *FIND_A_REPLACEMENT_SET* could be improved to a linear algorithm, *FIND_T($C_1, ..., C_m$)* is still an NP algorithm.

Let $T = T(C_1, ..., C_m)$ ($m = 2n - 3$); let $S_T = \{S_T^{(i)}|\ i = 1, ..., n_T\}$ be a set of edge-set trees; and let S be a list in which a record, S', holds $\{S_E \subset T; S'_E \not\subset T\}$, where $S_E \subset T$ is a set of edge sets that can be replaced by a set of edge sets, $S'_E \not\subset T$, to generate a new edge-set tree containing a set of MRSTs. Finding S_E and S'_E, which connect the same set of edge-set trees for n_T subsets of nodes, can be expressed as follows:

PROCEDURE *FIND_ALL_REPLACEMENT_SET(input: n_T, T;*
$$\text{output: } S)$$
{

    (1) Load T to a binary tree, t_T;

    (2) Find t_r by BREAK_TREE($t_T, t'_T, n_T, f; t_r$);

    (3) $S = S_E = S'_E = \emptyset$;
    traversing tree t_r,
    retrieve an edge set in each node of n_T-levels in t_r in terms of
    the "father-son" relation, and
    store it to $\{E_i| \ i = 1, ..., n_T - 1\}$;

    (4) $S_T = T - E_1$; for $i = 2$ to $n_T - 1$, do $S_T = S_T - E_i$;

    (5) Find a set of edge sets, S'_E, connecting edge-set trees in S_T
    by FIND_A_REPLACEMENT_SET(S_T, n_T, S'_E);

    (6) If $l_{S'_E} = \sum_{i=1}^{n_T} l_{E_i}$,
    where $l_{S'_E}$ is the total length of edge sets in S'_E,
    for $i = n_T$ down to 1, do $S_E = S_E + E_i$;

    (7) $S = S \cup \{S_E, S'_E\}$;

    (8) If not end of traversing on t_r, go to (3).

}

Using *FIND_ALL_REPLACEMENT_SET* in an edge-set tree, T, containing a set of MRSTs, all combinations of replacement sets of edge sets in T can be found (see Section 3.6). Let G be a list for storing the general solution. Initially, set $G = T$. An alternative algorithm finding the general solution by replacing a set of edge sets in edge-set trees stored in G with each replacement set of edge sets is expressed as follows:

PROCEDURE *FIND_GENERAL_SOLUTION(input: T; output: G)*
{

    (1) $G = T$; $S = \emptyset$;

    (2) For $i = 2$ to n, do FIND_ALL_REPLACEMENT_SET(i, T, S);

    (3) While not end of list S, do (a) and (b)
        (a) Get $S' = \{S_E, S'_E\} \in S$;
        (b) From the first to the last edge-set trees in G, do:
            (b1) Get $T_G \in G$;
            (b2) If $S_E \subset T_G$, or if a set of edges consisting of an edge
                  in each edge set in S_E is a subset of edges contained
                  in T_G,
                  generate an edge-set tree, T_{new}, by replacing $S_E \subset T_G$
                  by S'_E;
            (b3) $G = G \cup T_{new}$.

}

3.7.7 BINARY TREES STORING EDGE-SET TREES

An initial edge-set tree, T, generated by connecting p_1, p_7, p_6, p_2, p_5, p_4, and p_3, according to the ascendant priority of x-coordinates of nodes, is shown in Figure 3.28d. Figures 3.28a, 3.28b and 3.28c also demonstrate the connecting and breaking rules.

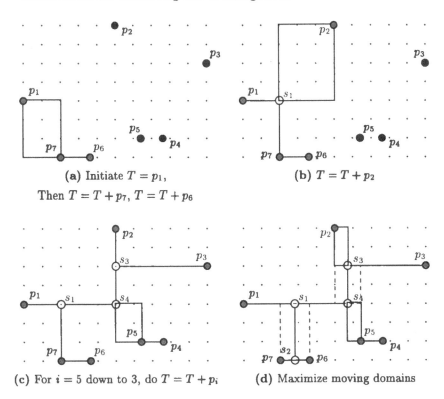

(a) Initiate $T = p_1$,

Then $T = T + p_7$, $T = T + p_6$

(b) $T = T + p_2$

(c) For $i = 5$ down to 3, do $T = T + p_i$

(d) Maximize moving domains

Figure 3.28 Find an initial tree

A binary tree holding an edge-set tree can be expressed as an x-coordinate or a y-coordinate oriented binary tree, or as other binary trees. An x-coordinate oriented binary tree obeys the following rules:

(1) *Select a vertex or a node that is an intersection of two edge sets in an edge-set tree as the root of the binary tree.*

(2) *Let p_1 be the father in $\cup_{i=1}^{3} E_{p_i \bullet}$. If $x_{p_2} < x_{p_3}$, or if $x_{p_2} = x_{p_3}$ and $y_{p_2} < y_{p_3}$, then p_2 is the left son and p_3 is the right son.*

(3) *Let p_2 be the only son of p_1. If $x_{p_1} < x_{p_2}$, or if $x_{p_1} = x_{p_2}$ and $y_{p_2} < y_{p_1}$, then p_2 is the left son; otherwise, it is the right son.*

In generating an initial edge-set tree in Figure 3.28d, a corresponding binary tree can be built in the following process:

(1) Set p_1 as the root.
(2) Link p_7 as a son of p_1.
(3) Link p_6 as a son of p_7.
(4) Lind p_2 to $E_{p_1 p_7}$: $s_1 = (mid(x_{p_1}, x_{p_2}, x_{p_7}), mid(y_{p_1}, y_{p_2}, y_{p_7}))$; break $E_{p_1 p_7}$ into $E_{p_1 s_1}$ and $E_{p_7 s_1}$ by s_1; update p_7 to be a son of s_1; link s_1 as a son of p_1; and link p_2 as a son of s_1.
(5) Link p_5 to $E_{p_2 s_1}$: break $E_{p_2 s_1}$ into $E_{s_1 s_4}$ and $E_{p_2 s_4}$; update p_2 to be a son of s_4; link s_4 as a son of s_1; and link p_5 as a son of s_4.
(6) Link p_4 as a son of p_5.
(7) Link p_3 to $E_{p_2 s_4}$: break $E_{p_2 s_4}$ into $E_{p_2 s_3}$ and $E_{s_3 s_4}$; update p_2 to be a son of s_3; link s_3 as a son of s_4; and link p_3 as a son of s_3.

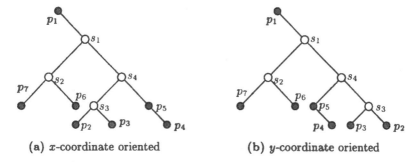

(a) x-coordinate oriented   (b) y-coordinate oriented

Figure 3.29 Binary trees for the initial tree

Rules for constructing y-coordinate and x-coordinate oriented binary trees are similar. Figure 3.29 shows two binary trees expressing the same edge-set tree, T, in Figure 3.28d.

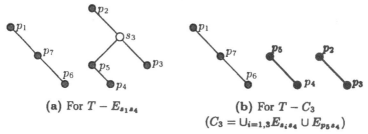

(a) For $T - E_{s_1 s_4}$ (b) For $T - C_3$
$(C_3 = \cup_{i=1,3} E_{s_i s_4} \cup E_{p_5 s_4})$

Figure 3.30 Binary trees expressing edge-set trees for subsets of nodes

Figure 3.30 shows x-coordinate oriented binary trees holding edge-set subtrees generated by removing an edge set or a clique from T in Figure 3.28d. When $E_{s_1 s_4}$ is removed from T, $E_{s_1 s_4}$ is maximized to $E_{s_1 s_4}^{max}$, and s_2 overlaps p_7. According to the breaking rule, p_6 and p_5 are adjacent to p_7 and s_3, respectively, after s_1 and s_4 are removed (Figure 3.30a). Removing a three-edge-set clique can be substituted for by removing an edge set and then removing a descendant edge set in an edge-set subtree. Edge-set trees in Figure 3.30b can be generated by removing $E_{p_5 s_3}$ from an edge-set tree in Figure 3.30a. Thus, removing $n - 1$ edge sets in turn from edge-set trees for subsets of nodes can generate a set of edge-set subtrees, in that each one becomes a node (see *BREAK_TREE*).

In finding an NR edge-set tree, each edge set is removed in turn from a binary tree to generate edge-set trees for two subsets of nodes (Figure 3.30a). Using *BREAK_TREE* on T generates t_r, holding each set of removal edge sets for generating edge-set trees for three subsets of nodes. The process can be described as follows:

(1) Let the root of t_r be empty; remove edge sets from T; and store each removal edge set as a son of the root of t_r in level 1:

    (1.1) Removing $E_{p_1 s_1}$ generates $S_{11} = E_{p_1 p_1}$, and $S'_{11} = E_{p_7 p_6} \cup E_{p_6 s_4} \cup E_{p_5 s_4} \cup E_{s_3 s_4} \cup E_{p_4 p_5} \cup E_{p_2 s_3} \cup E_{p_3 s_3}$;

    (1.2) Remove other edge sets in T (omitted).

(2) Remove edge sets from each pair of edge-set trees generated by removing an edge set in level 1, and store each removal edge set as a node in level 2 of t_r:

    (2.1) In turn, remove edge sets from S_{11} and S'_{11}, and store each removal edge set as a son of $E_{p_1 s_1}$ in level 2 of t_r:

        (2.1.1) Removing $E_{p_7 s_6}$ generates $S_{21} = E_{p_7 p_7}$, and $S'_{21} = E_{p_6 s_4} \cup E_{p_5 s_4} \cup E_{s_3 s_4} \cup E_{p_4 p_5} \cup E_{p_2 s_3} \cup E_{p_3 s_3}$;

        (2.1.2) Removing $E_{p_6 s_4}$ generates $S_{22} = E_{p_7 p_6}$, and $S'_{22} = E_{p_5 s_3} \cup E_{p_4 p_5} \cup E_{p_2 s_3} \cup E_{p_3 s_3}$;

        (2.1.3) Removing $E_{p_5 s_4}$ generates $S_{23} = E_{p_7 p_6} \cup E_{p_6 s_3} \cup E_{p_2 s_3} \cup E_{p_3 s_3}$, and $s'_{23} = E_{p_4 p_5}$;

        (2.1.4) Removing $E_{p_4 p_5}$ generates $S_{24} = E_{p_7 p_6} \cup E_{p_6 s_4} \cup E_{p_5 s_4} \cup E_{s_3 s_4} \cup E_{p_2 s_3} \cup E_{p_3 s_3}$, and $S'_{24} = E_{p_4 p_4}$;

        (2.1.5) Removing $E_{s_3 s_4}$ generates $S_{25} = E_{p_7 p_6} \cup E_{p_6 p_5} \cup E_{p_4 p_5}$, and $S'_{25} = E_{p_2 p_3}$;

        (2.1.6) Removing $E_{p_2 s_3}$ generates $S_{26} = E_{p_2 p_2}$, and $S'_{26} = e_{P_7 p_6} \cup E_{p_6 s_4} \cup E_{p_5 s_4} \cup E_{p_4 p_5} \cup E_{p_3 s_4}$;

        (2.1.7) Removing $E_{p_3 s_3}$ generates $S_{27} = E_{p_7 p_6} \cup E_{p_6 s_4} \cup E_{p_5 s_4} \cup E_{p_4 p_5} \cup E_{p_2 s_4}$, and $S'_{27} = E_{p_3 p_3}$;

    (2.2) Similar to the process in 2.1, remove edge sets from other pairs of edge-set trees to generate other nodes in level 2 of t_r.

In traversing tree t_r, a set of edge-set trees for i-subsets of nodes is generated by removing a set of $i - 1$ edge sets that have the relations of son, grandson, ..., $(i - 1)$-generation son starting from the root of T. A set of removal edge sets for generating three edge-set trees consists of two edge sets — for example, $\{E_{p_1 s_1}, E_{p_7 s_6}\}$, $\{E_{p_1 s_1}, E_{p_6 s_4}\}$, etc. — where $E_{p_1 s_1} \in T$ and $\{E_{p_7 s_6}, E_{p_6 s_4}\} \subset (T - E_{p_1 s_1})$.

3.8 ALGORITHM EXPANSIONS

The computational efficiency of an algorithm is mainly indicated by *time complexity*, that relates the running time and the size of a given problem. A problem that can be solved by an algorithm whose running time is less than $O(2^n = \sum_{i=0}^{n} C(n, i))$ is called a *polynomial problem (P problem)*; otherwise, it is a *non-deterministic polynomial problem (NP problem)*, where n is the size — for example, the number of given nodes in the Steiner's problem. A corresponding algorithm for a P problem or an NP problem is called a *P algorithm* or an *NP algorithm*. NP problems that can be transformed to each other in polynomial time are called *NP-complete problems (NPC problems)*.

3.8.1 DEFINITIONS FOR AN ALGORITHM EXPANSION

An NP problem is considered to be a problem that cannot be solved by an efficient algorithm. Current approaches for NP problems can be summarized as:

(1) heuristic;

(2) finding low-order P algorithms for special cases; or

(3) finding a P algorithm for an NPC problem.

Approaches (1) and (2) are used extensively in engineering, and it is not necessary to relate them with NPC problems. However, the margin between the solution and a heuristic approximation often cannot be determined. A special case is often too restricted for an engineering problem. Regarding approach (3), unfortunately, no P algorithm has been found, nor has a proposition that no P algorithm exists for an NPC problem been proven. Provided that a high-order P algorithm were found for an NPC problem, it might be an academic triumph; however, a high-order P algorithm is still inefficient for engineering purposes.

To solve an inefficient P or NP problem, we propose another approach called an *algorithm expansion*:

(1) Find an algorithm, A, for an exact solution measured by an index, I.

(2) Expand A into a series of P algorithms (subalgorithms), $A_1, A_2, ...,$ A_{n_a}, measured by indices, $I_1, I_2, ..., I_{n_a}$, respectively, under these requirements:

   (a) A_i is implemented before implementing A_j, where $1 \leq i < j \leq n_a$;

   (b) An exact solution is generated after implementing A_{n_a};

   (c) The time order of A_i must be equal to or less than the time order of A_j;

   (d) $I = I_0 + I_p$ for maximizations, or $I = I_0 - I_p$ for minimizations, where I_0 is an index for an initial state, $I_p = \sum_{i=1}^{n_a} I_i$, and $I_1 \geq I_2 \geq ... \geq I_{n_a}$.

According to the index series and the required accuracy, algorithm A can be replaced by a subseries, $A_1, A_2, ..., A_j$ $(j < n_a)$. An expansion from an algorithm, A, to subalgorithm series, $A_1, A_2, ..., A_{n_a}$, is an *inefficient expansion* if an inefficient subalgorithm, A_j, must be implemented for satisfying the required accuracy. If A_j is efficient, the expansion including $\{A_i|\ i = 1, ..., j\}$ is an *efficient expansion*.

A good heuristic algorithm can be considered as A_1 of an efficient expansion, regardless of the approximation accuracy. If no efficient expansion series exists for any algorithm of a problem, then no good heuristic algorithm exists for the problem. If $\{A_i|\ i = 1, 2, 3\}$ can run in $O(n^2)$ and if $I_p - \sum_{i=1}^{3} I_i$ meets the required accuracy, $\{A_i|\ i = 1, ..., n_a\}$ is called a *highly efficient series* of subalgorithms. In a highly efficient series, $\{A_i|\ i = 1, ..., n_a\}$, A_1, A_2, and A_3 must be low-order algorithms and $I_1 >> I_2 >> ... >> I_{n_a}$. The goal of the algorithm expansion is to generate a highly efficient series.

Let two different algorithms for the same problem be expanded into series $\{A_i|\ i = 1, ..., n_a\}$ and series $\{A'_i|\ i = 1, ..., n'_a\}$, and let series $\{I_i|\ i = 1, ..., n_a\}$ and series $\{I'_i|\ i = 1, ..., n'_a\}$ correspond to series $\{A_i|\ i = 1, ..., n_a\}$ and series $\{A'_i|\ i = 1, ..., n'_a\}$, respectively, and be computed by the same measurement. Series $\{A_i|\ i = 1, ..., n_a\}$ is considered to be more efficient than series $\{A'_i|\ i = 1, ..., n'_a\}$, if

- subalgorithm A_j $(j < n_a)$ is efficient;
- a subseries, $A_1, A_2, ..., A_j$, replacing series $A_1, A_2, ..., A_{n_a}$, satisfies the accuracy requirement; and
- $\sum_{i=1}^{j} I_i > \sum_{i=1}^{j} I'_i$.

An index, I_i, for subalgorithm A_i can be determined by the number, n_{e_i}, of events that occur in A_i; by the possibility, P_{ik} $(1 \leq k \leq e_i)$, of the kth event in A_i occurring; and by the contribution, $I_{c_{ik}}$, of the kth event:

$$I_i = \sum_{k=1}^{n_{e_i}} P_{ik} I_{c_{ik}} \tag{3.5}$$

An event that can occur in both A_i and A_j is called a *duplicate event* in A_i and A_j. An event occurring in A_i that contributes to I_i will no longer be considered in computing I_j $(i < j)$, even though it duplicates an event in A_j. If events occurring in A_i duplicate events occurring in A_j one by one, then $P_{jk} = 0$, where $i < j$ and $k = 1, 2, ..., n_{e_j}$. Thus, $I_j = 0$. If each event occurring in A_j $(j = k+1, k+2, ..., n_a)$ duplicates an event occurring in A_i $(i = 1, 2, ..., k)$, then an exact solution can be found by $A_1, ..., A_k$. Maximizing the number of duplicate events occurring in both A_i and A_j, where $i = 1, 2, 3$ and $j = i+1, ..., n_a$, if the major contribution for I_p can be converged to the indices for the low-order subalgorithms, an efficient expansion series can be generated, even though the algorithm for the exact solution becomes more inefficient.

3.8.2 EXPANSION INDICES FOR THE SOLUTION OF THE RECTILINEAR STEINER'S PROBLEM

According to Clique Theorem 10, Clique Corollary 2, and procedure *FIND_T* in Section 3.7.5, the algorithm finding a set of MRSTs can be expanded into a series of subalgorithms finding $\{T(C_{1n}, C_{3n}, ..., C_{in})|\ i = 1, ..., m\}$, or $\{T(C_1, C_3, ..., C_i)|\ i = 1, ..., m\}$, where $m = 2n - 3$ and n is the number of given nodes. However, one expansion series is more efficient than the other.

According to the clique evolutions (see Section 3.4), an i-edge-set clique (subtree) can be included in a j-edge-set clique (subtree), where $i < j$ and both i and j are odd. In contrast, an NN clique cannot be included in another NN clique. If $C_i \subset C_j$ and C_j is also an NN clique, the length reduction contributed by finding a $T(C_1, C_3, ..., C_i)$ from a $T(C_1, C_3, ..., C_{i-2})$ duplicates a part of that contributed by finding a $T(C_1, C_3, ..., C_j)$ from a $T(C_1, C_3, ..., C_{j-2})$. Because of these duplicate events, $I_i > I_i'$ and $I_j < I_j'$, where I_i, I_i', I_j and I_j' are indices for subalgorithms finding a $T(C_1, C_3, ..., C_i)$ from a $T(C_1, C_3, ..., C_{i-2})$, finding a $T(C_{1n}, C_{3n}, ..., C_{in})$ from a $T(C_{1n}, C_{3n}, ..., C_{(i-2)n})$, finding a $T(C_1, C_3, ..., C_j)$ from a $T(C_1, C_3, ..., C_{j-2})$, and finding a $T(C_{1n}, C_{3n}, ..., C_{jn})$ from a $T(C_{1n}, C_{3n}, ..., C_{(j-2)n})$, respectively. Thus, the series finding a $T(C_1, C_3, ..., C_i)$ from a $T(C_1, C_3, ..., C_{i-2})$ is more efficient than the series finding a $T(C_{1n}, C_{3n}..., C_{in})$ from a $T(C_{1n}, C_{3n}, ..., C_{(i-2)n})$, where $i = 3, 5, ..., m$. In this section, only the expansion finding $T(C_1, C_3, ..., C_m)$ is discussed.

Using each i-edge-set subtree of a clique as an i-edge-set clique in replacements can maximize the number of duplicate replacements both in finding a $T(C_1, C_3, ..., C_i)$ from a $T(C_1, C_3, ..., C_{i-2})$ and in finding a $T(C_1, C_3, ..., C_j)$ from a $T(C_1, C_3, ..., C_{j-2})$, where $j = i+2, i+4, ..., m$. In computing the index, I_k, for a subalgorithm finding a $T(C_1, C_3, ..., C_i)$ from a $T(C_1, C_3, ..., C_{i-2})$, the total length reduction contributed by events that occur in this subalgorithm and also duplicate the events occurring in algorithms measured by $I_1, I_2, ..., I_{k-1}$ equals zero. To avoid computing a contribution repeatedly, a replacement between an i-edge-set clique (or an i-edge-set subtree of a clique) and a set of edge sets must satisfy the *constraint rule*, consisting of two types of constraints:

(1) The replacement must not have been done by a subalgorithm finding a $T(C_1, C_3, ..., C_j)$ from a $T(C_1, C_3, ..., C_{j-2})$, where $j = 3, ..., i - 2$.

(2) It must contribute a length reduction in finding a $T(C_1, C_3, ..., C_i)$ from a $T(C_1, C_3, ..., C_{i-2})$.

Using constraint rule (1), a $T(C_1, C_3, ..., C_i)$ can be considered as a $T(C_i)$ generated by testing C_i in T_{NR} if $l_{T_{NR}} = l_{T(C_1, C_3, ..., C_{i-2})}$ is considered.

Figure 3.31a shows $C_3 = E_{as} \cup E_{bs} \cup E_{cs} \subset T_{NR}$, where thick lines represent edge-set subtrees. In finding a $T(C_1, C_3)$ from T_{NR} by replacing C_3 with $\{E_{pq}, E_{fg}\}$, the following constraints must be satisfied: $l_{E_{bs}} \le l_{E_{pq}}$, $l_{E_{cs}} \le l_{E_{pq}}$, $l_{E_{as}} \le l_{E_{fg}}$, and $l_{E_{cs}} \le l_{E_{fg}}$; otherwise, the replacement has been done in finding T_{NR}. Furthermore, $l_{E_{as}} + l_{E_{bs}} + l_{E_{cs}} > l_{E_{gf}} + l_{E_{pq}}$; otherwise, no length reduction is contributed from the replacement. Thus, the possibility for reducing length by replacing $C_3 \subset T_{NR}$ with $\{E_{pq}, E_{fg}\} \subset T(C_1, C_3)$ is $2^{-5} = 1/32 = 0.031$. This conclusion will be proven further.

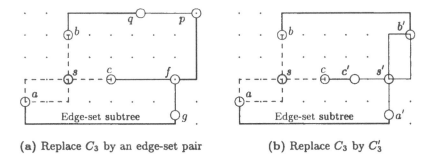

(a) Replace C_3 by an edge-set pair   (b) Replace C_3 by C_3'

Figure 3.31 Constraints for replacing C_3 by edge sets

Proposition 3.1 *Five independent constraints are required for replacing a three-edge-set clique with two edge sets or by replacing two edge sets with a three-edge-set clique in finding a $T(C_1, C_3)$ from a T_{NR}; the length reduction contributed by the replacement, at most, equals the length of an edge set in the three-edge-set clique.*

Proof: If an edge-set tree, $T = T(C_1, C_3)$ with $l_T < l_{T_{NR}}$, is generated by replacing $C_3 = E_{as} \cup E_{bs} \cup E_{cs} \subset T_{NR}$ with an edge-set pair, $\{E_{fg}, E_{pq}\} \subset T$ (Figure 3.31a), then there are two edge-set cycles in $T_{NR} \cup \{E_{fg}, E_{pq}\}$ or in $T \cup C_3$: C_{y1}, including E_{bs}, E_{cs}, and E_{pq}; and C_{y2}, including E_{as}, E_{cs}, and E_{fg}. To avoid repeating the length reduction contributed in finding an NR edge-set tree, two independent constraints, $l_{E_{bs}} \le l_{E_{pq}}$ and $l_{E_{cs}} \le l_{E_{pq}}$ in C_{y1}, and another two independent constraints, $l_{E_{as}} \le l_{E_{fg}}$ and $l_{E_{cs}} \le l_{E_{fg}}$, are required. To generate a length reduction, $l_{E_{gf}} + l_{E_{pq}} < l_{C_3} = l_{E_{as}} + l_{E_{bs}} + l_{E_{cs}}$ is also required.

After replacing C_3 with E_{fg} and E_{pq}, the following inequalities exist: $l_{E_{pq}} < l_{E_{bc}}$, $l_{E_{pq}} < l_{E_{ab}}$, $l_{E_{fg}} < l_{E_{ac}}$, and $l_{E_{fg}} < l_{E_{ab}}$. Summing these inequalities yields $2(l_{E_{fg}} + l_{E_{pq}}) < 2l_{C_3} + l_{E_{ab}}$, consistent with $l_{E_{fg}} + l_{E_{pq}} < l_{C_3}$. Similarly, an inequality generated by summing two or three arbitrary inequalities is also consistent with $l_{E_{fg}} + l_{E_{pq}} < l_{C_3}$. Therefore, the five constraints are independent.

Let $l_{E_{fg}} + l_{E_{pq}} \ge l_{E_{bs}} + l_{E_{cs}} = l_{C_3} - l_{E_{as}}$. Thus, $l_{C_3} - (l_{E_{fg}} + l_{E_{pq}}) \le l_{E_{as}}$, and the maximum length reduction equals $l_{E_{as}}$.

Similarly, if an edge-set tree, $T = T(C_1, C_3)$ with $(l_T < l_{T_{NR}})$, is generated by replacing an edge-set pair, $\{E_{fg}, E_{pq}\} \subset T_{NR}$, with $C_3 = E_{as} \cup E_{bs} \cup E_{cs} \subset T$, then $\{E_{ab}, E_{fg}, E_{pq}\}$, $\{E_{ac}, E_{fg}\}$, and $\{E_{bc}, E_{pq}\}$ are included in three edge-set cycles, respectively (Figure 3.31a). In addition to the constraint for generating a length reduction, the following five independent constraints must be satisfied: $l_{E_{fg}} \leq l_{E_{ab}}$, $l_{E_{fg}} \leq l_{E_{ac}}$, $l_{E_{pq}} \leq l_{E_{ab}}$, $l_{E_{pq}} \leq l_{E_{bc}}$, and $l_{E_{fg}} + l_{E_{pq}} > l_{E_{as}} + l_{E_{bs}} + l_{E_{cs}}$.

According to these inequalities, $2(l_{E_{fg}} + l_{E_{pq}}) \leq 2l_{C_3} + l_{E_{ab}}$. Thus, $l_{E_{fg}} + l_{E_{pq}} - l_{C_3} \leq l_{E_{ab}}/2 = (l_{E_{as}} + l_{E_{bs}})/2$. □

Proposition 3.2 *Seven independent constraints are required for replacing a three-edge-set clique with another three-edge-set clique, and the length reduction contributed by the replacement, at most, equals a half length of the replaced three-edge-set clique in finding a $T(C_1, C_3)$ from a T_{NR}.*

Proof: If an edge-set tree, $T = T(C_1, C_3)$ with $l_T < l_{T_{NR}}$, is generated by replacing $C_3 = E_{as} \cup E_{bs} \cup E_{cs} \subset T_{NR}$ with $C_3' = E_{a's'} \cup E_{b's'} \cup E_{c's'} \subset T$ (Figure 3.31b), then $\{E_{as}, E_{cs}, E_{a'c'}\}$, $\{E_{bs}, E_{cs}, E_{b'c'}\}$, and $\{E_{as}, E_{bs}, E_{a'b'}\}$ are included in three edge-set cycles, respectively. To avoid repeating the length reduction contributed in finding an NR edge-set tree, the following six independent constraints must be satisfied: $l_{E_{as}} \leq l_{E_{a'c'}}$, $l_{E_{cs}} \leq l_{E_{a'c'}}$, $l_{E_{bs}} \leq l_{E_{b'c'}}$, $l_{E_{cs}} \leq l_{E_{b'c'}}$, $l_{E_{as}} \leq l_{E_{a'b'}}$, and $l_{E_{bs}} \leq l_{E_{a'b'}}$. Summing these inequalities yields $2l_{C_3} \leq 4l_{C_3'}$. Furthermore, $l_{C_3'} < l_{C_3}$ is also required. Thus, $l_{C_3}/2 \leq l_{C_3'} < l_{C_3}$. When $l_{C_3'} = l_{C_3}/2$, $l_{C_3} - l_{C_3'} = l_{C_3}/2$ reaches the maximum.

If C_3 exists, the following relations exist: $l_{E_{a's'}} < l_{E_{ac}}$, $l_{E_{c's'}} < l_{E_{ac}}$, $l_{E_{b's'}} < l_{E_{bc}}$, $l_{E_{c's'}} < l_{E_{bc}}$, $l_{E_{a's'}} < l_{E_{ab}}$, and $l_{E_{b's'}} < l_{E_{ab}}$. Summing these inequalities yields $2l_{C_3'} < 4l_{C_3}$, which is consistent with $l_{C_3'} < l_{C_3}$. Thus, there are a total of seven independent constraints required for replacing a three-edge-set clique with another three-edge-set clique. □

Testing a clique or an edge-set subtree of a clique in an edge-set tree consists of three types of replacements (see Section 3.5):

- replacing a clique or an edge-set subtree of a clique in the edge-set tree with a set of edge sets;
- replacing a set of edge sets in the edge-set tree with a clique or an edge-set subtree; and
- replacing a clique or an edge-set subtree of a clique in the edge-set tree with another clique or edge-set subtree.

According to Propositions 3.1 and 3.2, the total length reduction contributed by testing a C_3 or a three-edge-set subtree in an NN clique of a T_{NR} equals $(2 \times 2^{-5} + 1.5 \times 2^{-7})\bar{l} = 2.4 \times 2^{-5}\bar{l} = 0.075\bar{l}$, where $\bar{l}$ is the average length of an edge set in the three-edge-set clique (or subtree) that replaces another one of an edge-set tree.

The average length reduction contributed by replacing a C_i with a C_i' is decreased when i is increased. As in Propositions 3.1 and 3.2, the total length reduction contributed by testing a $C_5 \subset T_{NR}$ equals $(2 \times 2^{-10} + 2.5 \times 2^{-17})\bar{l} = 2.02 \times 2^{-10}\bar{l} = 0.002\bar{l}$. The average length reduction contributed by replacing a $C_i \subset T_{NR}$ with a C_i' is equal to or less than $2.5 \times 2^{-17}\bar{l} \ll 2 \times 2^{-10}\bar{l}$ for $i \geq 5$, and thus can be ignored. Furthermore, replacing a C_i with a set of $(i+1)/2$ edge sets is equivalent to replacing a set of $(i+1)/2$ edge sets with a C_i in finding the length reduction. Thus, the common reference graph T_{NR} can always be represented by a $C_m^{(i)}$ that is an m-edge-set clique whose length equals the length of $T(C_1, C_3, ..., C_i)$, where $m = 2n - 3$ and $n \geq 7$ (see Section 3.9.4).

Therefore, three types of replacements can be simplified to one replacement that replaces an i-edge-set clique in $C_m^{(i-2)}$ with a set of $(i+1)/2$ edge sets with no vertices common at a Steiner point. However, it should be noted that the average length reduction contributed by testing $C_3 \subset T_{NR}$ is increased $1 - (2/2.4) = 17\%$ when the replacements between two three-edge-set subtrees are ignored.

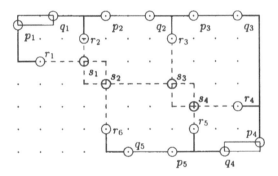

Figure 3.32 Independent constraints for replacing C_9 with five edge sets

As in the proof for the average length reduction contributed by testing a C_3 in a T_{NR}, the number of independent constraints required for generating a length reduction in finding a $T(C_1, C_3, ..., C_9)$ from a $T(C_1, C_3, ..., C_7)$ can be derived. Let $S_t = E_{r_1 s_1} \cup_{i=1}^{3} E_{s_i s_{i+1}} \cup E_{r_4 s_4}$ be the stem of $C_9 = S_t \cup E_{r_2 s_1} \cup E_{r_6 s_2} \cup E_{r_3 s_3} \cup E_{r_5 s_4} \subset T(C_1, C_3, C_5, C_7)$; and $E = \{E_{p_i q_i} \mid i = 1, ..., 5\} \not\subset T(C_1, C_3, C_5, C_7)$ be a set of edge sets that can replace C_9 to generate a new edge-set tree, where the total length, l_E, of edge sets in E is less than l_{C_9} (Figure 3.32). In $E \cup T(C_1, C_3, C_5, C_7)$, there are five edge-set cycles. Each edge-set cycle includes two or three edge sets of C_9 and one edge set of E. Furthermore, there are, at most, two edge-set cycles; each of them includes two edge sets in C_9.

Generalizing C_9 to $C_i \subset C_m^{(i-2)}$ $(i = 3, 5, ..., 2n-3)$ and using the constraint rule, each independent constraint needed for generating a length reduction in finding a $T(C_1, C_3, ..., C_i)$ from a $C_m^{(i-2)}$ is determined by one of the following conditions:

(1) Any edge set of C_i is equal to or shorter than an edge set in E within the same edge-set cycle in $E \cup C_i$, where $|E| = (i+1)/2$ and edge sets in E have the minimum length in replacing $C_i \subset C_m^{(i-2)}$. For example, $l_{E_{r_1 s_1}} \leq l_{E_{p_1 q_1}}$ and $l_{E_{r_2 s_1}} \leq l_{E_{p_1 q_1}}$ in the edge-set cycle including $E_{p_1 q_1}$, $E_{r_1 s_1}$ and $E_{r_2 s_1}$ (Figure 3.32).

(2) The length of any j-edge-set subtree ($j = 3, 5, ..., i - 2$) in C_i is equal to or less than the total length of the $(j + 1)/2$ edge sets in E that can be replaced by the j-edge-set subtree; otherwise, the j-edge-set subtree has been replaced by edge sets of E in finding $T(C_1, C_3, ..., C_j)$ from $C_m^{(j-2)}$. For example, when $i \geq 5$ in Figure 3.32, $l_{E_{r_1 s_1}} + l_{E_{r_2 s_1}} + l_{E_{s_1 s_2}} \leq l_{E_{p_1 q_1}} + l_{E_{p_2 q_2}}$; otherwise, $E_{r_1 s_1} \cup E_{r_2 s_1} \cup E_{s_1 s_2}$ has been replaced by $\{E_{p_1 q_1}, E_{p_2 q_2}\}$ in finding $T(C_1, C_3)$ from T_{NR}.

(3) $l_E < l_{C_i}$.

There are a total of $4 + 3(k - 2)$ independent constraints determined by condition (1). These independent constraints are determined by the conditions for finding a T_{NR}. In a C_i, there are $(i - 1)/2$ inner Steiner points. Each inner Steiner point determines a C_3; a pair of adjacent inner Steiner points determines a C_5; ... and a set of $(i - 3)/2$ adjacent inner Steiner points determines a C_{i-2}. Thus, the number of independent constraints determined by testing $C_3, C_5, ..., C_{i-2}$ in $C_i \subset C_m^{(i-2)}$ equals $n_{is} + (n_{is} - 1) + (n_{is} - 2) + ... + 2$, where $n_{is} = (i - 1)/2 = k - 1$ is the number of inner Steiner points in C_i, and k is the number of edge sets in E. Replacing an arbitrary C_3, C_5, ..., or C_{i-2} in C_i with edge sets in E cannot contribute a length reduction in finding $T(C_1, C_3, ..., C_i)$ from $C_m^{(i-2)}$. Therefore, there are n_k independent constraints in reducing the length of a $C_m^{(i-2)}$ by replacing a $C_i \subset C_m^{(i-2)}$ with $(i + 1)/2$ edge sets:

$$
\begin{aligned}
n_k &= 4 + 3(k - 2) + \sum_{j=1}^{k-1}(k - j) \\
&= \frac{k^2 + 5k - 4}{2} \quad (k > 1)
\end{aligned}
\tag{3.6}
$$

The possibility for contributing a length reduction by replacing a C_i in a $C_m^{(i-2)}$ with k edge sets in a $T(C_1, C_3, ..., C_i)$ is 2^{-n_k}, where $i = 2k - 1$.

According to conditions (2) and (3), the length reduction contributed by replacing a C_i or an i-edge-set subtree of a clique in a $C_m^{(i-2)}$ with a set of k edge sets, at most, equals the length of an edge set in C_i (Proposition 3.1). Furthermore, there are, at most, $(n - 1)/k$ i-edge-set cliques (or subtrees) that can be replaced by edge sets in finding a

$T(C_1, C_3, ..., C_i)$ from a $C_m^{(i-2)}$, because i-edge-set cliques (or subtrees) in $C_m^{(i-2)}$ can be replaced by the maximum of $(n-1)/k$ sets of edge sets, or by the maximum of $(n-1)/k$ i-edge-set subtrees. Let I_k be the index indicating the average length reduction contributed by finding a $T(C_1, C_3, ..., C_i)$ from a $C_m^{(i-2)}$, where $3 \leq i \leq 2n-3$ and $k = (i+1)/2$.

$$
\begin{aligned}
I_k &= \frac{n-1}{k} 2^{-n_k} \frac{\bar{l}_k}{2} \\
&= \frac{2(n-1)\bar{l}_k}{k} 2^{-k(k+5)/2} \quad (k > 1)
\end{aligned}
\tag{3.7}
$$

where $\bar{l}_k$ is an average length of edge sets in $C_m^{(i-2)}$. Thus, the ratio,

$$
\begin{aligned}
r_k &= \frac{I_{k+1}}{I_k} \\
&= \frac{k}{k+1} 2^{-(k+3)} \frac{\bar{l}_{k+1}}{\bar{l}_k} \quad (k > 1)
\end{aligned}
\tag{3.8}
$$

If the same ratio is applied to compute I_1,

$$
I_1 = 0.25(n-1)\bar{l}_1
\tag{3.9}
$$

Based on finding $T(C_1) = T_{NR}$ from a minimum spanning tree, the same I_1 can be derived. Let T be an edge-set tree generated by extending each edge in a minimum spanning tree into an edge set. The possibility for replacing an edge set in T with another edge set is $1/2$. When a total of $n-1$ edge sets of the minimum spanning tree are replaced with new edge sets, the length reduction reaches the maximum. Therefore, $I_1 = (1/2)(n-1)\frac{\bar{l}_1}{2} = 0.25(n-1)\bar{l}_1$, and T_0 is an edge-set tree extended from the minimum spanning tree. Denoting the average length of an edge set in T_0 as $\bar{l}_0$,

$$
\bar{l}_1 = \frac{(n-1)\bar{l}_0 - I_1}{2n-3}
\tag{3.10}
$$

According to Equations 3.9 and 3.10,

$$
\bar{l}_1 = \frac{4(n-1)}{9n-13} \bar{l}_0
\tag{3.11}
$$

Substituting Equation 3.11 into Equation 3.9 yields

$$
\begin{aligned}
I_1 &= \frac{(n-1)^2}{9n-13} \bar{l}_0 \\
&= \frac{n-1}{9n-13} l_{T_0}
\end{aligned}
\tag{3.12}
$$

where l_{T_0} is the length of T_0, and $I_1 = 11.1 \sim 12\%$. When nodes are properly distributed, using good heuristics to merge redundant edges from the minimum spanning trees can yield an average length reduction slightly less than I_1 (see Section 3.8.3). Note that I_1 does not affect the efficiency of the expansion. When $k > 1$, $\bar{l}_k$ is uniformly computed from $C_m^{(2k-3)}$. Thus,

$$\bar{l}_k = \bar{l}_{k-1} - \frac{I_k}{2n-3} \quad (k > 1) \tag{3.13}$$

According to Equation 3.8,

$$
\begin{aligned}
I_k &= r_{k-1} I_{k-1} \\
&= \frac{k-1}{k} 2^{-(k+2)} \frac{\bar{l}_k}{\bar{l}_{k-1}} I_{k-1} \quad (k > 1)
\end{aligned}
\tag{3.14}
$$

Eliminating $\bar{l}_k$ from Equations 3.13 and 3.14 yields

$$I_k = \frac{k-1}{k} 2^{-(k+2)} \frac{(2n-3)\bar{l}_{k-1}}{1 + (2n-3)\bar{l}_{k-1}} I_{k-1} \quad (k > 1) \tag{3.15}$$

Thus,

$$
\begin{aligned}
r_{k-1} &= \frac{I_k}{I_{k-1}} \\
&= \frac{k-1}{k} 2^{-(k+2)} \frac{(2n-3)\bar{l}_{k-1}}{1 + (2n-3)\bar{l}_{k-1}} \\
&\approx \frac{k-1}{k} 2^{-(k+2)} \quad (k > 1)
\end{aligned}
\tag{3.16}
$$

The average length reduction contributed by testing $C_3 \in T_{NR}$ has been increased 17%. Thus, $I_2 = 0.83 r_1 I_1$, and

$$
\begin{aligned}
I_k &= 0.83 \prod_{i=1}^{k-1} r_{k-1} I_{k-1} \\
&\approx \frac{0.83}{k 2^{(k-1)(k+6)/2}} I_1 \quad (1 < k < n)
\end{aligned}
\tag{3.17}
$$

According to Equation 3.16, $r_1 \approx (1/2)2^{-4}$, $r_2 \approx (2/3)2^{-5}$, $r_3 \approx (3/4)2^{-6}$, $r_4 \approx (4/5)2^{-7}$, ..., and $r_{n-2} \approx [(n-2)/(n-1)]2^{-(n+1)}$. A series of indices for measuring all subalgorithms, using Equation 3.17, is listed below:

$I_2 = 0.83 r_1 I_1 \approx 0.026 I_1$ for $T(C_1, C_3)$;
$I_3 = r_2 I_2 = 0.83 r_1 r_2 I_1 \approx 0.00054 I_1$ for $T(C_1, C_3, C_5)$;
$I_4 = r_3 I_3 = 0.83 r_1 r_2 r_3 I_1 \approx 0.0000112 I_1$ for $T(C_1, C_3, C_5, C_7)$;
...

Based on these indices, edge-set trees $T(C_1)$ and $T(C_1, C_3)$ contribute 97.9% and 99.9%, respectively, of the average length reduction that can be contributed by $T(C_1, C_3, ..., C_m)$ for $n = (m+3)/2$ nodes.

3.8.3 LENGTH REDUCTION CONTRIBUTED BY MERGING REDUNDANT EDGES

Reducing the length in the minimum spanning tree is often used to approximate the minimum Steiner tree. Let lengths of edges in the minimum spanning tree be 1, 2, 3, ..., m units, respectively. When the redundant sections of two edges with lengths l_a and l_b ($1 \leq a \leq b \leq m$) are merged into a Steiner point, the maximum length reduction is $l_a/2$. Similarly, merging the redundant sections of an edge whose length equals i and another one whose length is equal to or greater than i yields a maximum length reduction of $i/2$. The total number of combinations of edge pairs is $C(m, 2) + m = m(m+1)/2$, and the maximum total length reduction for $m(m+1)/2$ edge pairs is

$$
\begin{aligned}
\Delta L_{total\_max} &= \frac{m}{2} + \frac{2(m-1)}{2} + \frac{3(m-2)}{2} + ... + \frac{(m-1)2}{2} + \frac{m}{2} \\
&= \frac{1}{2} \sum_{i=1}^{m} i(m+1-i) \\
&= \frac{1}{2}[\frac{(m+1)m(m+1)}{2} - \frac{m(m+1)(2m+1)}{6}] \\
&= \frac{m(m+1)(m+2)}{12} \tag{3.18}
\end{aligned}
$$

if the distributions of all edges with a length from 1 to m units are uniform. Thus, the maximum length reduction, Δl_{max}, per edge-set pair is

$$
\Delta l_{max} = \frac{(m+2)}{6} = \frac{2\bar{l}+1}{6} \tag{3.19}
$$

where $\bar{l}$ is the average length of an edge in the minimum spanning tree. Because a Steiner point is generated by a merged edge-set pair, Δl_{max} is also the maximum length reduction per Steiner point. The average length reduction per Steiner point is $\Delta l_{max}/2 = (m+2)/12$. Since there are n nodes per tree, the maximum length reduction per tree is

$$
\Delta L_{max} = (n-2)\Delta l_{max} = \frac{(n-2)(2\bar{l}+1)}{6} \approx \frac{(n-2)\bar{l}}{3} = 0.333(n-2)\bar{l} \tag{3.20}
$$

Letting the average number of Steiner points be $(n-2)/2$, the average length reduction per tree is

$$
\Delta L_{ave} = \frac{\Delta L_{max}}{4} \approx 0.083(n-2)\bar{l} \tag{3.21}
$$

and the average reduction rate without "global refinement" from minimum spanning trees to Steiner trees is

$$\frac{\Delta L_{ave}}{(n-1)\bar{l}} \approx 0.083 \qquad (3.22)$$

When a Steiner tree is generated by merging redundant sections of each pair of adjacent edge sets into Steiner points, and when the average number of edges in the Steiner tree is $(n-1) + (n-2)/2 = (3n-4)/2$, the average edge length of the Steiner tree is

$$
\begin{aligned}
\bar{l}' &= \frac{(n-1)\bar{l} - \Delta L_{ave}}{(3n-4)/2} \\
&= \frac{(n-1)\bar{l} - (n-2)(2\bar{l}+1)/24}{(3n-4)/2} \\
&\approx \frac{11}{18}\bar{l} = 0.61\bar{l} \qquad (3.23)
\end{aligned}
$$

Provided that the lengths of most edges range between u and v, the maximum total length reduction for $C(v-u+1,2) + (v-u+1) = (v-u+1)(v-u+2)/2$ edge pairs is

$$
\begin{aligned}
\Delta L_{total\_max} &= \frac{1}{2}u(v-u+1) + \frac{1}{2}(u+1)(v-u) + \\
&\quad \frac{1}{2}(u+2)(v-u-1) + \dots + \frac{1}{2}(v-1)(2) + \frac{1}{2}v \\
&= \frac{1}{2}\sum_{i=u}^{v} i(v+1-i) \\
&= \frac{1}{12}[v(v+1)(v+2) - u(u-1)(3v-2u+4)]
\end{aligned}
$$
$$(3.24)$$

Thus, the average length reduction per Steiner tree is

$$
\begin{aligned}
\Delta L_{ave} &= \frac{(n-2)\Delta L_{total\_max}}{(v-u+1)(v-u+2)/2} \\
&= \frac{(n-2)[v(v+1)(v+2) - u(u-1)(3v-2u+4)]}{6(v-u+1)(v-u+2)}
\end{aligned}
$$
$$(3.25)$$

When $u = m/3$ and $v = 2u$, $\Delta L_{ave} = (n-2)u/6 = (n-2)m/18 = (1/9)(n-2)\bar{l} = 0.111(n-2)\bar{l}$. When $u = v = m/2$, $\Delta L_{ave} = 0.0625(n-2)m = 0.125(n-2)\bar{l}$ reaches the maximum. However, $u = v = m/2$ is not realized for the analysis of a large number of trees. Based on the above analysis, the length reduction rate is dominated by both the distribution of nodes and the merging method. Using good heuristic algorithms, it is 8% to 10%, on average (see Chapters 1 and 2).

3.9 LINEAR AND NEARLY LINEAR ALGORITHMS FOR SPECIAL CASES

This section further proves linear algorithms for engineering purposes by applying the theory for the general solution of the rectilinear Steiner's problem. Linear algorithms are presented for the general solution of the switch-box problem, and for the solution of the problem when nodes are lying on the boundary of a polygon. A nearly linear algorithm finding an NR edge-set tree and the number of nodes in an NR edge-set tree containing exact solutions are also discussed.

3.9.1 A LINEAR ALGORITHM FOR THE GENERAL SOLUTION OF THE SWITCH-BOX PROBLEM

The switch-box problem is a special case of the rectilinear Steiner problem, in that all given nodes are located on the perimeter of a rectangle. The solution of the switch-box problem can be used in switch-box routing, because all signal terminals are located on the boundary of a switch-box.

Special MRST Problem Theorem 1 *A solution of the switch-box problem is contained in one or more edge-set trees:*

(1) an edge-set tree tested by C_{1n} and C_{3n}; and/or

(2) an edge-set tree including an NN clique whose main vertices lying on two opposite sides of the boundary of a switch-box.

Proof: Let the solution of the switch-box problem be edge-set tree T that connects nodes lie on the perimeter of rectangle R.

If an arbitrary NN clique, $C_{in} \subset T$, connects nodes on two or three sides of R, then its main vertices must lie on two sides of R. If the main vertices of C_{in} lie on two opposite sides of R, then i must be 1 or 3 (Figure 3.33a). If the main vertices of C_{in} lie on two adjacent sides of R, then an edge set whose vertices are Steiner points in the stem of C_{in} is a parallel moving edge set, and it can be moved to the boundary (Figure 3.33b). Thus, i must also be 1 or 3. If an arbitrary NN clique, $C_{in} \subset T$, connects nodes on four sides of R, the main vertices must lie on two opposite sides of R (Clique Theorems 5 and 9).

According to Clique Theorem 10, the solution of the switch-box problem is either an edge-set tree tested by one- and three-edge-set NN cliques, or an edge-set tree tested by an NN clique whose main vertices lie on two opposite sides of R.   □

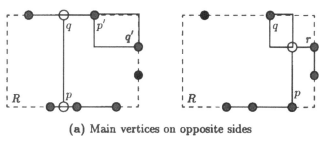

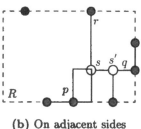

(a) Main vertices on opposite sides

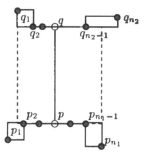

(b) On adjacent sides

Figure 3.33 An NN clique connects nodes on two or three sides (p and q are main vertices; $E_{ss'}$ can be moved to the lower side of R)

Figure 3.34 Extended parallel moving edge set

When the moving domains of two vertices of a parallel moving edge set, E_{pq}, are extended to two point sets on a pair of parallel edges $e_{f_1 g_1}$ and $e_{f_2 g_2}$, E_{pq} becomes an *extended parallel moving edge set*, where $e_{f_1 g_1}$ and $e_{f_2 g_2}$ occupy the projections of $\cup_{i=1}^{n_1-1} E_{p_i p_{i+1}}$ on $\cup_{j=1}^{n_2-1} E_{q_j q_{j+1}}$ and $\cup_{j=1}^{n_2-1} E_{q_j q_{j+1}}$ on $\cup_{i=1}^{n_1-1} E_{p_i p_{i+1}}$, and connect $\{p_i | \ i = 2, ..., n_1 - 1\}$ and $\{q_j | \ j = 2, ..., n_2 - 1\}$, respectively (Figure 3.34). Extended parallel moving edge set E_{pq} can replace a set of parallel moving edge sets connecting pairs of edge sets in $\{E_{p_i p_{i+1}} | \ i = 1, ..., n_1 - 1\}$ and $\{E_{q_j q_{j+1}} | \ j = 1, ..., n_2 - 1\}$; and all these parallel moving edge sets are enclosed by the same rectangle whose two opposite sides are $e_{f_1 g_1}$ and $e_{f_2 g_2}$. If a vertex, p or q, of an extended parallel moving edge set becomes a 2-D Steiner point, E_{pq} is called an extended moving edge set. Similarly, a clique whose moving edge sets are extended to extended moving edge sets is called an extended clique. An edge-set tree in which each moving edge set is extended to an extended moving edge set is called an *extended edge-set tree*. Using extended edge-set trees can reduce the number of edge-set trees in the general solution.

According to the Special MRST Problem Theorem 1, the solution of the switch-box problem can be found by comparing the lengths of the following three extended edge-set trees that include:

- an extended NN cliques whose main nodes lie on two horizontal sides,
- an extended NN cliques whose main nodes lie on two vertical sides, and
- one- or three-edge-set extended NN cliques only, respectively.

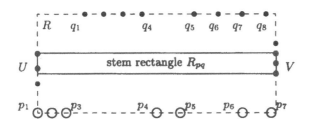

(a) Point sets P and Q

$P_1 = \{p_1, ..., p_3\}, Q_1 = \{q_1, ..., q_4\},$

$P_2 = \{p_4, p_5\}, Q_2 = \{q_5, q_6, q_7\},$

$P_3 = p_6, Q_3 = q_8, P_4 = p_7$

(b) A sorted sequence, S^*_{PQ}, from $P \cup Q$

Figure 3.35 A sorted sequence for nodes
on two opposite sides of a rectangle

A set with a specified precedence of elements is called a *sequence*, and is notated as $\{...\}^*$. Let $P = \{p_i | i = 1, ..., n_p\}^*$ and $Q = \{q_i | i = 1, ..., n_q\}^*$ be point sets lying on two opposite sides of a rectangle. Sorting elements in $P \cup Q$ in terms of the ascendent priority of x- or y-coordinates generates a *sorted sequence*, $S^*_{PQ} = \{P_1, Q_1, ..., P_{n_P}, Q_{n_Q}\}^*$, of subsequences of nodes, where $P_i \subset P$ and $Q_i \subset Q$ (Figure 3.35). Similarly, points on the other two sides of the rectangle can be considered to be two point sets, U and V. Determining the stem rectangle by point sets U and V and connecting each subsequence in S^*_{PQ} to a side of the stem rectangle generates an NN clique, C_{mn}, where $m = 2|S^*_{PQ}| + 1$.

The length, $l_{C_{mn}}$, of a possible NN clique can be directly computed from the coordinates of each subsequence in S^*_{PQ}. Let R_{pq} be the stem rectangle, the minimum rectangle connecting U and V or the rectangle whose opposite sides occupy the projections of U on V and of V on U (Figure 3.35a); let n_P and n_Q be the numbers of point sets belonging to P and Q in S^*_{PQ}, respectively; and let l_d be the length of the shortest degenerated edge set in C_{mn}. If $n_P > n_Q$, l_d equals the distance between p_i and R_{pq}. If $n_Q > n_P$, l_d equals the distance between q_i and R_{pq}. If $n_P = n_Q$, $l_d = 0$. In the case that P and Q lie on two horizontal sides,

$$l_{C_{mn}} = l_h + min(n_P, n_Q)l_v + (|n_P - n_Q|)l_d \qquad (3.26)$$

where l_h and l_v are the lengths of a horizontal and a vertical side of the rectangle.

Let the total length of edges connecting nodes in all subsequences be l_S. The length, l_T, of edge-set tree T, including C_{mn}, is $l_{C_{mn}} + l_S$. If $l_T > l_{T'}$, T is not a solution, where $l_{T'}$ is the length of edge-set tree T', which includes one- or three-edge-set NN cliques only. If $l_T = l_{T'}$, both T and T' are solutions.

Let $p_f^{(i)}$ and $p_g^{(i)}$ be the first and the last nodes in $P_i \subset P$; let $q_f^{(i)}$ and $q_g^{(i)}$ be the first and the last nodes in $Q_i \subset Q$; let $u_f^{(i)}$ and $u_g^{(i)}$ be the first and the last nodes in $U_i \subset U$; and let $v_f^{(i)}$ and $v_g^{(i)}$ be the first and the last nodes in $V_i \subset V$. Furthermore, let S^*_{PQ} and S^*_{UV} be sorted from $P \cup Q$ and $U \cup V$, respectively. Because only one- and three-edge-set NN cliques exist in edge-set tree T', there is no edge set with two Steiner points as vertices in T'. Thus, T' can be sequentially generated by connecting a node to its nearest node or to its nearest edge set whose vertices are nodes. An algorithm finding T' can be expressed as:

(1) Implement FIND_FOR_SWITCH_BOX(S^*_{PQ}, U, V, T_{PQ}) *to generate an edge-set tree, T_{PQ}, in terms of S^*_{PQ}.*

(2) Implement FIND_FOR_SWITCH_BOX(S^*_{UV}, P, Q, T_{UV}) *to generate an edge-set tree, T_{UV}, in terms of S^*_{UV}.*

(3) If $l_{T_{PQ}} \le l_{T_{UV}}$, then $T' = T_{PQ}$, else $T' = T_{UV}$.

(4) Maximize the moving domain of each Steiner point.

The procedure *FIND_FOR_SWITCH_BOX(S^*_{PQ}, U, V, T_{PQ})* includes three subprocesses: connecting nodes in U to the first and/or the second subsequences of S^*_{PQ}, then sequentially connecting each subsequence in S^*_{PQ} to edge sets generated in the previous process, and finally connecting point set V to the last two subsequences in S^*_{PQ}. The first and the last subprocesses are expressed by *CONNECT_FIRST_POINT_SET* and *CONNECT_LAST_POINT_SET*, respectively. In these procedures, a variable, l_{max}, is used to specify an edge set whose length is the maximum. This edge set may be removed in the final step.

PROCEDURE *CONNECT_FIRST_POINT_SET(input: S_{PQ}^*, U;*
$$\textit{output: } T, l_{max})$$

{

    (a) *Connect U by straight-edge sets as an initial edge-set tree, T; specify the straight-edge set whose length is the maximum and is stored in l_{max};*
    notate the first and the last nodes of U as u_1 and u_2, respectively.

    (b) *If $l_{E_{p_f^{(1)}u_1}} < l_{E_{q_f^{(1)}u_2}}$, then*
        {
            $T = E_{p_1 q_1}$, where $p_1 = p_f^{(1)}$ and $q_1 = u_1$;
            $S = (mid(x_{p_1}, x_{q_1}, x_{u_2}), mid(y_{p_1}, y_{q_1}, y_{u_2}))$;
        }
        else
        {
            $T = E_{p_1 q_1}$, where $p_1 = q_f^{(1)}$ and $q_1 = u_2$;
            $S = (mid(x_{p_1}, x_{q_1}, x_{u_1}), mid(y_{p_1}, y_{q_1}, y_{u_1}))$.
        }

    (c) *Find one of the following sets of edge sets: $\{E_{q_f^{(1)}s}, E_{q_g^{(1)}p_f^{(2)}}\}$, $\{E_{q_f^{(1)}s}, E_{p_g^{(1)}p_f^{(2)}}\}$, and $\{E_{q_f^{(1)}s'}, E_{p_g^{(1)}s'}, E_{p_f^{(2)}s'}\}$, in terms of the total minimum length of edge sets, and store it in E; find the edge set whose length, l, is the maximum from E; if $l_{max} < l$, then specify the edge set whose length is l and $l_{max} = l$.*

    (d) *Connect edge sets in P_1 and Q_1 with straight-edge sets and store these straight-edge sets and E, found in (c), to T.*

}

PROCEDURE *FIND_FOR_SWITCH_BOX(input: S_{PQ}^*, U, V;*
$$\textit{output: } T_{PQ}, l_{max})$$

{

    (a) *CONNECT_FIRST_POINT_SET(S_{PQ}^*, U, T, l_{max});*
        $i = 2$.

    (b) *Compute $d = min(d_1, d_2, d_3)$, where $d_1 = dist(p_f^{(i)}, p_g^{(i-1)})$, $d_2 = dist(p_f^{(i)}, q_g^{(i-1)})$, $d_3 = dist(p_f^{(i)}, E_{p_g^{(i-1)}q_f^{(i-1)}})$, and dist(a, b) is the distance between a and b.*

(c) If $d = d_1$, connect $p_f^{(i)}$ and $p_g^{(i-1)}$ by $E_{p_f^{(i)} p_g^{(i-1)}}$;

   else if $d = d_2$, connect $p_f^{(i)}$ and $q_g^{(i-1)}$ by $E_{p_f^{(i)} q_g^{(i-1)}}$;

   else connect $p_f^{(i)}$ to $E_{p_g^{(i-1)} q_f^{(i-1)}}$.

   (note that $E_{p_g^{(i-1)} q_f^{(i-1)}} + p_f^{(i)} = E_{p_g^{(i-1)}, s} \cup E_{q_f^{(i-1)}, s} \cup$
   $E_{p_f^{(i)}, s}$, where $x_s = mid(x_{p_g^{(i-1)}}, x_{q_f^{(i-1)}}, x_{p_f^{(i)}})$ and $y_s = mid(y_{p_g^{(i-1)}}, y_{q_f^{(i-1)}}, y_{p_f^{(i)}})$).

(d) If $d > l_{max}$, then $l_{max} = d$;

   else if $d = d_2$ and $d > l_{max}$, then $l_{max} = d_2$;

   else reset $l_{max} = max(l_{E_{p_f^{(i)}, s}}, l_{E_{q_f^{(i-1)}, s}})$;

   specify the edge set whose length is l_{max}.

(e) Compute $d' = min(d_1', d_2', d_3')$, where $d_1' = dist(q_g^{(i-1)}, q_f^{(i)})$,
   $d_2' = dist(q_f^{(i)}, p_g^{(i)})$. and $d_3' = dist(q_f^{(i)}, E_{q_g^{(i-1)} p_f^{(i)}})$.

(f) If $d' = d_1'$, connect $p_g^{(i-1)}$ and $q_f^{(i)}$ by $E_{p_g^{(i-1)} q_f^{(i)}}$;

   else if $d' = d_2$, connect $q_f^{(i)}$ and $p_g^{(i)}$ by $E_{q_f^{(i)} p_g^{(i)}}$;

   else connect $q_f^{(i)}$ to $E_{q_g^{(i-1)} p_f^{(i)}}$.

(g) If $d' = d_1'$ and $d' > l_{max}$, then $l_{max} = d'$;

   else if $d' = d_2'$ and $d' > l_{max}$, then $l_{max} = d'$;

   else reset $l_{max} = max(l_{E_{p_f^{(i)}, s'}}, l_{E_{q_f^{(i)}, s'}})$, where

   $x_{s'} = mid(x_{p_f^{(i)}}, x_{q_f^{(i)}}, x_{q_g^{(i-1)}})$ and
   $y_{s'} = mid(y_{p_f^{(i)}}, y_{q_f^{(i)}}, y_{q_g^{(i-1)}})$;

   specify the edge set whose length is l_{max}.

(h) If not end of S_{PQ}^*, then $i = i + 1$ and go to (b).

(i) Connect points in V to edge-set tree T generated by (a) to (h) by implementing CONNECT_LAST_POINT_SET(T, V, l_{max}, T_{PQ}).

}

PROCEDURE *CONNECT_LAST_POINT_SET(input: T, V, l_{max};*
                                            output: T_{PQ})

{

(a) Connect V by straight-edge sets; specify the straight-edge set whose length is the maximum and is stored in l_{max};
   notate the first and the last nodes of V as v_f and v_g, respectively;
   notate the last node of the second to last subsequence and the first and last nodes of the last subsequence in S_{PQ}^* as g_2, f_1, and g_1, respectively.

(b) If $E_{f_1 g_2}$ has been generated in *(c)* or *(f)* in
FIND_FOR_SWITCH_BOX, *then*
{

    if $l_{E_{f_1 g_2}} + (x_{f_1} - x_{v_1}) < dist(v_1, g_1) + dist(v_2, g_2)$, *then*
    connect v_1 to $E_{f_1 g_2}$,
        where $\{v_1, v_2\} = \{v_f, v_g\}$, v_1 is closest to g_1 and v_2
    is closest to g_2;
        else remove $E_{f_1 g_2}$ and generate $E_{g_1 v_1}$ and $E_{g_2 v_2}$.

}
(c) Generate $C_3 = E_{v_1 s''} \cup E_{f_1 s''} \cup E_{g_2 s''}$ and $\{E_{g_1 v_1}, E_{g_2 v_2}\}$,
    where s'' is 1-D and its moving domain occupies an edge from
    (x_{f_1}, y_{v_f}) to (x_{f_1}, y_{v_g});
    if $l_{C_3} < l_{E_{g_1 v_1}} + l_{E_{g_2 v_2}}$, *then*
    {

        remove $E_{g_1 v_1}$ and $E_{g_2 v_2}$;
        if $l_{E_{f_1 s''}^{max}} < l_{max}$ or $l_{E_{g_2 s''}^{max}} < l_{max}$,
            remove the edge set specified by the length of l_{max};
        else if $l_{E_{f_1 s''}^{max}} > l_{max}$, then remove $E_{f_1 s''}$;
        else remove $E_{g_2 s''}$;

    }
    else
    {

        remove C_3;
        remove the edge set whose length is the maximum over
        $E_{g_1 v_1}$, $E_{g_2 v_2}$, and edge sets specified by l_{max} and l'_{max}.

    }

}

According to step (c) of procedure CONNECT_LAST_POINT_SET, when edge sets $E_{g_1 v_1}$ and $E_{g_2 v_2}$ and edge sets specified by l_{max} and l'_{max} have the same length, there are, at most, four different extended edge-set trees, in that only one- and three-edge-set NN cliques exist. Including the extended edge-set trees having an NN clique whose main end vertices lie on two opposite sides of the rectangle, there are, at most, six extended edge-set trees for the general solution of the switch-box problem.

The linear algorithm proposed by Cohoon et al. for the switch-box problem was considered to be the best achievement in applying the theoretical research on the rectilinear Steiner's problem. In their algorithm, ten topologies should be computed [23]. The computation in our algorithm, however, is equivalent to the computation of one topology.

Example: The configuration of Figure 3.36 is used to explain the implementation of the algorithm finding the solution for the switch-box problem. The point sets on four sides, P, Q, U, V, of rectangle R are

$P = \{p_1, p_2\}$, $Q = \{q_i | i = 1, ..., 4\}$, $U = u$, and $V = \{v_1, v_2\}$, respectively.

Sorting nodes in $P \cup Q$ in terms of x-coordinates generates a sorted sequence, $S_{PQ}^* = \{Q_1, P_1, Q_2\}$, where $Q_1 = \{q_1, q_2, q_3\}$, $P_1 = \{p_1, p_2\}$, and $Q_2 = q_4$. Sorting nodes in $U \cup V$ in terms of y-coordinates generates a sorted sequence, $S_{UV}^* = \{V_1, U_1, V_2\}$, where $V_1 = v_1$, $U_1 = u$, and $V_2 = v_2$.

The finding starts with connecting U to S_{PQ}^*. Connecting the first node, $q_1 \in Q_1 \in S_{PQ}^*$, to the last node, $u \in U$, generates $T' = E_{q_1 u}$. Connecting nodes in the same subsequence, Q_1, with straight-edge sets generates $T' = T' \cup \bigcup_{i=1}^{2} E_{q_i q_{i+1}}$. Comparing the total length of edge sets in $\{E_{p_1 s}, E_{p_2 q_4}\}$, the total length of edge sets in $\{E_{p_1 q_3}, E_{p_2 q_4}\}$, and the length of clique $C_3 = E_{p_1 s_1} \cup E_{q_3 s_1} \cup E_{q_4 s_1}$ determines that the length of the clique is the minimum. Thus, $T' = T' \cup C_3$. Furthermore, $E_{p_1 s_1}$ and the maximum length, $l_{max} = l_{E_{p_1 s_1}}$, are specified for detecting the redundant length in the edge-set cycle completed by connecting edge sets in V to T'. *FIND_FOR_SWITCH_BOX* is not necessary in this example because connecting S_{PQ}^* has been finished.

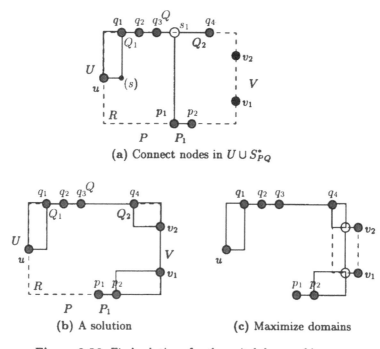

(a) Connect nodes in $U \cup S_{PQ}^*$

(b) A solution   (c) Maximize domains

Figure 3.36 Find solutions for the switch-box problem

Because q_4 is the last node in the last subsequence in S_{PQ}^*, $T' = T' \cup E_{q_4 v_2}$. Connecting v_1 to v_2 generates $T' = T' \cup E_{v_1 v_2}$. Comparing the distance between p_2 and v_1 and the distance between p_2 and $E_{q_4 v_2}$ establishes that $l_{E_{p_2 v_1}}$ is shorter. Note that there is an edge-set cycle in $T' \cup E_{p_2 v_1}$ and that $l_{max} = l_{E_{p_1 s_1}} > l_{E_{p_2 v_1}}$. Thus, $T' = (T' - E_{p_1 s_1}) \cup E_{p_2 v_1}$ (Figure 3.36b). Maximizing the moving domain of each Steiner point generates T' in Figure 3.36c.

The NN clique $C_{mn} \subset T$, whose main vertices, p and q, lie on U and V, respectively, is generated by determining a rectangle, R_{pq}, connecting U and V, where $p = u$, $q = (x_v, y_u)$ in this example. Thus, $l_{C_{mn}} = l_h + min(1,2)l_v + (2-1)l_d$, where l_d is the distance between side Q and e_{pq}. Then, $l_T = l_{C_{mn}} + l_S$, where $l_S = l_U + l_V + l_{P_1} + l_{Q_1} + l_{Q_2}$ and l_X is the total length of edge sets in X. Similarly, another NN clique, C'_{mn}, whose main vertices, p' and q', lie on P and Q, respectively, can be found for computing the length of another edge-set tree.   ◇

3.9.2 LINEAR ALGORITHMS FOR MRSTS CONNECTING NODES ON THE BOUNDARY OF A POLYGON

The terminals of a circuit board or a multichip module (MCM) are often located on the boundary of a polygon, where the signal wires are routed. Finding MRSTs connecting terminals on the boundary of a polygon is significant for routing. A side adjacent to two 270° inner angles of a polygon is called a *concave side* (Figure 3.37). A polygon with a concave side is called a *concave polygon*; a polygon with no concave side is called a *convex polygon*. The switch-box problem is a special problem in finding an MRST connecting nodes on the boundary of a convex polygon.

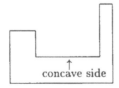

Figure 3.37 A concave side in a concave polygon

Special MRST Problem Theorem 2 *MRSTs for a set of nodes on the boundary of a convex polygon are confined by the boundary.*
Proof: Let s be a Steiner point in an NR edge-set tree, T_{NR}, connecting a set of nodes on the boundary of a convex polygon, G. Assuming s to be located outside G, s must be adjacent to three points (Steiner points or nodes) on three sides of the polygon (see Property 1). Because no concave side exists in G, a redundant length must exist in an edge set with s as its vertex (Figure 3.38). However, this conflicts with the condition that s is a Steiner point in an NR edge-set tree. Thus, it is

impossible to locate s outside G. Because an MRST must be contained in an NR edge-set tree, an arbitrary subtree of an MRST connecting a set of nodes on the boundary of G cannot exist outside G. □

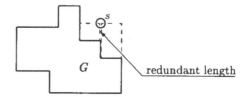

Figure 3.38 A redundant length in one of
three edge sets intersecting at s

According to Special MRST Problem Theorem 2, MRSTs connecting nodes on the boundary of a convex polygon can be found in the polygon. By projecting nodes on the boundary of a convex polygon to the boundary of a rectangle, Special MRST Problem Theorem 1 for the switch-box problem can be used in finding MRSTs connecting terminals on the boundary of a convex polygon. Thus, linear algorithms exist for this problem.

An algorithm finding MRSTs connecting nodes on the boundary of a convex polygon also consists of three major steps: sorting nodes in terms of x- and y-coordinates, respectively, finding an edge-set tree including one- and three-edge-set NN cliques, and computing the length of each possible NN clique connecting nodes on four or more sides.

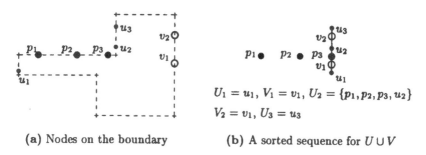

$U_1 = u_1$, $V_1 = v_1$, $U_2 = \{p_1, p_2, p_3, u_2\}$
$V_2 = v_1$, $U_3 = u_3$

(a) Nodes on the boundary   (b) A sorted sequence for $U \cup V$

Figure 3.39 Nodes on the boundary of a polygon

A sorted sequence is slightly different from that used in finding the solution for the switch-box problem. For example, a sorted sequence, S_{UV}^*, in terms of vertical (horizontal) coordinates, includes nodes lying on horizontal (vertical) sides (Figure 3.39). Thus, both horizontal and vertical coordinates can be considered in connecting a node. According to Figure 3.39b, the connection of p_3 is determined by $dist(p_3, p_2)$,

$dist(p_3, v_1)$, and $dist(p_3, E_{p_1 u_1})$, where $dist(X, Y)$ indicates the distance between X and Y.

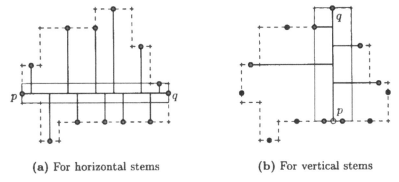

(a) For horizontal stems (b) For vertical stems

Figure 3.40 Computing the length of each NN clique

If a solution is contained by an edge-set tree including an NN clique connecting nodes on four or more sides, the main vertices of the NN clique must lie on two sides whose projections on each other are not empty (Figure 3.40). For example, if the stem of the NN clique is horizontal (Figure 3.40a), two vertical sides connected by a rectangle constitute a side pair holding main vertices, p and q, selected for computing the length of an NN clique and then the length of an edge-set tree. The solution is contained in the edge-set tree whose length is the minimum.

If all terminals lie on the boundary of a concave (maze) polygon (Figure 3.41), and if all wires are confined by this polygon, sorted sequences similar to that shown in Figure 3.39b can be generated. Selecting main vertices lying on two sides whose projections on each other are not empty, the length of each NN clique connecting nodes on four or more sides can be computed. There may be more than one NN clique in an edge-set tree containing a solution. Thus, a linear algorithm also exists for finding MRSTs confined in a concave polygon. However, an obstacle formed by the area outside the concave polygon has been considered in the solution.

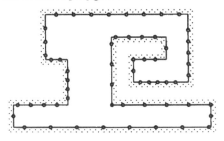

Figure 3.41 A maze concave polygon

3.9.3 A NEARLY LINEAR ALGORITHM
FINDING AN NR EDGE-SET TREE

An edge-set tree can be expressed as a binary tree for computer data processing. Edge sets in the edge-set tree correspond to the lines of a binary tree, one by one. A line, pq, in a binary tree, t_b, is expressed as $pg \in t_b$; removing pq from t_b is expressed as $t_b - pq$; and connecting a line, fg, to t_b is expressed as $t_b \cup fg$.

Connecting two nodes in a binary tree generates a *cycle*. A cycle that does not embed other branches is called a *no-embedded-branch cycle (NEB cycle)*. Correspondingly, an edge-set cycle that does not embed other edge-set subtrees is called a *no-embedded-branch edge-set cycle (NEB edge-set cycle)*.

To relate NEB cycles generated from an edge-set tree to NEB cycles generated from a corresponding binary tree one by one, we propose a *vector binary tree* in which the coordinates of sons and their father and grandfather determine which son is on the left (right). Letting two sons, the father and the grandfather be s_1, s_2, f, and g, respectively, a vector binary tree, t_{vb}, is built by the rule as follows:

(1) Fix g and swing edge set E_{fg} clockwise or counterclockwise in an edge-set tree.

(2) If E_{fs_1} is pointed by E_{fg} prior to E_{fs_2} in the edge-set tree, then s_1 is the left son of f in t_{vb}; otherwise, s_2 is the left son of f.

(3) If f has only one son, and g, f, and s lie on the same line in the edge-set tree, then s is the left (right) son of f in t_{vb} if f is the left (right) son of g.

If f has only one son, s, and if s, f, and g do not lie on the same line, then set $s = s_1 = f$ or $s = s_2 = f$ in (2) of the rule. According to the swinging direction, a vector binary tree is either clockwise or counterclockwise. Only counter-clockwise binary trees are used in this section.

Special MRST Problem Theorem 3 *If an NEB edge-set cycle, C_y, exists in $T \cup E_{pq}$, there must exist an NEB cycle, $c_y \subset t_{vb} \cup pq$, where t_{vb} is a vector binary tree corresponding to edge-set tree T, and pq corresponds to E_{pq}.*

Proof: Let $E_{fg} \cup E_{fs_1} \cup E_{fs_2} \subset T$ correspond to branch $\{fg, fs_1, fs_2\} \subset t_{vb}$, where $E_{fs_1} \cup E_{fs_2} \subset C_y \subset T \cup E_{pq}$ and $\{fs_1, fs_2\} \subset c_y \subset t_{vb} \cup pq$.

Assuming that fg is embedded in c_y, g must be a son of f. Let f, s_2 and p be the left son of s_1, the left son of f, and the lth generation son of s_2, respectively, where $l \geq 1$. Then, g, s_1, and q must be the right son of f, the l'th generation son of s_0 that must have a right son, s_3, and the rth generation son of s_3, respectively, where $l' \geq 0$, $r \geq 0$, and $s_0 s_3 \in c_y$ (Figure 3.42a).

Using the rule of counterclockwise swinging, no Steiner point exists in the edge-set subtree from s_0 to f. Moreover, E_{fg} must be on the right of E_{fs_2}, and $E_{s_0s_3}$ must be on the right of $E_{s_0s_1'}$, where s_1' is a son of s_0 (Figure 3.42b). Because f is the left son of s_1, and because q is the rth generation son of s_3, p is an end vertex of the edge-set subtree from s_2 to p, and q is an end vertex of the edge-set subtree from s_3 to q. Thus, E_{fg} is embedded in C_y. This conflicts with the given condition. Therefore, it is impossible to embed fg in c_y.   □

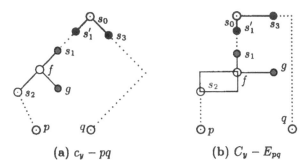

(a) $c_y - pq$ (b) $C_y - E_{pq}$

Figure 3.42 An NEB cycle must correspond to an NEB edge-set cycle

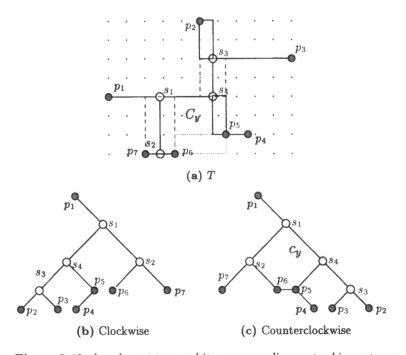

(a) T

(b) Clockwise   (c) Counterclockwise

Figure 3.43 An edge-set tree and its corresponding vector binary trees

Example: Figure 3.43 shows an edge-set tree and its corresponding vector binary trees. A binary tree shown in Figure 3.43c is built in the following process:

(1) Select p_1 (or other node) as the root; s_1 is either the left son or the right son.

(2) Swinging $E_{p_1 s_1}$ counterclockwise, $E_{s_1 s_2}$ is pointed by $E_{p_1 s_1}$ prior to $E_{s_1 s_4}$; thus, s_2 is the left son and s_4 is the right son of s_1.

(3) Following subprocess (2), other branches can be determined.

If an NEB edge-set cycle can be formed by connecting a pair of nodes or Steiner points in the edge-set tree in Figure 3.43a, then there must exist a corresponding NEB cycle formed by connecting two corresponding nodes in a vector binary tree in Figure 3.43b or 3.43c. Cycle $c_y = \{p_6 s_2, s_2 s_1, s_1 s_4, s_4 p_5, p_5 p_6\}^*$ in Figure 3.43c corresponds to edge-set cycle $C_y = \{E_{p_6 s_2}, E_{s_2 s_1}, E_{s_1 s_4}, E_{s_4 p_5}, E_{p_5 p_6}\}^*$ in Figure 3.43a.   ◇

A node adjacent to only one line in a branch of a binary tree is called a *tip* of the branch. A left (right) son having only one right (left) son in a binary tree is a *flex point*. A flex point that is the left (right) son and has a right (left) son is a *left (right) flex point*. Tips and flex points can be found by traversing a binary tree. Based on the traversing precedence, tips form a *tip list*, and flex points between two tips form a *flex point list*. Connecting any pair of adjacent tips in the tip list generates an NEB cycle.

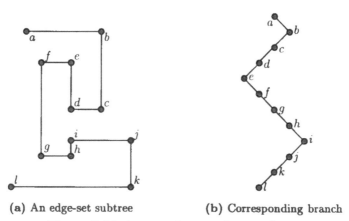

(a) An edge-set subtree   (b) Corresponding branch

Figure 3.44 A maze edge-set subtree and
its corresponding branch in a vector binary tree

Example: Traversing the binary tree t_{vb} in Figure 3.43c generates a tip list: $\{p_1, p_7, p_6, p_4, p_3, p_2\}^*$. Connecting any pair of adjacent tips in the tip list generates an NEB cycle. Node p_5 is a right flex point in Figure 3.43b and is a left flex point in Figure 3.43c.

An NEB edge-set cycle formed by connecting a and l in Figure 3.44a corresponds to an NEB cycle formed by connecting a and l of the vector binary tree in Figure 3.44b. In a branch of a vector binary tree in Figure 3.44b, a flex point list between a and l is $\{b, e, i\}$, where b and i are right flex points and e is a left flex point. ⋄

An NEB cycle, $c_y \subset t_{vb} \cup pq$, is called a *triangular cycle* if no more than one flex point exists between p and q. A cycle, $c_y \subset t_{vb} \cup pp' \cup qq'$, embedded in an NEB cycle is called a *convex cycle*, if $\{p, p'\}$ is a pair of left (right) flex points, $\{q, q'\}$ is a pair of right (left) flex points, and two triangular cycles exist in $t_{vb} \cup pp'$ and $t_{vb} \cup qq'$, respectively.

Let T be an edge-set tree, and $C_y \subset T \cup E_{pq}$ be an NEB edge-set cycle corresponding to a convex cycle. In order to find a linear-time algorithm to detect redundant lengths of edge sets in an NEB edge-set cycle, we use double pointers, i and j, moving backward in the corresponding convex cycle. The algorithm for advancing these pointers to detect redundant lengths is as follows:

(1) Initially, i and j point to p and q, respectively; and the distance, $d(E_{ii+}, E_{jj+})$, between E_{ii+} and E_{jj+} is set to be the minimum distance d_{kl} for connecting two points, k and l, to form an NEB edge-set cycle, where $k = i$, $l = j$, and i^+ and j^+ point to the fathers of i and j, respectively, in a vector binary tree.

(2) If $l_{E_{ii+}} > d_{kl}$, then replace E_{ii+} with E_{kl}, and exit.

(3) If $l_{E_{jj+}} > d_{kl}$, then replace E_{jj+} with E_{kl}, and exit.

(4) If $d(E_{i+i++}, E_{jj+}) \leq d(E_{j+j++}, E_{ii+})$, then $i = i^+$.
   If $d(E_{i+i++}, E_{jj+}) < d_{kl}$,
      then $d_{kl} = d(E_{i+i++}, E_{jj+})$, $k = i^+$ and $l = j$.

(5) If $d(E_{j+j++}, E_{ii+}) \leq d(E_{i+i++}, E_{jj+})$, then $j = j^+$.
   If $d(E_{j+j++}, E_{ii+}) < d_{kl}$,
      then $d_{kl} = d(E_{j+j++}, E_{ii+})$, $k = i$ and $l = j^+$.

(6) If i and j meet each other in the backward movement, then exit; otherwise, go to (2).

In the algorithm for detecting redundant lengths in a convex cycle, the edge set connecting E_{i+i++} and E_{jj+} or connecting E_{j+j++} or E_{ii+} is called a *step edge set*. It determines which pointer should be advanced.

Example: In NEB cycle $c_y = \{p_6s_2, s_1s_2, s_1s_4, s_4p_5, p_5p_6\}^*$, shown in Figure 3.43c, the search starts by setting i and j pointing to p_5 and p_6, respectively, where p_5p_6 corresponds to an edge set connecting p_5 and p_6 in Figure 3.43a. In the backward search in t_{vb}, i and j are moved to their fathers by the following: (1) Initially set $d_{kl} = d(E_{p_6s_2}, E_{s_4p_5})$ because $d(E_{p_6s_2}, E_{s_4p_5}) < d(p_5p_6)$. (2) If $d(E_{s_1s_2}, E_{p_5s_4}) < d(E_{p_6s_2}, E_{s_1s_4})$, i is moved to s_2; otherwise, j is moved to s_4. However, $d(E_{p_6s_2}, E_{s_4p_5}) <$

$l_{E_{s_1 s_2}}$ is found and $E_{s_1 s_2}$ is replaced with an edge set connecting $E_{p_6 s_2}$ and $E_{s_4 p_5}$. In this example, the backward progression in c_y is terminated in (2); otherwise, it is not terminated until $i = j$.   ◇

Let $r_1, r_2, ..., r_z$, be tips between tips p and q in the tip list of t_{vb}. If $c_y \subset t_{vb} \cup pq$ is not an NEB cycle, and if tip r_h ($1 \le h \le z$) is the first one hit by a step edge set generated in moving pointers i and j backward starting from p and q, then two branches traversed by i and j are called a pair of *tails*, with tips p and q. If there are a total of m tips in the tip list for t_{vb}, there are, at most, $C(m, 2) - (m - 1) = (m - 1)(m - 2)/2$ pairs of tails. Therefore, finding redundant lengths in tails spends $O(m^2)$.

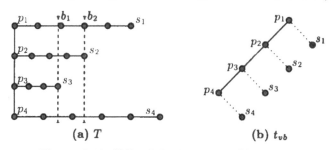

(a) T (b) t_{vb}

Figure 3.45 Tails of tips in a vector binary tree

Example: Vector binary tree t_{vb} in Figure 3.45b corresponds to edge-set tree T in Figure 3.45a. A tip list for t_{vb} is $S^* = \{s_i | i = 1, ..., 4\}^*$. Thus, there are a total of three pairs of tails for tip pairs $\{s_1, s_3\}$, $\{s_1, s_4\}$ and $\{s_2, s_4\}$. The tail pair with tips s_1 and s_3 is empty because s_2 meets the step edge set $E_{s_1 s_3}$ in the initial step. The tail pair with tips s_2 and s_4 includes edge-set subtrees between s_2 and line b_1 and between s_4 and b_1, respectively.   ◇

Using Special MRST Problem Theorem 3 and the algorithm for detecting redundant lengths from both convex cycles and tails, the algorithm finding an NR edge-set tree can be expressed as:

(1) Generate a vector binary tree, t_{vb}, from T.

(2) Traverse t_{vb} to find a tip list and flex point lists, determine each triangular cycle and each convex cycle formed by connecting tips and flex points.

(3) Detect redundant lengths from each triangular cycle and each convex cycle.

(4) Detect redundant lengths from each pair of tails.

(5) If no redundant length is found, then exit; otherwise, go to (2).

Because redundant lengths can be detected in each convex cycle and in each pair of tails, individually, this algorithm satisfies the requirement for parallel computing.

Using the vector binary tree, an edge-set tree, $T(C_1, C_3)$, tested by three-edge-set cliques from an NR edge-set tree, can be found by a similar algorithm:

(1) Generate a vector binary tree, t_{vb}, from T.

(2) Traverse t_{vb} to find a tip list and flex point lists.

(3) Find a pair of step edge sets from two adjacent NEB cycles formed by connecting three adjacent tips.

If their total length is less than that of the three-edge-set subtree whose inner vertex is a common Steiner point of the adjacent NEB cycles, then replace the three-edge-set subtree of T with the step edge sets.

(4) Using three pointers in three tails, find a three-edge-set subtree formed by a step edge set connecting two tails; then find an edge set that connects the step edge set and an edge set in the third tail into a three-edge-set subtree.

Connect the three-edge-set subtree to T to form two cycles (not necessarily an NEB cycle).

In two cycles, if the total length of a pair of edge sets or a three-edge-set subtree of T is less than that of the new three-edge-set subtree, replace the edge set pair or the three-edge-set subtree in T with the new three-edge-set subtree.

(5) If no redundant length is found, then exit; otherwise, go to (2).

The time complexity of this algorithm is dominated by steps (3) and (4), and is less than $O(n^2)$.

3.9.4 THE MINIMUM NUMBER OF NODES CONNECTED BY AN NR EDGE-SET TREE CONTAINING A SET OF MRSTS

An NR edge-set tree may contain a set of MRSTs when the number of nodes is limited. Thus, finding an exact solution in simple cases can be transferred to finding NR edge-set trees.

Clique Lemma 5 *If there exists a pair of edge sets that can replace a three-edge-set clique (subtree) in an NR edge-set tree to generate a new edge-set tree whose length is shorter than that of the NR edge-set tree, then at least seven nodes exist in the NR edge-set tree.*

Proof: Let six nodes be connected by three edge-set subtrees that are connected by a three-edge-set clique (subtree), C_3, to form an NR edge-set tree, T_{NR}. There are three cases for connecting nodes in three edge-set subtrees:

(1) Every edge-set subtree connects two nodes.

(2) There are 3, 2, and 1 nodes in the three edge-set subtrees, respectively.

(3) There are 4, 1, and 1 nodes in the three edge-set subtrees, respectively.

For case (1), each edge-set subtree is a degenerated clique (Figure 3.46a). Moving r' to the location of q (the lowest location in $M_{r'}$), the relation among $l_{E_{f_s}}$, $l_{E_{p_q}}$ and l_{C_3} can be expressed by Figure 3.46b, where l_X indicates the length of X and $C_3 = E_{r_s} \cup E_{r'_s} \cup E_{p_s}$. Because $\{E_{r'_s}, E_{r_s}\} \subset T_{NR}$, we have $l_{E_{r'_s}} < l_{E_{f_s}}$ and $l_{E_{r_s}} \leq l_{E_{f_s}}$. Because $l_{E_{p_q}} = l_{E_{r'_s}} + l_{E_{p_s}}$, we have $l_{C_3} = l_{E_{r_s}} + l_{E_{r'_s}} + l_{E_{p_s}} \leq l_{E_{f_s}} + l_{E_{p_q}}$. Thus, it is impossible to generate a new edge-set tree whose length is less than $l_{T_{NR}}$ by replacing $C_3 \subset T_{NR}$ with two edge sets if there are less than seven nodes in T_{NR}. Similarly, the same conclusion can be proven for cases (2) and (3).   □

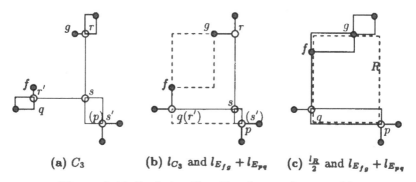

(a) C_3 (b) l_{C_3} and $l_{E_{f_s}} + l_{E_{p_q}}$ (c) $\frac{l_R}{2}$ and $l_{E_{f_s}} + l_{E_{p_q}}$

Figure 3.46 Replacing $C_3 \subset T_{NR}$ (or an edge set pair)
with an edge-set pair (or $C_3 E_{r_s} \cup E_{r'_s} \cup E_{s_s'}$)

Clique Lemma 6 *If there exists a three-edge-set clique (subtree) that can replace a pair of edge sets in an NR edge-set tree to generate a new edge-set tree whose length is shorter than that of the NR edge-set tree, then there exist, at least, seven nodes in the NR edge-set tree.*

Proof: Assume $\{E_{fg}, E_{pq}\} \subset T_{NR}$, and assume there exists a three-edge-set clique (subtree), C_3, that can connect edge-set trees for three subsets of nodes in $T_{NR} - E_{fg} - E_{pq}$ to generate a new edge-set tree whose length is less than $l_{T_{NR}}$. Thus, there must exist a minimum rectangle, R, that connects edge-set trees in $T_{NR} - E_{fg} - E_{pq}$ and encloses E_{fg} and E_{pq} (Figure 3.46c). Because a half perimeter, $l_R/2$, of R equals l_{C_3}, we have $l_R/2 \leq l_{E_{f_g}} + l_{E_{p_q}}$.

However, $l_{E_{f_g}}$ is equal to or less than the distance, $dist(g, E_{pq})$, between g and E_{pq}, due to E_{fg} in T_{NR}. Thus, $l_R/2 = dist(g, E_{pq}) + l_{E_{p_q}} \geq l_{E_{f_g}} + l_{E_{p_q}}$. Therefore, C_3 cannot replace E_{fg} and E_{pq} to generate a new edge-set tree whose length is less than $l_{T_{NR}}$.   □

Special MRST Problem Theorem 4 *An NR edge-set tree connecting six or fewer nodes contains a set of MRSTs.*

Proof: According to Clique Lemmas 5 and 6, an NR edge-set tree for six nodes is always an edge-set tree, $T(C_1, C_3)$, tested by C_1 and C_3. As in the proofs of Clique Lemmas 5 and 6, an NR edge-set tree for six nodes is always an edge-set tree, $T(C_1, C_3, ..., C_9)$, for six nodes. According to Clique Theorems 3 and 10, this theorem is proven. $\square$

3.10 THE MRST ROUTING

An *MRST routing* is a routing technique using MRSTs, rather than finding a wire connecting two points. This section introduces the figure of merit for wire congestion to explain the benefits contributed by MRST routing. Furthermore, MRST routing techniques based on the solution of Steiner's problem are presented.

3.10.1 FIGURE OF MERIT FOR WIRE CONGESTION

Let a single-layer workspace be gridded into squares; the size of a grid (square) is significantly larger than that of a pin or a via. The number of nets crossing a horizontal or vertical grid edge is called the *wire density* of the vicinity of that grid edge, and is denoted as d_w. The *wire capacity*, c_w, of a grid edge is the allowed wire density. Because the diameter of a via is much less than the length of a grid edge, the effect of vias can be ignored in computing the wire density. Placing all nets on one gridded layer generates a *density topography*, in which wire density is considered to be a topographical height. A two-dimensional map recording the density of each grid edge is called a *topography map (T-map)*. The vicinity of a grid edge on which the density is the maximum (minimum) is called the *top (valley)* of the topography. Slicing the topography yields an individual layer.

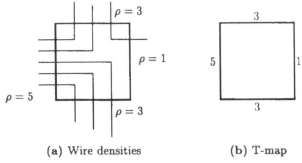

(a) Wire densities   (b) T-map

Figure 3.47 Wire densities in a T-map

Example: Figure 3.47a shows the wire density in each grid edge. The effect of vias is ignored. A wire density of $d_w = 5$ means that the wire density in the vicinity of the left vertical grid edge is 5. Figure 3.47b shows the T-map corresponding to the wire density of each grid edge in Figure 3.47a.   ◇

Indicating the maximum wire density and the minimum wire density in a T-map as ρ_{max} and ρ_{min}, respectively, the minimum number of layers required for routing all nets, N_{layer}, can be computed by

$$N_{layer} = even(\frac{2\rho_{max}}{c_w}) \tag{3.27}$$

where $even(x) = 2(i+1)$, when $2i < x \le 2(i+1)$ for integer i. Horizontal wires and vertical wires are routed on different sandwiching layers. According to Equation 3.27, the maximum wire density, δ_{max}, in each layer is computed by

$$\delta_{max} = \frac{\rho_{max}}{N_{layer}/2} \tag{3.28}$$

According to Equations 3.27 and 3.28, $\delta_{max} \approx c_w$.

Example: If, at most, 10 wires are allowed to cross a grid edge on a layer, then $c_w = 10$. If all nets are placed on a single layer and the maximum wire density is 92, then $\rho_{max} = 92$, the minimum number of layers required for routing is $N_{layer} = even(2 \times 92/10) = even(18.4) = 20$ and $\delta_{max} = 92/10 = 9.2 \approx c_w$.   ◇

Let N_ρ be the number of grid edges with wire density ρ in a T-map. The total routing length, L_{total}, can be computed by

$$L_{total} = l_{edge} \sum_{\rho=\rho_{min}}^{\rho_{max}} \rho N_\rho \tag{3.29}$$

where l_{edge} is the length of a grid edge. The relation between N_ρ and the total number, N_{edge}, of grid edges in a T-map can be expressed as

$$N_{edge} = \sum_{\rho=\rho_{min}}^{\rho_{max}} N_\rho \tag{3.30}$$

For a workspace with $N_x \times N_y$ grids, $N_{edge} = (N_x - 1)N_y + N_x(N_y - 1)$ when all grid edges are available. Let a wire segment and its neighboring clearance be accommodated by a track segment whose length equals that of a grid edge. The number, n_{track}, of all total available track segments on one layer can be computed by Equation 3.31:

$$n_{track} = \frac{1}{2}c_w N_{edge} = \frac{L_{track}}{l_{edge}} \tag{3.31}$$

where L_{track} is the total length of all available track segments on one layer and is a constant for a layer.

According to Equations 3.29 and 3.30, the average wire density, ρ_{ave}, in a T-map can be expressed as

$$\rho_{ave} = \frac{L_{total}}{l_{edge} N_{edge}} \tag{3.32}$$

If $\rho_{ave} \ll \rho_{max}$, then wires are concentrated in the vicinity of a grid edge whose wire density equals or is nearly equal to ρ_{max}; thus, wire congestion occurs in the vicinity of that grid edge. Minimizing $\rho_{max} - \rho_{ave}$ will digest or eliminate the wire congestion. Therefore, the wire congestion can be measured by a *figure of merit for congestion*, defined as

$$m_c = \frac{\rho_{max} - \rho_{ave}}{\rho_{max}} = 1 - \frac{\sum_{\rho=\rho_{min}}^{\rho_{max}} \rho N_\rho}{\rho_{max} \sum_{\rho=\rho_{min}}^{\rho_{max}} N_\rho} \tag{3.33}$$

Because ρ_{max} is dominated by the distributions of wires, the figure of merit is a function in which wire distributions are major variables and the total wire length is a secondary variable. According to Equation 3.33, the figure of merit is not sensitive to the size of the grids.

Example: Figure 3.48 shows m_c for two different T-maps corresponding to the same terminal (node) configuration. In Figure 3.48a, $\rho_{min} = 0$, $\rho_{max} = 3$, $N_0 = 15$, $N_1 = 7$, $N_2 = N_3 = 1$, the total wire length is $\sum_{\rho=0}^{3} \rho N_\rho = 7 + 2 + 3 = 12$ units, total number of grid edges $\sum_{\rho=0}^{3} N_\rho = 15 + 7 + 1 + 1 = 24$ and $m_c = 1 - \sum_{\rho=1}^{3} \rho N_\rho / (\rho_{max} \sum_{\rho=0}^{4} N_\rho) = 1 - (1 \times 7 + 2 \times 1 + 3 \times 1)/(3 \times 24) = 0.833$. Figure 3.48b shows that ρ_{max} is reduced from 3 to 1 by selecting alternative nets; the congestion is also improved: $m_c = 1 - \rho N_\rho / \sum_{\rho=0}^{1} N_\rho = 1 - 12/24 = 0.5$. ◇

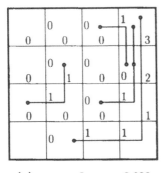 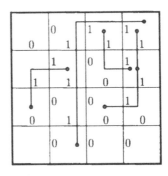

(a) $\rho_{max} = 3$, $m_c = 0.833$ (b) $\rho_{max} = 1$, $m_c = 0.5$

Figure 3.48 Wire density and m_c

According to Equations 3.32 and 3.33, m_c can be expressed as

$$m_c = 1 - \frac{L_{total}}{l_{edge}\rho_{max}N_{edge}} = 1 - \frac{c_w L_{total}}{2\rho_{max}L_{track}} \qquad (3.34)$$

Let δ_{ave} be the average wire density of a layer. Then

$$\delta_{ave} = \frac{\rho_{ave}}{N_{layer}/2} \qquad (3.35)$$

and

$$m_c = 1 - \frac{\delta_{ave}}{\delta_{max}} \qquad (3.36)$$

If $\delta_{ave} \approx \delta_{max}$, all nets are uniformly routed in every layer and almost no available track is wasted. If $\delta_{ave} \ll \delta_{max}$, most of the available tracks are wasted. According to Equations 3.28, 3.35 and 3.36, $m_c \approx 1 - \delta_{ave}/c_w$. Therefore, m_c represents approximately the average ratio of the wasted tracks to all available tracks, or represents the average ratio of the wiring capacity wasted to the wire congestion in every layer. In most current routing techniques, $m_c \geq 50\%$. Thus, there is still a large margin for routing improvement.

According to Equation 3.34, the relationship between m_c and ρ_{max} can be expressed as

$$\rho_{max} = \frac{c_w L_{total}}{2(1 - m_c)L_{track}} \qquad (3.37)$$

According to Equations 3.27 and 3.37, the relationship among the number of layers, the total wire length, and the area occupied by wires can be expressed as

$$N_{layer} = even(\frac{L_{total}}{(1 - m_c)L_{track}}) \qquad (3.38)$$

For convenience, the function of $even(x)$ can be removed and Equation 3.38 can be expressed as

$$N_{layer} = \frac{L_{total}}{(1 - m_c)L_{track}} \qquad (3.39)$$

When chips on a board and the area of the board are fixed, L_{track} is a constant, and N_{layer} is determined by $L_{total}/(1 - m_c)$. Because m_c is dominated by the wire positions, both the minimization of L_{total} and the position selection of each wire must be considered in the minimization of N_{layer}. The general solution of Steiner's problem provides both the minimum wire length and different solutions for route selections (Figure 3.49).

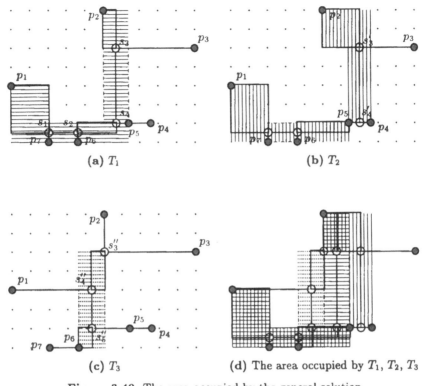

(a) T_1 **(b)** T_2

(c) T_3 **(d)** The area occupied by T_1, T_2, T_3

Figure 3.49 The area occupied by the general solution

When L_{total} reaches the minimum, N_{layer} can be minimized further by selecting wire positions. If m_c is decreased to $m_c - \Delta m_c$ by increasing L_{total} to $L_{total} + \Delta L_{total}$, and if N_{layer} is also reduced, then

$$\frac{L_{total} + \Delta L_{total}}{1 - (m_c - \Delta m_c)} < \frac{L_{total}}{1 - m_c} \tag{3.40}$$

Equation 3.40 can be simplified as

$$\frac{\Delta L_{total}}{L_{total}} < \frac{\Delta m_c}{1 - m_c} \tag{3.41}$$

Therefore,

$$\Delta L_{total} < \Delta m_c N_{layer} L_{track} = 2\Delta m_c \rho_{max} L_{track}/c_w \tag{3.42}$$

Reducing m_c to $m_c - \Delta m_c$ is equivalent to reducing a length, $L_{total} + 2\Delta m_c \rho_{max} L_{track}/c_w$, to the total length L_{total}. If the increase in the total routing length is less than $2\Delta m_c \rho_{max} L_{track}/c_w$, the number of layers is reduced.

3.10.2 AN MRST ROUTING

Commercial routers are mainly based on pin-to-pin and net-by-net sequential routing that must follow obstacles created by previously routed wires [114]. However, the shortest network to connect a signal set is an MRST, which cannot be generated by pin-to-pin connection techniques. Thus, current commercial routers cannot minimize the total wire length of a PCB. Furthermore, net-by-net routing on a board cannot properly assess the local congestion. Consequently, the result is more incomplete wires or more layers needed to accommodate the nets.

Although commercial routers are supposedly state-of-the-art, most made less than 50% routing area available. According to the "20/80 rule" in PCB design, 80% of the wires on a board may be updated or rerouted to make space for 20% incomplete wires.

To speed up routing, parallel computing based on a "divide and conquer" strategy is used in hardware acceleration (micro-processor level). In estimating wire congestion in global routing, a multiterminal net is divided into two-terminal nets in terms of the connection of a minimum spanning tree, whose length is 12% longer than that of an MRST, on average. In consequent detailed routing, the routing area is divided into subareas. Because the routing length is longer, because local congestion cannot be properly estimated, and because nets may cross several subareas, the result is worse than that generated by a sequential routing.

The pin-to-pin and net-by-net techniques have a thirty-five year history and were improved in various aspects, such as algorithm, programming techniques, and hardware accelerations. It is difficult to improve them further. To achieve a breakthrough in auto-router design, a new approach is needed. Based on the theory for Steiner's problem, the algorithm expansions, and the efficient algorithm for the general solution, the total routing length and wasted routable area can be minimized. Thus, we propose an MRST router using the following process:

(1) Find the general MRST solution for each signal set (a set of signal terminals).

(2) Construct a density topography and its corresponding T-map by assigning nets for each signal set to one layer.

(3) Find one net from the nets for each signal set to minimize the maximum density.

(4) Compact the valley (low-density area) and depress the top (high-density area) to equalize density in the T-map.

(5) Minimize the number of layers and vias by determining nets or subsets on each layer.

In steps (1) and (3), an MRST solution and the best net selection for each signal set can be computed independently. Thus, these steps satisfy the requirement of parallel computing. Moreover, parallel computing can be implemented starting from "valleys" and "tops", respectively, in step (4). In subset assignments of step (5), parallel processing can be continued in each subarea. The consequent detailed routing is completed by a process similar to step (4).

Example: Assigning the MRST solutions for the three signal sets in Figure 3.50 to one layer generates a T-map corresponding to the topography constructed by the solutions (Figure 3.51). The total density on an edge of an edge set is called the *edge weight* of the edge. The maximum weight of an edge in Figure 3.51a is $2 + 3 + 2 + 2 = 9$ (Figure 3.51b).

Selecting an edge whose weight is the minimum from each bent-edge set or each moving edge set generates a density topography in which the total routing length is the minimum and the maximum density is reduced. Figure 3.51c shows three edges whose weights are the minimum in edge sets $E_{s_2 s_3}$, E_{ij}, and E_{jk}. Replacing three bent-edge sets in the the topography shown in Figure 3.51a by three edges shown in Figure 3.51c generates the topography shown in Figure 3.52.

Based on the T-map corresponding to the density topography, the center line of a valley surrounding the top can be found by scanning the T-map starting from the top (Figure 3.52b). The process is similar to the forward search in maze searching [114]. The whole valley can be found by scanning the T-map starting from the center line. Moving wires to their vicinity to increase (decrease) the density of that area is called a *compaction (depression)*. The process consisting of both compaction and depression is called *local congestion elimination*. Connecting the top and a grid of the valley generates a guide line that guides the local congestion elimination from the valley to the top (Figure 3.52b). Thus, the difference between the density of the top and the density of the valley can be minimized iteratively (Figure 3.53). In local congestion elimination, the routing length increase is controlled by wire movements, and the wire movement is constrained by via minimization. ◇

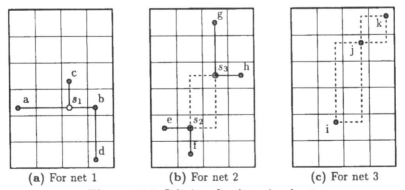

(a) For net 1   (b) For net 2   (c) For net 3

Figure 3.50 Solutions for three signal nets

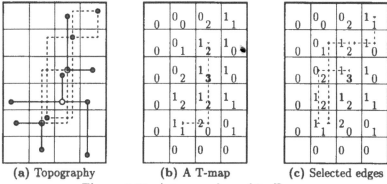

| (a) Topography | (b) A T-map | (c) Selected edges |

Figure 3.51 A topography and its T-map

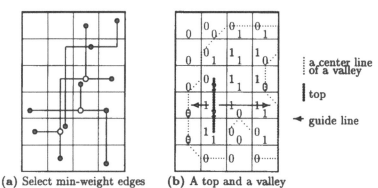

| (a) Select min-weight edges | (b) A top and a valley |

Figure 3.52 A topography on which the total length is minimized

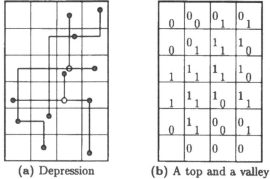

| (a) Depression | (b) A top and a valley |

Figure 3.53 Density is equalized (m_c is minimized)

In a T-map, w_{pq} is used to indicate a wire with terminals p and q; a wire can consist of a sequence of connected wires in which a terminal of a wire is, at most, common with a terminal of another wire. Wires that are neighboring, but separate on the same layer are called *nested wires* (Figure 3.54a). Wire nesting is important in via minimization because nested wires can be routed on the same layer. The Nested-Wire Theorem is provided to nest wires in local congestion elimination for via minimization.

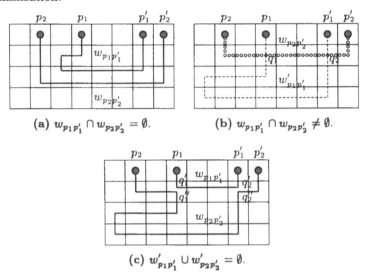

(a) $w_{p_1 p_1'} \cap w_{p_2 p_2'} = \emptyset$.   (b) $w_{p_1 p_1'} \cap w_{p_2 p_2'} \neq \emptyset$.

(c) $w_{p_1 p_1'}' \cup w_{p_2 p_2'}' = \emptyset$.

Figure 3.54 Nesting $w_{p_1 p_1'}$ by $w_{p_2 p_2'}$

Nested-Wire Theorem *If edge $e_{p_1 p_1'}$ is a part of straight edge $e_{p_2 p_2'}$, and if wires $w_{p_1 p_1'}$ and $w_{p_2 p_2'}$ intersect at grids q_1 and q_2 in a T-map, then wire $\{w_{p_1 q_1'}, w_{q_1' q_2'}, w_{q_2' p_1'}\}$ is nested by wire $\{w_{p_2 q_1''}, w_{q_1'' q_2''}, w_{q_2'' p_2'}\}$, where q_1' and q_1'' are two separate points in q_1, and q_2' and q_2'' are two separate points in q_2.*

Proof: Because $w_{p_1 p_1'}$ and $w_{p_2 p_2'}$ intersect at grids p and q in a T-map, there must exist a cycle, c_y, including $w_{q_1 q_2}'$ and $w_{q_1 q_2}''$, where $w_{q_1 q_2}'$ is a part of $w_{p_1 p_1'}$, and $w_{q_1 q_2}''$ is a part of $w_{p_2 p_2'}$ (Figure 3.54b). Let $\Delta w_{p_1 p_1'} = w_{p_1 p_1'} - w_{q_1 q_2}' = \{w_{p_1 q_1}, w_{p_1' q_2}\}$ and $\Delta w_{p_2 p_2'} = w_{p_2 p_2'} - w_{q_1 q_2}'' = \{w_{p_2 q_1}, w_{p_2' q_2}\}$. Because both $\Delta w_{p_1 p_1'}$ and $\Delta w_{p_2 p_2'}$ are outside of c_y, there must exist two nested wires, $\{\Delta w_{p_1 p_1'}, w_{q_1 q_2}''\}$ and $\{\Delta w_{p_2 p_2'}, w_{q_1 q_2}'\}$, if two wires are separate in grids q_1 and q_2. Selecting separate points q_1' and q_1'' in q_1 and separate points q_2' and q_2'' in q_2 generates $w_{p_1 p_1'}' = \{w_{p_1 q_1'}, w_{q_1' q_2'}, w_{q_2' p_1'}\}$ and $w_{p_2 p_2'}' = \{w_{p_2 q_1''}, w_{q_1'' q_2''}, w_{q_2'' p_2'}\}$, where $w_{q_1' q_2'} = w_{q_1 q_2}'$ and $w_{q_1'' q_2''} = w_{q_1 q_2}''$ (Figure 3.54c). □

According to the proof of the nested-wire theorem, the conclusion is still correct when edge $e_{p_2 p_2'}$ is extended to a staircase curve (Figure 3.55a). If $w_{p_1 p_1'}$ and $w_{p_2 p_2'}$ are shown as in Figure 3.55b, in which $w_{p_1 p_1'} \cap w_{p_2 p_2'} \neq \emptyset$, then the nested wires are as shown in Figure 3.55c.

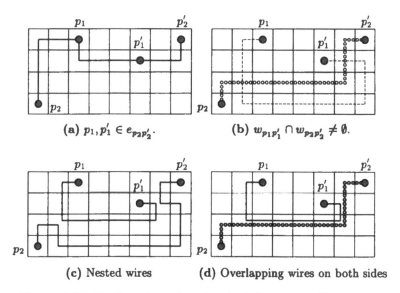

(a) $p_1, p_1' \in e_{p_2 p_2'}$.   (b) $w_{p_1 p_1'} \cap w_{p_2 p_2'} \neq \emptyset$.

(c) Nested wires   (d) Overlapping wires on both sides

Figure 3.55 Nesting wires whose terminals lie on a rectilinear curve

In the local congestion elimination for two layers, $w_{p_1 p_1'}$ in Figure 3.55b can be allocated as that in Figure 3.55d, because the area on the opposite side of $w_{p_2 p_2'}$ is available. However, a net-by-net routing technique cannot realize the routing in Figure 3.55d, because an obstacle representing a wire segment on one side occupies both sides. This is a major cause that more than 50% routing area is wasted by an commercial router.

3.11 CONCLUSIONS

In this chapter, we have presented a theory to solve the infamous rectilinear Steiner's problem. This theory is built on an edge-set tree generated by extending each edge and each Steiner point of a rectilinear tree into an edge set and dynamic Steiner point, respectively. Based on an edge-set tree without redundant length, dynamic length-conservation edge-set subtrees called cliques are proven for arithmetics of generating new edge-set trees by replacing cliques with other edge sets. Eliminating each redundant length found by clique and edge-set replacements generates the general solution of the rectilinear Steiner's problem, which is independent of an initial edge-set tree.

According to the algorithm generating the general solution, we propose an algorithm expansion that expands an algorithm into a series of subalgorithms measured by a series of corresponding indices. An index for a subalgorithm is determined by the number of events occurring in the subalgorithm and by the total contribution of the events. An event that has contributed to an index for a subalgorithm will never contribute to an index for other subalgorithms. Thus, all different events have a common reference from which indices for different subalgorithms can be computed. Maximizing the number of events that occur in efficient subalgorithms and duplicate events occurring in inefficient subalgorithms will converge the major contribution to indices for efficient subalgorithms.

If the sum of all indices is highly converged in indices for efficient subalgorithms, the expansion is called an efficient expansion. An accuracy-estimable solution can be found efficiently by an efficient expansion generated from an algorithm for a solution of an inefficient P or NP problem.

In combinatorial graph theory, an objective graph is often generated by improving a graph by replacing subgraphs in the graph with new subgraphs. In a graph, the possibility of a complex subgraph existing is much less than that of a simple subgraph existing. Furthermore, a complex subgraph is often formed by simple subgraphs. Maximizing the total contribution by replacing simple subgraphs with new simple subgraphs, the reference graph can be efficiently improved to a graph very close to or exactly equal to the objective graph. In finding a solution for the rectilinear Steiner's problem starting from a minimum spanning tree, 97.9% length reduction can be achieved by implementing a nearly linear subalgorithm, and 99.9% length reduction can be achieved by implementing this subalgorithm and an $O(n^2)$ subalgorithm.

The theory for the rectilinear Steiner's problem can also be applied in proving linear algorithms to find MRSTs connecting nodes on the boundary of a polygon. These algorithms will simplify switch-box routing and complex channel routing.

To avoid the drawbacks generated by traditional pin-to-pin connection and net-by-net routing techniques, we present an algorithm for MRST routing, based on the general solution that can provide many selections for an MRST connecting a signal set. Properly selecting a solution from each general solution can eliminate or reduce the local congestion that often produces more incomplete wires after routing. The algorithm for MRST routing also fits the requirements of parallel computing and considers via minimization. Thus, a router based on parallel MRST routing combining local congestion elimination, via minimization, and high-speed simulation can be realized.

REFERENCES

1. P. K. Agarwal and M. T. Shing, "Algorithms for the Special Cases of Rectilinear Steiner Trees: I. Points on the Boundary of a Rectilinear Rectangle," *Networks*, vol. 20, 1990, pp. 453-485.

2. A. Agrawal, P. Kein, and R. Ravi, "When Trees Collide: an Approximation Algorithm for the Generalized Steiner Tree Problem on Networks," *Proc. 23rd Annu. ACM Symp. Theory of Comput.*, 1991, pp. 134-144.

3. A. V. Aho, M. R. Garey, and F. K. Hwang, "Rectilinear Steiner Trees: Efficient Special Case Algorithms," *Networks*, vol. 7, 1977, pp. 37-58.

4. J. M. Basart and L. Huguet, "An Heuristic Algorithm for Rectilinear Steiner Minimal Trees," working paper, Department d'Informàtica, Facultat de Ciències, Universitat Autònoma de Barcelona, 08193 Bellaterra, 1987.

5. J. E. Beasley, "A Heuristic for Euclidean and Rectilinear Steiner Problem," *Eur. J. Oper. Res.*, vol. 58, 1992, pp. 284-292.

6. S. Bhaskaran, "Optimal Design of Gas Pipeline Networks," Ph.D. Dissertation, University of Adelaide, Australia, 1978.

7. S. Bhaskaran and F. J. M. Salzborn, "Optimal Design of Gas Pipeline Networks," *J. Oper. Res. Soc.*, vol. 30, 1979, pp. 1047-1060.

8. M. W. Bern, "Network Design Problems: Steiner Trees and Spanning *k*-Trees," Ph.D. Dissertation, Computer Science Division, University of California at Berkeley, 1987.

9. M. W. Bern, "Two Probabilistic Results on Rectilinear Steiner Trees," *Algorithmica*, vol. 3, 1988, pp. 191-204.

10. M. W. Bern, "Faster Exact Algorithms for Steiner Trees in Planar Networks," *Networks*, vol. 20, 1990, pp. 109-120.

11. M. W. Bern and M. de Carvalho, "A Greedy Heuristic for the Rectilinear Steiner Tree Problem," technical report, Computer Science Division, University of California at Berkeley, 1985.

12. J. H. Camin and R. R. Sokal, "A Method for Deducing Branching Sequences in Phylogeny," *Evolution*, vol. 19, 1965, pp. 311-326.

13. L. L. Cavalli-Sforza and A. W. F. Edwards, "Phylogenetic Analysis: Models and Estimation Procedures," *Am. J. Hum. Genet.*, vol. 19, 1967, pp. 233-257.

14. R. J. Cedergren, D. Sankoff, B. LaRue, and H. Grosjean, "The Evolving *t*RNA Molecule," *Crit. Rev. Biochem*, vol. 11, 1981, pp. 35-104.

15. S. Chang, "The Generation of Minimal Trees with a Steiner Topology," *J. Assoc. Comput. Mach.*, vol. 1, 1972, pp. 699-711.

16. T-H. Chao and Y-C. Hsu, "Rectilinear Steiner Tree Construction by Local and Global Refinement," *Proc. IEEE Int. Conf. CAD*, 1990, pp. 432-435.

17. T.-H. Chao and Y.-C. Hsu, "Rectilinear Steiner Tree Constructions by Local and Global Refinement," *IEEE Trans. CAD*, vol. 13, no. 3, 1994, pp. 303-309.

18. C. Chiang, M. Sarrafzadeh, and C. K. Wong, "A Powerful Global Router: Based on Steiner Min-Max Trees," *IEEE Int. Conf. CAD*, 1989, pp. 2-5.

19. C. Chiang, M. Sarrafzadeh, and C. K. Wong, "Global Routing Based on Steiner Min-Max Trees," *IEEE Trans. CAD*, vol. 9, no. 12, 1990, pp. 1318-1325.

20. F. R. K. Chung, M. Gardner, and R. L. Graham, "Steiner Trees on a Checkerboard," *Math. Mag.*, vol. 62, 1989, pp. 83-96.

21. F. R. K. Chung and H. K. Hwang, "The Largest Minimal Rectilinear Steiner Trees for a Set of n Points Enclosed in a Rectangle with Given Perimeter," *Networks*, vol. 9, 1979, pp. 19-36.

22. E. J. Cockayne and D. G. Schiller, "Computation of Steiner Minimal Trees," in *Combinatorics*, ed. by D. J. A. Welsh and D. R. Woodall, Institute for Mathematics and Applications, Southend-on-Sea, Essex, England, 1972, pp. 53-71.

23. J. P. Cohoon, D. S. Richards, and J. S. Schowe, "An Optimal Steiner Tree Algorithm for a Net Whose Terminals Lie on the Perimeter of a Rectangle," *IEEE Trans. CAD*, vol. 9, no. 4, 1990, pp. 398-407.

24. R. Courant and H. Robbins, *What is Mathematics?*, Oxford University Press, New York, 1941.

25. W. H. E. Day, "Computationally Difficult Parsimony Problem in Phylogenetic Systematics," *J. Theor. Biol.*, vol. 103, 1983, pp. 429-438.

26. W. H. E. Day, "Computational Complexity of Inferring Phylogenies from Dissimilarity Matrices," *Bull. Math. Biol.*, vol. 49, 1987, pp. 461-467.

27. W. H. E. Day, D. S. Johnson, and D. Sankoff, "Computational Complexity of Inferring Phylogenies by Compatibility," *Syst. Zool.*, vol. 35, 1986, pp. 224-229.

28. W. H. E. Day and D. Sankoff, "Computational Complexity of Inferring Phylogenies from Chromosone Inversion Data," *J. Theor. Biol.*, vol. 124, 1987, pp. 213-218.

29. L. L. Deneen, G. M. Shute, and C. D. Thomborson, "A Propably Fast, Provably Optimal Algorithm for Rectilinear Steiner Trees," in *Random Structure and Algorithms*, vol. 5, no. 4, John Wiley & Sons, Inc., New York, 1994, pp. 535-557.

30. S. E. Dreyfus and R. A. Wagner, "The Steiner Problem in Graphs," *Networks*, 1972, pp. 195-207.

31. A. M. Farley, S. T. Hedetniemi, and S. L. Mitchell, "Rectilinear Steiner Trees in Rectangle Trees," *SIAM J. Alg. Disc. Methods*, vol. 1, 1980, pp. 70-81.

32. J. S. Farris, "Methods for Computing Wagner Trees," *Syst. Zool.*, vol. 19, 1970, pp. 83-92.

33. J. Felsenstein, "Numerical Methods for Inferring Evolutionary Trees," *Q. Rev. Biol.*, vol. 57, 1982, pp. 379-404.

34. J. Felsenstein, "Phylogenies from Molecular Sequences: Inference and Reliability," *Annu. Rev. Genet.*, vol. 22, 1988, pp. 521-565.

35. L. R. Foulds, "Maximum Savings in the Steiner Problem in Phylogeny," *J. Theor. Biol.*, vol. 107, 1984, pp. 471-474.

36. L. R. Foulds and V. J. Rayward-Smith, "Steiner Problems in Graphs: Algorithms and Applications," *Eng. Optimiz.*, vol. 7, 1983, pp. 7-16.

37. Y. Fu, "Application of Linear Graph Theory to Printed Circuits," *Proc. Asilomar Conf. Syst. Circuits*, 1967, pp. 721-728.

38. M. R. Garey and D. S. Johnson, "The Rectilinear Steiner Tree Problem is NP-Complete," *SIAM J. Appl. Math.*, vol. 32, 1977, pp. 826-834.

39. E. N. Gilbert and H. O. Pollak, "Steiner Minimal Trees," *SIAM J. Appl. Math.*, vol. 16, no. 1, 1968, pp. 1-29.

40. R. L. Graham and L. R. Foulds, "Unlikelihood that Minimal Phylogenies for a Realistic Biological Study Can Be Constructed in Reasonable Computation Time," *Math. Biosci.*, vol. 60, 1982, pp. 133-142.

41. J. Griffith, G. Robins, J. S. Salowe, and T. Zhang, "Closing the Gap: Near-Optimal Steiner Trees in Polynomial Time," *IEEE Trans. CAD*, vol. 13, no. 11, 1994, pp. 1351-1365.

42. S. L. Hakimi, "Steiner's Problem in Graphs and Its Implications," *Networks*, vol. 1, 1972, pp. 113-135.

43. S. Hambrusch and L. TeWinkel, "Parallel Heuristics for the Steiner Tree Problem in Images Without Sorting or Routing," Technical Report 89-048, Computer Science Dept., Purdue University, 1989.

44. M. Hanan, "Net Wiring for Large Scale Integrated Circuits," *IBM Res. Rep., RC 1375*, 1965.

45. M. Hanan, "On Steiner's Problem with Rectilinear Distance," *SIAM J. Appl. Math.*, vol. 14, No. 3, 1966, pp. 255-265.

46. M. Hanan, "A Counter Example to a Theorem of Fu on Steiner's Problem," *IEEE Trans. Circuit Theory*, CT-19, 1972, pp. 74.

47. M. Hanan and J. M. Kurtzberg, "Placement Techniques," in *Design Automation of Digital Systems, Volume One: Theory and Techniques*, ed. by M.A. Breuer, Prentice-Hall, Englewood Cliffs, New Jersey, 1972, ch. 5, pp. 213-282.

48. M. Hanan, P. K. Wolff Sr., and B. J. Agule, "Some Experimental Results on Placement Techniques," *Proc. 13th Des. Autom. Conf.*, 1976, pp. 214-224.

49. J. A. Hendrickson, "Clustering in Numerical Cladistics: a Minimum-length Directed Tree Problem," *Math. Biosci.*, vol. 3, 1968, pp. 371-381.

50. J. M. Ho, G. Vijayan, and C. K. Wong, "A New Approach to the Rectilinear Steiner Tree Problem," *Proc. 26th Des. Autom. Conf.*, 1989, 161-166.

51. J. M. Ho, G. Vijayan, and C. K. Wong, "Constructing the Optimal Rectilinear Steiner Tree Derived from a Minimum Spanning Tree," *Dig. Tech. Pap., ICCAD-89*, 1989, pp. 6-9.

52. J. M. Ho, G. Vijayan, and C. K. Wong, "New Algorithm for the Rectilinear Steiner Tree Problem," *IEEE Trans. CAD*, vol. 9, no. 2, 1990, pp. 185-193.

53. Y. C. Hsu, Y. Pan, and W. J. Kubitz, "A Path Selection Global Router," *24th ACM/IEEE Des. Autom. Conf.*, 1987, pp. 641-644.

54. F. K. Hwang, "On Steiner Minimal Trees with Rectilinear Distance," *SIAM J. Appl. Math.* vol. 30, no. 1, 1976, pp. 104-114.

55. F. K. Hwang, "The Rectilinear Steiner Problem," *Des. Autom. Fault Tolerant Comput.*, vol. 3, 1978, pp. 303-310.

56. F. K. Hwang, "An $O(n \log n)$ Algorithm for Suboptimal Rectilinear Steiner Trees," *IEEE Trans. Circuits Syst.*, vol. CAS-26, no. 1, 1979, pp. 75-77.

57. F. K. Hwang, "An $O(n \log n)$ Algorithm for Rectilinear Minimal Spanning Trees," *J. ACM*, vol. 26, 1979, pp. 177-182.

58. F. K. Hwang and D. S. Richards, "Steiner Tree Problems," *Networks*, vol. 22, 1992, pp. 55-89.

59. F. K. Hwang, D. S. Richards, and P. Winter, *The Steiner Tree Problem*, North-Holland, New York, 1992.

60. F. K. Hwang and Y. C. Yao, "Comments on Bern's Probabilistic Results on Rectilinear Steiner Trees," *Algorithmica*, vol. 5, 1990, pp. 591-598.

61. A. Iwainsky, "Some Notes on the Steiner Tree Problem in Graphs," in *Optimization of Connection Structures in Graphs*, ed. by A. Iwainsky, Centr. Inst. of Cyb. a. Inform. Processes, Berlin, 1985, 57-73.

62. A. Iwainsky, E. Canuto, O. Taraszow, and A. Villa, "Network Decomposition for the Optimization for Connection Structures," *Networks*, vol. 16, 1986, pp. 205-235.

63. V. Jarnik and O. Kössler, "O Minimálnich Gratech Obsahujicich n Daných Bodu [in Czech]," *Casopis Pesk. Mat. Fyr.*, vol. 63, 1934, pp. 223-235.

64. J. W. Jiang and P. S. Tang, "An Algorithm for Generating Rectilinear Steiner Trees," [in Chinese] *J. Fudan Univ. (Natural Science)*, vol. 25, 1986, pp. 343-349.

65. A. B. Kahng and G. Robins, "A New Class of Steiner Tree Heuristic with Good Performance: The Iterated 1-Steiner Approach," *Proc. IEEE Int. Conf. CAD*, 1990, pp. 428-431.

66. A. B. Kahng and G. Robins, "A New Class of Iterative Steiner Tree Heuristics with Good Performance," *IEEE Trans. CAD*, vol. 11, no. 7, 1992, pp. 893-902.

67. A. B. Kahng and G. Robins, "On Performance Bounds for a Class of Rectilinear Steiner Tree Heuristics in Arbitrary Dimension," *IEEE Trans. CAD*, vol. 11, no. 11, 1992, pp. 1462-1465.

68. J. Komlos and M. T. Shing, "Probabilistic Partitioning Algorithms for the Rectilinear Steiner Problem," *Networks*, vol. 15, 1985, pp. 413-423.

69. J. B. Kruskal, "On the Shortest Spanning Subtree of a Graph and the Traveling Salesman Problem," *Proc. Am. Math. Soc.*, vol. 7, 1956, pp. 48-50.

70. E. S. Kuh and M. Marek-Sadowska, "Global Routing," in *Layout Design and Verification*, ed. by T. Ohtsuki, North-Holland, Amsterdam, 1986, pp. 169-198.

71. D. H. Lee, "Low Cost Drainage Networks," *Networks*, vol. 6, 1976, pp. 351-371.

72. D. T. Lee and C. K. Wong, "Voronoi Diagrams in L_1 (L_∞) Metrics with 2-Dimensional Storage Applications," *SIAM J. Comput.*, vol. 9, 1980, pp. 200-211.

73. J. H. Lee, N. K. Bose, and F. K. Hwang, "Use of Steiner's Problem in Suboptimal Routing in Rectilinear Metric," *IEEE Trans. Circuits Syst.*, vol. CAS-23, no. 7, 1976, pp. 470-476.

74. K. W. Lee and C. Sechen, "A New Global Router for Row-based Layout," *IEEE Int. Conf. CAD*, 1988, pp. 180-183.

75. R. R. L. Matos, "Rectilinear Arborescence and Rectilinear Steiner Tree Problems," Ph.D. Dissertation, University of Birmingham, 1980.

76. Z. A. Melzak, "On Problem of Steiner, " *Can. Math. Bull.*, vol. 4, 1961, pp. 143-148.

77. Z. Miller and M. Perkel, "The Steiner Problem in the Hypercube," *Networks*, vol. 22, 1992, pp. 1-19.

78. A. P-C. Ng, P. Raghavan, and C. D. Thompson, "A Language for Describing Rectilinear Steiner Tree Configurations," *Proc. 23rd Des. Autom. Conf.*, 1986, pp. 659-662.

79. M. Pecht, *Placement and Routing of Electronic Modules*, Marcel Dekker, New York, 1993.

80. R. C. Prim, "Shortest Connection Networks and Some Generalizations," *Bell Syst. Tech. J.*, vol. 36, 1957, pp. 1389-1401.

81. J. S. Provan, "Convexity and the Steiner Tree Problem," *Networks*, vol. 18, 1988, pp. 55-72.

82. J. S. Provan, "Two New Criteria for Finding Steiner Hulls in Steiner Tree Problems," *Algorithmica*, vol. 7, 1992, pp. 289-302.

83. S. K. Rao, P. Sadayappan, F. K. Hwang, and P. W. Shor, "The Rectilinear Steiner Arborescence Problem," *Algorithmica*, vol. 7, 1992, pp. 277-288.

84. R. L. Rardin, R. G. Parker, and M. B. Richey, "A Polynomial Algorithm for a Class of Steiner Tree Problems on Graph," Indus. and Syst. Engr. Report Series J-82-5, Georgia Institute of Technology, 1982.

85. V. J. Rayward-Smith, "The Computation of Nearly Minimal Steiner Trees in Graphs," *Int. J. Math. Ed. Sci. Technol.*, vol. 14, 1983, pp. 15-23.

86. V. J. Rayward-Smith and A. Clare, "On Finding Steiner Vertices," *Networks*, vol. 16, 1986, pp. 283-294.

87. D. Richards, "Fast Heuristic Algorithms for Rectilinear Steiner Trees," *Algorithmica*, vol. 4, 1989, pp. 191-207.

88. D. Richards and J. S. Salowe, "Special Convex Cases for Rectilinear Steiner Minimal Trees," *Proc. NATO Workshop Topol. Network Des.*, 1989.

89. D. Richards and J. S. Salowe, "A Simple Proof of Hwang's Theorem for Rectilinear Steiner Minimal Trees," *Ann. Oper. Res.*, 1992.

90. D. Richards and J. S. Salowe, "A Linear-time Algorithm to Construct Rectilinear Steiner Minimal Trees for k-extremal Point Sets," *Algorithmica*, vol. 7, 1992, pp. 247-276.

91. F. J. Rohlf, "A Note on Minimum Length Trees," *Syst. Zool.*, vol. 33, 1984, pp. 341-343.

92. M. H. Rudowski, "Hypothesis on the Length of Minimal Steiner Tree in Multidimensional Spaces with a Rectilinear Metric," [in Polish] *Arch. Autom. Telemech.*, vol. 29, 1984, pp. 353-358.

93. D. Sankoff and P. Rousseau, "Locating the Vertices of a Steiner Tree in an Arbitrary Metric Space," *Math. Program*, vol. 9, 1975, pp. 240-246.

94. M. Sarrafzadeh and C. K. Wong, "Hierarchical Steiner Tree Construction in Uniform Orientations," *IEEE Trans. CAD*, vol. 11, no. 9, 1992, pp. 1095-1101.

95. M. Sarrafzadeh and D. Zhou, "Global Routing of Short Nets in Two Dimensional Arrays," *Int. J. Comput. Aid. VLSI Des.*, vol. 2, no. 2, 1990, pp. 197-211.

96. M. Servit, "Heuristic Algorithm for Rectilinear Steiner Trees," *Digit. Process.*, vol. 7, 1981, pp. 21-32.

97. J. M. Smith and J. S. Liebman, "Steiner Trees, Steiner Circuits and the Interference Problem in Building Design," *Eng. Optimiz.*, vol. 4, 1979, pp. 15-36.

98. J. M. Smith and M. Gross, "Steiner Minimal Trees and Urban Service Networks," *J. Socio. Econ. Plan.*, vol. 16, 1982, pp. 21-38.

99. J. M. Smith, D. T. Lee, and J. S. Liebman, "An $O(N log N)$ Heuristic Algorithm for the Rectilinear Steiner Minimal Tree Problem," *Eng. Optimiz.*, vol. 4, 1980, 179-192.

100. T. Snyder, "On Minimal Rectilinear Steiner Trees in All Dimensions," *Proc. 6th ACM Symp. Comput. Geom.*, 1990, pp. 311-320.

101. J. Soukup and W. F. Chow, "Set of Test Problems for the Minimum Length Connection Networks," *ACM/SIGMAP Newsl.*, vol. 15, 1973, pp. 48-51.

102. J. Soukup, "On Minimum Cost Networks with Non-linear Costs," *SIAM J. Appl. Math.*, vol. 25, 1975, pp. 571-581.

103. C. D. Thomborson (A. K. A. Thompson), L. L. Deneen, and G. M. Shute, "Computing a Rectilinear Steiner Minimal Tree in $n^{O(n^{1/2})}$ Time," in *Parallel Algorithms and Architectures [LNCS 269]*, ed. by Albrecht et al., Akademie-Verlag, Berlin, 1987, pp. 176-183.

104. E. A. Thompson, "The Method of Minimum Evolution," *Ann. Hum. Genet. London*, vol. 36, 1973, pp. 333-340.

105. V. A. Trubin, "Subclass of the Steiner Problem on a Plane with Rectilinear Metric," *Cybernetics*, vol. 21, 1985, pp. 320-324.

106. T. Watanabe and Y. Sugiyama, "A New Routing Algorithm and Its Hardware Implementation," *Proc. 23rd ACM/IEEE Des. Autom. Conf.*, 1986, pp. 574-580.

107. B. M. Waxman, "Routing of Multipoint Connections," *IEEE J. Select. Areas Comm.*, vol. 6, 1988, pp. 1617-1622.

108. B. M. Waxman, "Probable Performance of Steiner Tree Algorithms," Technical Report WUCS-88-4, Dept. of Computer Science, Washington University, St. Louis, 1988,

109. B. M. Waxman and M. Imase, "Worst-case Performance of Rayward-Smith's Steiner Tree Heuristics," *Inf. Process. Lett.*, vol. 29, 1988, pp. 283-287.

110. P. Winter, "Steiner Problem in Networks: A Survey," *Networks*, vol. 17, 1987, pp. 129-167.

111. Y. T. Wong and M. Pecht, "Approximating the Steiner Tree in the Placement Process," *ASME J. Electron. Packag.*, Sept. 1989, pp. 228-235.

112. Y. T. Wong, G. Li, and M. Pecht, "Characterization and Generation of Trees," in *Placement and Routing of Electronic Modules*, ed. by M. Pecht, Marcel Dekker, New York, 1993, pp. 29-58.

113. Y. T. Wong, M. Pecht, M. D. Osterman, and G. Li, "Placement for Routability," in *Placement and Routing of Electronic Modules*, ed. by M. Pecht, Marcel Dekker, New York, 1993, pp. 139-180.

114. Y. T. Wong, M. Pecht, and G. Li, "Detailed Routing," in *Placement and Routing of Electronic Modules*, ed. by M. Pecht, Marcel Dekker, New York, 1993, pp. 181-220.

115. Y. T. Wong and M. Pecht, "A Solution for Steiner's Problem," in *Placement and Routing of Electronic Modules*, ed. by M. Pecht, Marcel Dekker, New York, 1993, pp. 261-304.

116. Y. T. Wong and M. Pecht, "Introducing a Theory for the General Solution of Steiner's Problem in Rectilinear Space," *1st Int. Symp. Microelectron. Package & PCB Technol.*, Beijing, P. R. China, 1994.

117. J. G. Xiong, "Algorithms for Global Routing," *Proc. 23rd ACM/IEEE Des. Autom. Conf.*, 1986, pp. 824-830.

118. Y. Y. Yang, "Optimal and Suboptimal Solution Algorithm for the Wiring Problem," Ph.D. Dissertation, Columbia University, 1972.

119. Y. Y. Yang and O. Wing, "An Algorithm for the Wiring Problem," *Dig. IEEE Int. Symp. Electron. Networks*, 1971, pp. 14-15.

120. Y. Y. Yang and O. Wing, "Suboptimal Algorithm for a Wire Routing Problem," *IEEE Trans. Circuit Theory*, vol. CT-19, 1972, pp. 108-511.

121. Y. Y. Yang and O. Wing, "Optimal and Suboptimal Solution Algorithms for the Wiring Problem," *Proc. IEEE Int. Symp. Circuit Theory*, vol. 19, 1972, pp. 154-158.

122. Y. Y. Yang and O. Wing, "On a Multinet Wiring Problem," *IEEE Trans. Circuit Theory*, vol. CT-20, 1973, pp. 250-252.

123. A. Yao, "An $O(|E|\,loglog\,|E|)$ Algorithm for Finding Minimum Spanning Trees," *Inf. Process. Lett.*, vol. 4, 1975, pp. 21-23.

Chapter 4

PERFORMANCE-DRIVEN GLOBAL ROUTING AND WIRING RULE GENERATION FOR HIGH-SPEED PCBs AND MCMs

Paul D. Franzon and Sharad Mehrotra

Once a system's clock frequency exceeds 25 MHz, it is often necessary to control and restrict its placement and routing to ensure correct functionality. These restrictions arise from the need to control delay and noise on interchip interconnects. The purpose of this chapter is to describe to the reader the different ways in which electrical design requirements arise and the different computer-aided approaches used to ensure that the placement and routing constraints ensure a good electrical design.

The chapter is structured as follows: The first section will describe how the timing design requirements and signal integrity (noise control) requirements lead to the needs for restrictions. The following sections will describe the different computer-aided approaches to managing these requirements in physical design.

4.1 SIGNAL INTEGRITY MANAGEMENT

Signal integrity refers to the delay and noise properties of the electrical signals on the MCM. Signal integrity management refers to the modeling, design, and simulation of the phenomena that determine delay and noise; primarily the following: (1) propagation delay and reflection noise in signal interconnect; (2) crosstalk between neighboring signal lines; and (3) simultaneous switching noise and other common mode noise sources on power and ground planes.

Good signal integrity management is critical to good product design. A design with poor signal delay and noise requirements will either simply not work, will suffer from a high transient failure rate, or will suffer from a low yield (only a portion of the systems manufactured will work).

Fortunately, signal integrity management is usually easier in an MCM than in the corresponding PCB. Shorter distances and generally lower inductances conspire to help the designer. However, even in relatively low speed (33 MHz to 50 MHz) designs, signal integrity can not be ignored – there are many simple errors that can be made. In this section, signal integrity requirements will be presented before discussing each of the affecting phenomena in turn. This discussion is brief in nature. For further information, the reader is referred to reference [1] for a basic discussion, and to references [2, 4, 5, 6] for more detail.

4.1.1 SIGNAL INTEGRITY REQUIREMENTS

Consider the digital circuits, and associated waveforms shown in Figure 4.1. In order for the circuit to function correctly, the following conditions must be satisfied for all possible combinations of transistor and MCM or PCB characteristics:

(1) The sum t_{total} of all the delays shown must add up to less than the clock period. Note that for signal integrity purposes, the interconnect delay is NOT best defined as the delay between the 50% points in the waveform. Here it is defined as the *settling delay*, or the delay from the 50% point at the output pin until that time that the waveform at the input pin goes above, and stays above, V_{IH} (for the low to high transition here). Noise in the waveform is acceptable (and is in fact common) as long as the noise is settled out in sufficient time. The difference between the nominal high voltage and V_{IH} defines the high noise margin of the digital circuit. A similar definition is used for the low noise margin. As long as the total noise is less than the noise margin around the latch or flip-flop setup and hold period, the signal will be correctly latched.

(2) In contrast, noise in the clock waveform must be severely controlled. Any potential for false clock edges can lead to system failure. Figure 4.2 gives some examples of noisy waveforms that can lead to false clock edges.

(3) Additionally, on clocks, the *clock skew* must be well controlled. The clock skew is defined as any difference in the arrival times of the clock edges at the clock inputs of the flip-flops or latches. The portion of the clock period allowed for signal delay is directly reduced by any amount of clock skew.

In order to ensure good signal integrity it is necessary to meet these conditions across all possible variations in manufacturing and operating conditions. The *parametric variation* of the CMOS transistors can often be +/- 20%. This means that in any manufacturing run, a driver's strength might vary by 20%. Variations are accounted for in the design by the manufacturer or circuit designer providing best and worst case delays, noise margins and driver and receiver simulation circuit models. Usually the largest source of clock skew is driver-to-driver variation. Special clock chip sets are provided that minimize this variation. Transistor properties also vary considerably with temperature. Again the best and worst cases are guaranteed across a temperature range.

Noise can come from a number of sources. *Ringing noise* results from reflections from unmatched terminations if the interconnect is acting as a transmission line, or can result from resonance if the line is acting as an RLC line (Figure 4.3). *Crosstalk noise* results from neighboring lines inducing signals on each other. *Simultaneous Switching Noise* (SSN) appears on ground and power connections as a result of drivers switching simultaneously (Figure 4.4). SSN is one form of *common mode* noise; i.e., noise that results from the common action of different circuits. In an MCM-D (thin film MultiChip Module), it is also necessary to check the resistive voltage drop on the power and ground planes or rails.

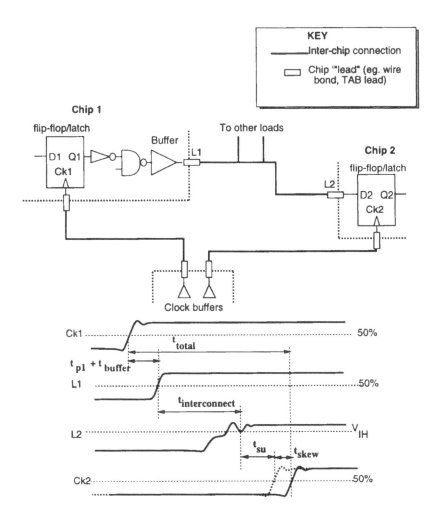

Figure 4.1 Example showing some digital circuits and associated waveforms

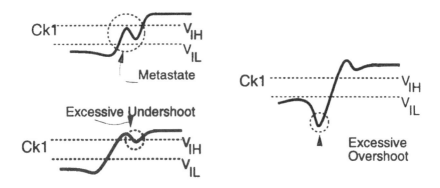

Figure 4.2 Examples of poor clock signals

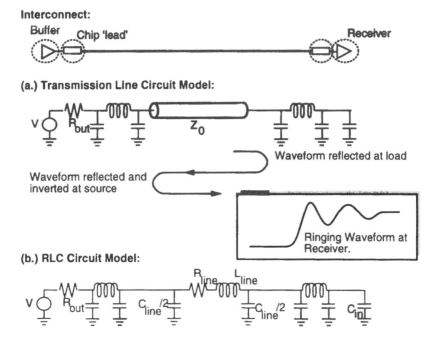

Figure 4.3 Sources of ringing noise

These same noise sources can also be found in the analog connections in a mixed signal MCM. However, there are a couple of important differences between noise control for analog and digital circuits. There is no concept of a 'noise margin' for analog signals; all noise is generally bad. Many analog signals have very small swings and are thus very sensitive to noise. In such cases, crosstalk noise from a neighboring digital signal can be devastating. Most analog circuits are also very sensitive to noise appearing on their power and ground connections. It is very important to control digital-circuit-induced common mode noise on these connections, particularly simultaneous switching noise.

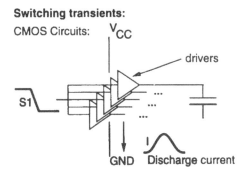

Switching transients:
CMOS Circuits:

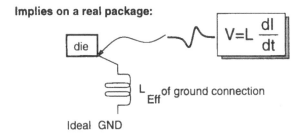

Implies on a real package:

Figure 4.4 Source of simultaneous switching noise

Next, we give an intuitive understanding of signal propagation basics and present a rule of thumb that allows the designer to select between transmission line and RLC circuit representations of interconnect structures.

4.1.2 DELAY AND NOISE BASICS

The basics of delay and noise control for digital circuits can be understood by considering how signals are actually propagated in interconnect circuits, and then understanding the different models used in interconnect circuits.

Signal propagation can always be understood by considering the *Cardinal Rule* of package-level interconnect design:[1]

---

[1] With thanks to Michael Steer for sharing the concept.

The Cardinal Rule. *The signal return path always exists whether you provide for it or not.*

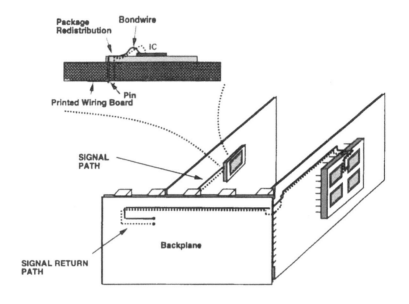

Figure 4.5 The Cardinal Rule

The signal return path is usually provided by the power and ground circuits in a system. For example, consider the signal and return path shown in Figure 4.5. The signal path is taken through the wires determined by the designer. The signal return path tries to stay as close as possible to signal path but is confined to the power and ground circuits. In this case, it flows on the power and ground planes in the PCBs and MCMs and through the power and ground pins on the connectors and the MCM packages. The capacitance per unit length of this circuit is the capacitance between the signal and signal returns. The inductance is determined by the loop inductance between the signal and return currents. The resistance is determined mainly by the cross section of the signal wires (the signal return current is more 'spread out' and only contributes to about 10% of the resistance). In an MCM-L or MCM-C, the resistance is generally small as the signal wire cross sectional area is generally large. In an MCM-D, the resistance can be 1,000 Ω per meter or more, depending on the signal wire cross section and materials. Though, such a high resistance does slow the signal down, the change in speed is generally not important except in the fastest clocked circuits. Usually the so-called 'skin effect' can be ignored in MCM-D circuits if CMOS drivers are being used.

A couple of observations can be made from this illustration:

(1) When the signal return path geometry changes with respect to the signal path geometry, the inductance per unit length and capacitance per unit length changes, possibly distorting the signal. When the two geometries do not change, a transmission line structure has been built. More on this aspect below.

(2) It is important to provide a high density of ground and power pins in the connectors and packages. If the density is low, the inductance of the circuits increase and the concentration of current in the few pins increases the common mode noise on the power and ground circuits. Generally, one power or ground is required for every one to four signal pins, the higher density being desired at higher speeds.

The first observation above leads to a couple of questions: (1) When do you need to build a transmission line structure; and (2) how do you model the elements in the interconnect circuit?

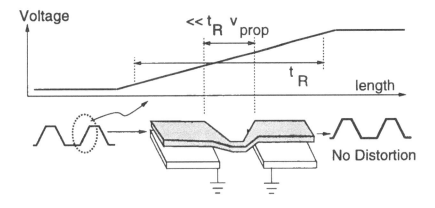

Figure 4.6 Short signal paths do not need to be built nor modeled as transmission lines

A useful rule of thumb that can be used to answer these questions is based on a comparison of the signal path length of each interconnect structure with the length over which the rising edge of the signal is spread out ($t_r c_0 / \sqrt{\epsilon_r}$, where t_r is the rise time, c_0 is the speed of light in a vacuum and and ϵ_r is the dielectric constant of the material). If the distance $t_r c_0 / \sqrt{\epsilon_r}$ is much shorter than the signal path length, then it is necessary to build the structure as a transmission line (usually by using power and ground planes), and model it as a transmission line (Figure 4.3a). Otherwise, the signal edges will be distorted as they try to 'jump' the gap in the uniform transmission line. If the distance $t_r c_0 / \sqrt{\epsilon_r}$ is much longer than the signal path length (three to five times longer), then it is not necessary to build a transmission line structure, and, whether a transmission line structure is built or not, the circuit can be modeled as a lumped RLC circuit (Figure 4.3b). In this case, a physical interpretation might be that the long edge of the rising signal can 'jump' the non-uniform gap without distortion (Figure 4.6).

For example, a high-speed BiCMOS driver might have a rise time of 0.5 ns on an MCM. On an MCM-L (FR4, ϵ_r=4.2), $t_r c_0/\sqrt{\epsilon_r} = 2.88$ inches. Signal wires shorter than one-third of this distance, 0.96 inches, can be safely modeled as RLC circuits.

4.1.3 HIGH-SPEED INTERCONNECT DESIGN

High-speed system interconnect design reduces to the problem of constraining the layout so that delay, reflection noise, and crosstalk noise constraints are met. (SSN is largely layout independent.) The intention of this section is not to equip you with all the techniques used in interconnect design but to explain how interconnect design affects layout. The section starts with a treatment of delay and reflection noise before turning to crosstalk noise.

The interconnect model type determines how the designer needs to constrain the physical circuit design in order to meet delay and noise requirements. If a transmission line model is required (e.g., 0.5 ns rise time fast CMOS driver, line length > 1 in) and the permitted interconnect delay is less than five times the speed-of-light delay on the line (e.g., on a three inch long line, the speed of light delay is about 0.5 ns; here any interconnect delay requirement of less than 2.5 ns requires attention) then the following steps are necessary:

- Reflection noise must be minimized by providing a matching termination at one or both ends of the transmission line (Figure 4.7).
- Reflection noise must be minimized by reducing the lengths of stubs off the main interconnect line. This is done by controlling the topology of the interconnection. The different topologies that arise if short stubs are required are shown in Figure 4.8. *It is important to note that a minimum spanning Steiner tree is NOT one of these topologies.* The allowed maximum stub length is best determined through simulation.
- When the allowed interconnect delay is very tight it might be necessary to control the relationships between the different branch lengths in the interconnection. Some combinations of branch lengths result in resonances and excessive ringing.

If the interconnect model has an RLC nature then it is often not necessary to control ringing noise by minimizing stub length and using restricted topologies. However, it might be necessary to control branch lengths in order to control delay ad ringing. Often, additional resistance is used to control the ringing noise. The resistance might be placed in series with the driver or provided (in an MCM-D only) by changing the line width (and thus its resistance). Analytic techniques to determine suitable line widths for simple cases are described in [7].

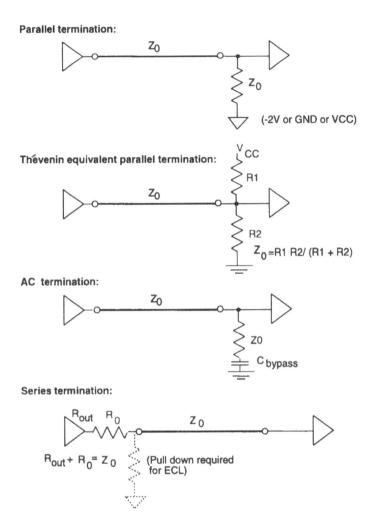

Figure 4.7 Different schemes for providing a terminating resistor

The constraints on wire length, stub length, topology, branch lengths and line widths are often referred to as *wiring rules*. Except for certain RLC cases with few loads, circuit simulation is generally required to determine the wiring rules. There are too many factors to consider to come up with satisfactory analytic answers.

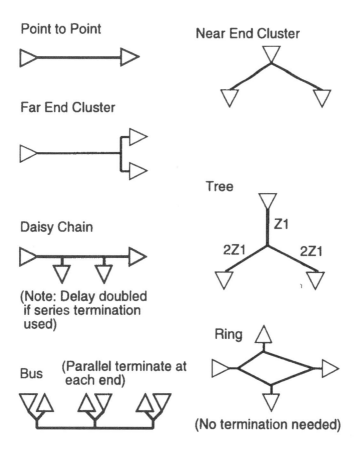

Figure 4.8 Standard topologies used to control reflection noise
in transmission lines

On most Printed Circuit Boards and MCM-Ds, crosstalk control is rarely an issue. However, on MCM-Ls, and MCM-Cs, crosstalk control measures are often required. Crosstalk control will become more important in the future as rise times decrease (crosstalk noise levels increase with decreasing rise time). Crosstalk is controlled by enforcing adjacency rules (forbidding certain nets from running parallel) and enforcing minimum spacing rules between some other neighboring nets. Most CAD tools allow the input of fixed spacing rules. Fixed spacing rules are suboptimal. The true minimum spacing rule depends on the coupled line length and how many nets are producing the crosstalk on the net being considered. Currently, only tools used within the IBM mainframe computer group permit such considerations [9]. Other approaches to managing crosstalk noise in CAD tools are described in [19].

4.2 CAD APPROACHES TO CONTROLLING DELAY
AND REFLECTION NOISE

In order to control delay and reflection noise it is necessary to conduct the following steps:

(1) Determine the minimum and maximum delay and reflection noise requirements for each data net and the maximum skew and reflection noise requirements for each clock net. These are obtained from the timing design for the system and the noise budget (see [5]) for each type of net.

(2) Obtain best and worst case circuit models for the on-chip drivers, on-chip receivers and between-chip interconnect circuits.

(3) Determine constraints on the net topology, termination resistance, stub lengths, branch lengths and widths, etc. These constraints must meet the delay and noise requirements while ensuring that routability and other requirements are met.

To give an example, consider the case where a two-terminal net must have a total settling delay of less than, say, 2 ns. The general configuration of a two terminal net is shown in Figure 4.9. The branch lengths are l_1 and l_2 and the stub length is l_3.

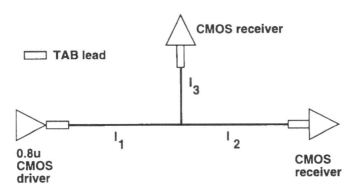

Figure 4.9 Two terminal net used in example

To illustrate this example, a large set of circuit simulations were conducted to obtain the response surface associated with the circuit. The response surface, in this case, represents the settling delay to the farthest receiver as a function of the variables l_1, l_2 and l_3. As the response surface here is a four-dimensional surface, only three-dimensional 'cuts' can be drawn. If the interconnect is a lossy MCM-D interconnect, then a typical response surface 'cut' for this case is presented in Figure 4.10. If the interconnect is built in a lossless MCM-L or PCB technology, then a typical response surface 'cut' for this case is presented in Figure 4.11.

5cm branch : 0.1V noise budget

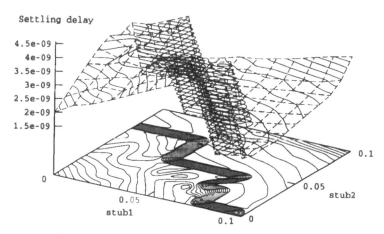

Figure 4.10 Response surfaces characterizing settling delay
over a range of lengths for a lossy net on an MCM-D
in the net class given in Figure 4.9
('stub1' is labelled 'l3' in Figure 4.9 and 'stub2' is labelled as 'l2')

"stableac" ——
"stableac" ——·

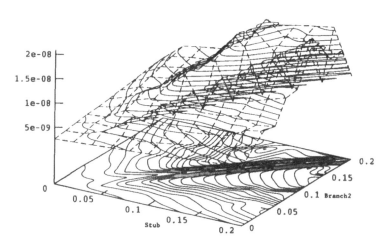

Figure 4.11 Response surfaces characterizing settling delay
over a range of lengths for a lossless net on an PCB
in the net class given in Figure 4.9
('stub' is labeled as 'l3' in Figure 4.9 and 'branch2' is labelled as 'l2')

It is the non-linearity of these response surfaces that makes the third step above ('Determine constraints') particularly difficult. The non-linearities arise from the non-linear characteristics of the digital drivers, from the broadband nature of digital signals (digital signals often have equivalent analog bandwidths of 1 GHz or more) and from the non-linear frequency response of the interconnect circuits.

In general, the non-linear nature of the circuit response makes the job of translating delay and noise requirements into physical constraints difficult. As a result, there are a number of computer-aided approaches used, in conjunction with appropriate placement and routing tools, to control delay and reflection noise. These approaches are the subject of the remainder of this chapter.

4.2.1 DELAY EQUATIONS

Though, in general, the response is non-linear, it is possible to accurately fit linear or quadratic equations to portions of it. For example, Davidson and Katopis [3,4] describe how they fit delay equations to portions of the response surface and then use these delay equations to drive the placement and routing tools. For the two-terminal example described in the previous section, they would use a delay equation of the form [4],

$$T_{D1} = a_0 + a_1 l_1 + a_2 l_2 + a_3 C_1 + a_4 C_2 + a5C_5 \qquad (4.1)$$

where
a_i are the fitting coefficients,
l_1 and l_2 are as given in Figure 4.9,
C_1 is the output capacitance of the driver,
C_2 is the input capacitance of the first receiver + the capacitance of stub l_3, and
C_3 is the input capacitance of the second receiver.

T. Mikazuki and N. Matsui describe a similar approach in which quadratic equations are used [16].

The biggest limitation with using delay equations comes about in deciding which region to fit the equation to. The resulting placement and routing tools must also be constrained to operate in the same region. As a result, it is very easy to overconstrain the tools, preventing any useful results from being gained. In practice, it is necessary to iterate on the fitted region, and resulting equation, until a placement and route are complete.

S. Simovich et al. describe a third alternative in which a piecewise linear equation is fitted over a larger region [21]. However, such a fitting is very computationly complex and would require excessive run times for problems with more than two receivers.

4.2.2 ON-LINE SIMULATION APPROACH

As the on-line simulation approach is partially an extension of the current mainstream non-automated approaches, the latter will be described first.

4.2.2.1 NON-AUTOMATED APPROACH

The current mainstream non-automated approach to generating physical constraints (wiring rules) for placement and routing tools is described in Figure 4.12. The process starts with the electrical designer using the available analytic expressions, rules-of-thumb, etc. to obtain an initial estimate for each wiring rule [5]. The wiring rule is then passed to the placement and routing engines which then attempts to generate a placement and route that meets the rule. The resulting nets are then extracted and simulated. If any circuit responses are not within the timing and noise budgets, the electrical designers uses his or her years of experience to adjust the wiring rules. Similarly, the wiring rules for unroutable nets are adjusted in order to make them routable. The process is iterated until the design is 100% electrically correct and 100% routed.

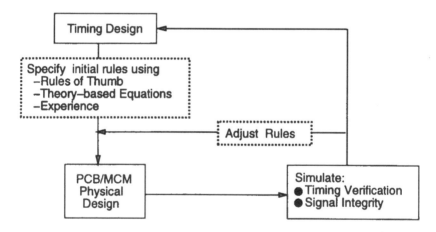

Figure 4.12 Current approach to generating wiring rules

4.2.2.2 AUTOMATED APPROACHES

More automated approaches to improving this process rely on automating one or more aspects of the cycle shown in Figure 4.12. For example, Simoudis [20] designed an expert system to replace the human designer in the 'adjustment' phase of this cycle. However, as the expert system relied on qualitative, rather than quantitative, means of improvement, its utility was limited. For example, one rule stated that if a delay was too long, then the rule 'shorten by one-third' was to be applied. Unfortunately, such arbitrary improvement increments might make the net unroutable instead. Quantitative improvements are needed.

One approach to quantitatively directing the 'adjustment' phase of the cycle, and thus automate the entire process, has been developed by the company Interconnectix (based in Portland, Oregon, USA). For each net that fails to meet delay and noise requirements after post-layout simulation, their tool conducts a sensitivity analysis. The tool not only simulates the combination of branch and stub lengths determined by the router, but also simulates the circuit response for small variations in these lengths. Thus the tool locally obtains a portion of the response surface. By fitting a linear equation to these points, the tool can provide quantitative guidance to the router. Using this technique, the router and simulator should be able to converge on a good solution, automatically, and in only a few cycles of the route-simulate-adjust loop.

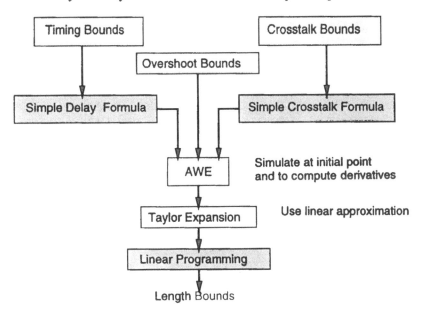

Figure 4.13 Replacement steps for 'specify initial rules' phase in the Lee and Shragowitz approach

A similar technique to producing better initial rules has been developed by Lee and Shragowitz [11]. Instead of automating the adjustment phase of the design loop shown earlier, they replace the 'specify initial rules' phase with the steps shown in Figure 4.13. Just as in the manual approach, they first use a simple analytic expression (such as the time-of-flight delay equation, delay $t_d = l/(c/\sqrt{\epsilon_r})$, where l is the length, c is the velocity of light in a vacuum and ϵ_r is the dielectric constant of the insulator) to arrive at an initial estimate for suitable lengths. They then use Asymptotic Waveform Evaluation (AWE) to conduct a local sensitivity analysis for small variations of these lengths. A linear equation is fitted to these points and by solving the equation against a set of delay and noise constraints, a wiring rule is obtained.

One problem with this approach to generating a wiring rule is that there is no guarantee that it is routable. If there is no satisfactory route, human intervention is required in the 'adjustment' phase to come up with a more routable wiring rule. (Due to the non-linear nature of the response surface, practical wiring rules are not unique.)

The two previous approaches represent opposite ends of the spectrum. In the Lee and Shragowitz approach, routability concerns are not automated at all; human intervention is required. In the Interconnectix approach, the router and wiring rule generation/adjustment engine are very tightly coupled; thus requiring that the router be written from scratch. A third approach is desired if one is constrained to work with existing routers but wish to take account of routability needs. Such an approach is described next.

4.2.2.3 USING A GLOBAL ROUTER

Mehrotra et al. describe an approach that uses a global router to quickly determine routability while using on-line simulation (or, as actually described in the paper, a predictor function based on precharacterized simulation results [15]; see Section 4.2.3) to ensure that delay and noise requirements are met. The result is a set of high-quality wiring rules that can be fed to conventional routers.

The steps used in this approach are outlined in Figure 4.14. From the placement, the routing channels are identified in the form of a channel intersection graph (for an example of a set of channels, please see Figure 4.15). The capacity of each channel is also determined. Simulation is then used to determine a set of feasible trees for each net. A feasible tree is a tree that spans the required connected nodes in the channel intersection graph but also meets signal integrity requirements (see Figure 4.16 for an example).

The global routing problem is then formalized as follows [12]: An instance of the global routing problem consists of a routing graph $G = (V, E)$, with *vertices* V and *edges* E, and a set of *supernets* N, where each supernet is a subset of V. A supernet is a collection of identical nets (same termination nodes) that are treated as a single net for simplicity.

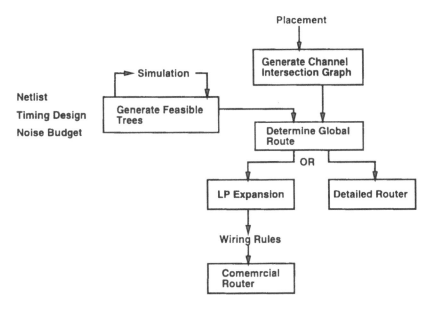

Figure 4.14 Using a global routing phase to obtain routability
and signal integrity simultaneously

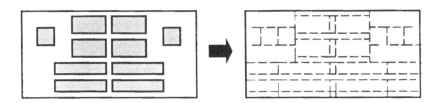

Figure 4.15 The global router determines a route
through a set of channels

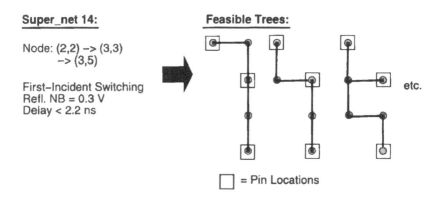

Super_net 14:

Node: (2,2) -> (3,3)
      -> (3,5)

First–Incident Switching
Refl. NB = 0.3 V
Delay < 2.2 ns

Feasible Trees:

etc.

☐ = Pin Locations

Figure 4.16 Example showing how a set of feasible trees are generated for a net (or a collection of identical nets, or supernet, in this case)

Each edge is labeled with a capacity $c : E \to R^+$ and edge lengths $l : E \to R^+$. Each net has a multiplicity $k_n \geq 1$. In addition, for each supernet $i \in N$, there is a set of admissible routes, or *trees* $T_i^1, \ldots, T_i^{t_i}$. A solution to the global routing problem is a set of admissible routes, one or more for each net, such that the capacity $c(e)$ on each edge is not exceeded by the *traffic* on that edge. The *traffic* on an edge is defined by the weighted sum of all the routes that contain edge e:

$$U(e) \quad = \quad \Sigma_{i \in N, t \in i_l, e \in T_i^t} w(i, t) \tag{4.2}$$

The weights $w(i, t)$ denote the number of wires in supernet i that are routed using tree t. The objective function minimized over all such feasible solutions varies, depending on the design problem. Some formulations try to minimize wire length, or the maximum ratio of the traffic on an edge to its capacity. Wire length minimization objective is used mostly for chip design. The global routing objective in this paper is to satisfy as many of the electrical constraints as possible, subject to the wiring capacity constraints. A benefit function $b : T \to R$ is associated with each tree. $b(i, j)$ reflects the likelihood of satisfying the electrical constraints associated with net i when routed using tree T_i^j. Hence the objective of the global routing is to maximize $B(T)$:

$$B(T) \quad = \quad \Sigma_{i \in N, j \in i_l} b(i, j) \tag{4.3}$$

The global routing problem is optimally solved by finding a set of routing trees for each net in the design with high probability of meeting the electrical constraints, and then maximizing the routing objective function while satisfying the edge capacity constraints.

4.2.2.4 INTEGER PROGRAMMING FORMULATION

The routing graph definition, the nets, routing trees for each net, and the associated benefit function fully specify a global routing problem. The routing problem can be formulated as an integer program, by associating an integer variable y_{ij} with tree i of supernet j. y_{ij} is the number of wires in supernet j routed with tree i. Then the global routing problem is given by the following integer program:

$$
\begin{aligned}
\text{Maximize} \quad & \sum_{i=1}^{N} \sum_{j=1}^{n_i} b_{ij} y_{ij} \\
\text{subject to} \quad & \sum_{j=1}^{n_i} y_{ij} = d_i && i = 1, \ldots, N \\
& \sum_{i=1}^{N} \sum_{j=1}^{n_i} a_{ij}^k y_{ij} \leq c_k && k = 1, \ldots, m \\
& 0 \leq y_{ij} \\
\text{integer} \quad & y_{ij}
\end{aligned}
$$

Here b_{ij} is the benefit for the tree i of net j, N is to total number of supernets, d_i is the number of wires in supernet i, n_i is the number of routing trees for net i, c_k is the capacity of edge k. a_{ij}^k is a $(0, 1)$ matrix that specifies whether or not tree i of net j uses the edge k.

Solution Methods: Integer programming is, in general, NP-hard. There are numerous ways of solving integer programs, e.g., cutting plane algorithms, branch and bound and Lagrangian relaxation [12]. One method that has been shown to be very effective for solving the global routing problem is a randomized rounding technique to the linear relaxation of the integer program [17]. The basic idea is to relax the integer constraint in the formulation, which make it a linear program, and to solve the linear programming problem. If the solution of the linear program is integral, then we have an optimal global routing. If not, then we need to transform it into an integer solution by rounding the non-integral values. Carden has shown how to correct the solution if some capacity constraints are violated after the rounding [8]. In our experience, almost all solutions turned out to be integer after solving the linear relaxation. This is primarily due to the high multiplicity of the nets. Hence there was usually no need to round to an integer solution. In the few cases with non-integer solutions, a simple rounding led to very few constraint violations, although the objective function might be sub-optimal.

For more details and examples, the reader is referred to reference [14].

4.2.3 PRE-CHARACTERIZATION

All of the above techniques assume the existence of an on-line simulator that is very fast. To be practical, the simulator should be able to simulate each net in less than one second. This goal is possible if behavioral models are used for the driver and receivers and the simplest, lossless, transmission line model is used. If the simulation model is more complex, for example a lossy MCM-D interconnect structure is present, then the simulation becomes too time-consuming with current simulation technologies. One solution is to conduct a preliminary characterization step and replace on-line simulation with evaluation of a predictor function. A separate characterization is performed for each unique circuit topology (for example, those listed in Figure 4.8).

The objective of the characterization is to obtain a *predictor function* that can be used to accurately and quickly obtain circuit responses of interest (e.g., delay and reflection noise) of a generalized interconnect circuit topology, over a range of certain *design variables*, such as length.

To obtain a characterization, we conduct a computer experiment [18] in which *samples* are taken at different points in the 'design space', the dimensions of which are the design variables. The sampling method is a heuristic, multistage experiment (see Figure 4.17) that uses Latin Hypercube Sampling [13] in each iteration. Resampling after the first iteration depends on the evaluation of the prediction error, measured by *cross-validation*.

The predictor function is a data interpolant. Interpolation is performed using Moving Least Square Interpolation [10], on the simulated points. The predictor function is given in the following form:

$$\phi(z) = \Sigma_{j=1}^{n} a_j b_j(z) \tag{4.4}$$

where z is the vector of design variables, $b_1(z), \ldots, b_n(z)$ are n linearly independent polynomials in z supplied by the user and the a_j's are constants to be determined. Whenever $\phi(z)$ is evaluated, Moving Least Squares are used to determine the a_j. The a_j's are chosen so that a weighted sum of the error of prediction at all sample points is minimized. The square error is given as

$$E_z(\phi) = \Sigma_{i=1}^{N} w_i(z)(\phi^*(z_i) - \phi(z_i))^2 \tag{4.5}$$

where $w_i(z)$ is the weight assigned to the error at z_i, and $z_1, \ldots, z_N$ are the N distinct sample points in the design space. The error $E_z(\phi)$ is minimized by solving the system of equations

$$BW(z)B^T a(z) = BW(z)\phi \tag{4.6}$$

where B is an $n \times N$ matrix whose jth column is $[b_j(z_1), \ldots, b_j(z_n)]$, ϕ is the $N \times 1$ vector of responses at the sample points, and $W(z)$ is a diagonal matrix

$$W(z) = diag[w_1(z), \ldots, w_N(z)] \tag{4.7}$$

Note that when $W(z)$ is the identity matrix, the equations are the same as those for the usual least square minimization. The weighting function has the form

$$w_i(z) = w(d(z, z_i)) \tag{4.8}$$

where d is the Euclidean distance between two points. To achieve exact interpolation at the sampled points, the function w should go to infinity at the sampled points z_i's. Functions of the form

$$w(d) = e^{-\alpha d^2}/(d^2) \tag{4.9}$$

have this behavior. These functions also attenuate rapidly and hence minimize the influence of remote data values, while smoothing the response.

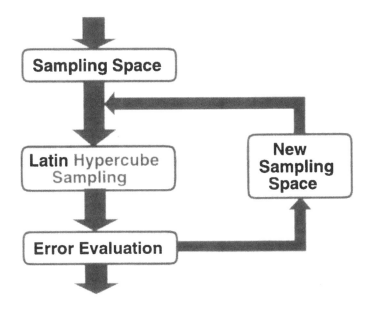

Figure 4.17 Steps in sequential sampling

4.3 CONCLUSIONS

In order to meet a specified timing design and to prevent excessive noise from upsetting the system it is necessary to control the placement and routing during physical design. In particular, it is necessary to constrain the topology, branch and stubs lengths of many nets. These constraints, often referred to as *wiring rules* allow delay and ringing noise to be controlled. Often, minimum spanning Steiner trees are unacceptable for controlling noise. Sometimes, it is also necessary to control crosstalk noise by specifying rules about net adjacency and spacing.

Specifying wiring rules is a non-trivial task. The available analytic expressions for estimating the appropriate rules are not good enough to guarantee success. Simulation must be used to obtain many rules. Additionally, it is very easy to arrive at a rule set that overconstrains the router.

Current approaches to specifying good wiring rules are usually iterative, as shown in Figure 4.12 above. Automating this cycle involves improving and automating either the 'Specify initial rules' step or the 'Adjust Rules' step. In this chapter, we discussed several different alternatives to automating these steps. The first alternative used empirically derived linear or quadratic delay-equations to drive the initial rule generation. The problem with this approach is that valid delay equations can only be determined over small ranges of applicability. Often, the electrical designer must iterate on the range over which the equation is fitted in order to obtain a routed solution.

The second alternative used a simulation-based sensitivity analysis to guide the 'Adjust Rules' step. This alternative required that the rules engine interface closely with the placement and routing tools. The latter tools must be rewritten for this task. The third alternative used a simulation-based sensitivity analysis to help generate better initial rules. However, until the routing is attempted the designer does not know if these rules are routable or not. Human intervention might be needed. The fourth alternative used a simulation-based rule generator with a global router to try to ensure that the specified rules are both electrically correct and routable. Unlike the second approach, this approach can be used with currently available routers.

In the last three of these alternatives, on-line simulation is needed. If the simulation model is more complex than a simple behavioral model for the digital circuits and a lossless transmission line for the interconnect, the simulation might be too slow. In these cases, a better alternative would be to conduct a pre-characterization phase and use a predictor function instead of a simulator.

REFERENCES

1. H.B. Bakoglu, *Circuits, Interconnections, and Packaging for VLSI*, Addison-Wesley, 1990.

2. J. E. Buchanan, *BiCMOS/CMOS Systems Design*, McGraw-Hill, 1990.

3. E.E. Davidson, "Electrical Design of a High Speed Computer Package," *IBM J. Res. Dev.*, vol. 26(3), May 1982, pp. 349–361.

4. E.E. Davidson and G.A. Katopis, "Package Electrical Design," In *Microelectronics Packaging Handbook*, ed. by R.R. Tummala and E.J. Rymaszewski, chapter 3. Van Nostrand Reinhold, 1989.

5. P.D. Franzon, "Electrical design," In *Multichip Module Technologies and Alternatives: The Basics*, edited by D.A. Doane and P.D. Franzon, chapter 11, Van Nostrand Reinhold (New York), 1992.

6. B. Gilbert and L.G. Salmon, *Interconnection Design and Process Characteristics for Digital Multichip Modules Operating at High Clock Frequencies*, ISHM, 1993, pp. 165–218.

7. R. Gupta and L.T. Pillage, "Otter: Optimal Termination of Transmission Lines Excluding Radiation," In *Proc. 31st ACM/IEEE Design Automation Conference*, 1994, pp. 640–645.

8. R. C. Carden IV and C.-K. Cheng, "A Global Router Using an Efficient Approximate Multicommodity Multiterminal Flow Algorithm," In *Proc. of 28th Design Automation Conference*, 1991, pp. 322–327.

9. G.A. Katopis and H. Smith, "Coupled Noise Predictors for Lossy Interconnects," In *IEEE 2nd Topical Meeting on Electrical Performance of Electronic Packaging*, 1993, pp 65–68.

10. Peter Lancaster and Kestutis Salkauskas, *Curve and Surface Fitting : An Introduction*, Academic Press, 1986.

11. J. Lee, E. Shragowitz, and D. Poli, "Bounds on Net Lengths for High-Speed PCBs," In *Int. Conf. on Computer-Aided Design*, 1993, pp. 73–76.

12. Thomas Lengauer, *Combinatorial Algorithms for Integrated Circuit Layout*, John Wiley and Sons, 1990.

13. M. D. McKay, R. J. Beckman, and W. J. Conover, "A Comparison of Three Methods for Selecting Values of Input Variables in the Analysis of Output From a Computer Code," *Technometrics*, vol. 21(2), May 1979, pp. 239–245.

14. Sharad Mehrotra, *Automated Synthesis of High Speed Digital Circuits and Package-Level Interconnect*, Ph.D. Dissertation, North Carolina State University, 1994.

15. Sharad Mehrotra, Paul Franzon, Griff Bilbro, and Michael Steer, "CAD Tools for Managing Signal Integrity and Congestion Simultaneously," In *Proc. Third Topical Meeting on Electrical Performance of Electronic Packaging*, 1994.

16. T. Mikazuki and N. Matsui, "Statistical Design Techniques for High-Speed Circuit Boards," In *IEEE/CHMT'90 IEMT Symposium*, number IEEE no. CH2864-7/90/-0000-0185, 1990, pp. 185–191.

17. P. Raghavan and C. D. Thompson, "Multiterminal Global Routing: A Deterministic Approximation," *Algorithmica*, vol. 6, 1991, pp. 73–82.

18. J. Sacks, W. J. Welch, T. J. Mitchell, and H. P. Wynn, "Design and Analysis of Computer Experiments," *Statistical Science*, vol. 4(4), 1989, pp. 409–435.

19. M. Sengupta, S. Lipa, P. Franzon, and M. Steer, "Control of Crosstalk Noise," In *Proc. 44th Electronic Components and Technology Conference*, 1994.

20. E. Simoudis, "A Knowledge-Based System for the Evaluation and Re-design of Digital Circuit Networks," *IEEE Trans on CAD*, vol. 8(3), March 1989, pp. 302–315.

21. Slobodan Simovich, Sharad Mehrotra, Paul Franzon, and Michael Steer, "A Priori Generation of Delay and Reflection Noise Macromodels and Wiring Rules for PCBs and MCMs," *IEEE Trans. CHMT, Part B. Advanced Packaging*, vol. 17(1), February 1994, pp. 15–21.

Chapter 5

ROUTING FOR ANALOG AND MIXED CIRCUITS

Enrico Malavasi

A key issue in the usability of CAD systems for analog and mixed circuits is their capability to enforce a large variety of simultaneous constraints. During the definition of layout details, the values of all performance functions (such as bandwidth, offset, noise, gain etc.) are subject to degradation with respect to their nominal values defined by the designer, due to the non-idealities of devices and interconnections. Routing in particular is responsible for the introduction of parasitics in the interconnections, which can have a significant impact on the electrical behavior of the circuit, although their exact value is difficult to predict before the layout is defined in all its details.

In digital systems, parasitics are important for speed and signal integrity considerations. The total capacitance between a wire and ground and with the surrounding nets determines the maximum speed of a signal on that wire. The total capacitance between a wire and the surrounding switching nets determines the crosstalk and therefore the maximum amount of noise injected in the wire itself. However, as long as the total noise is kept within the immunity margins, it can be neglected. Hence digital systems are characterized by an intrinsic robustness to parasitic effects, which allows automatic layout tools to use simplified parasitic models, neglect second order effects, and so achieve considerable efficiency and speed. With analog circuits, on the contrary, parasitics affect all performance functions, from bandwidth to phase margin, distortion, offset, linearity etc. The effects of stray resistances and capacitances can not be cumulated as they are with digital systems, but must be considered on a net-by-net basis. Furthermore, crosstalk can seldom be neglected; there are no thresholds or immunity margins, but rather a set of "tolerances" provided by the designer on the maximum degradation that performance functions are allowed to sustain. Therefore analog circuit layout tools need more accurate parasitic models and often they trade off computational efficiency for accuracy. On the other hand, in the analog systems the number of circuit elements (i.e., transistors, gates, interconnections) is often orders of magnitude lower than in the digital systems.

Analog routing is needed in different types of environments, in particular in device-level routing for full-custom cell design, and in system-level routing for macro-cell design with mixed circuits. Examples of the former case are the cell-level module generators for op-amps and comparators [20, 23]. Examples of the latter case are the module-generators for data converters and switched-capacitor circuits, often involving pre-defined floor-plans for large hierarchical mixed circuits [39, 21]. Although the separation between "digital" and "analog" still holds for most commercial applications, very high-speed integrated circuits on silicon are becoming the state of the art, and VLSI systems with clocks of 100 to 500 MHz are already commercially available. At these speeds, rising and falling delays must be kept within less than 100 ps, and clock skew constraints are extremely tight. As a consequence, it is becoming more and more important to model correctly all the effects related to crosstalks and to stray capacitances. CAD systems for digital design are introducing parasitic control features inherited from the analog CAD.

Beyond parasitic control, analog routing is characterized by a number of special features, not usually provided by the tools targeted toward digital design. The first of such features is **symmetric routing**, namely the capability to route symmetric interconnections. This requirement stems from the need to enforce the best possible matching in fully differential architectures, in order to achieve high performance in terms of gain, offset, and robustness to common-mode and supply noise. Another useful feature is the capability to build **electrostatic shields**, that is interconnection structures able to decouple effectively critical net pairs. This involves decoupling between noisy and sensitive signal nets, and between analog and digital signals and supply. At the system level, routing must provide the capability to **distribute** cells and devices **to different supply pins**, in order to avoid coupling between noise sources and sensitive devices, due to voltage drops on power lines and to substrate currents.

The complexity of parasitic control, noise decoupling, performance constraints, symmetries, power and ground sizing etc. make analog and mixed routing a formidable problem. So far this problem has not been solved satisfactorily, and no commercial products are available today, except for a few point tools targeted toward a limited set of circuit classes. The lack of tools can also be ascribed to the limited size of the market for analog integrated circuits, which is smaller than the digital market by at least an order of magnitude. However, although analog CAD so far has not reached the maturity of digital place and route tools, yet it has contributed to the knowledge of techniques and algorithms which are becoming more and more important also for the digital world.

5.1 OVERVIEW OF CURRENT APPROACHES

The first approaches to the automation of analog design appeared in the mid-'80s. They were mostly works on module generation for specific circuit classes, such as switched-capacitor circuits [1], or attempts to store knowledge about complex analog design in expert systems [2]. The need for specific routers to deal with the complexity of analog constraints has been apparent since the beginning. The first routers were based on a **net classification** scheme, as a simple way to capture the complex pattern of interrelations between nets due to crosstalk, matching, and mutual sensitivity to noise. A second generation of routers tried to remove the rigidity of net classification, and introduced a **weight-controlled** cost function. The router became an optimizer trying to minimize the value of a cost function containing all the non-idealities to be controlled in the form of a weighted sum, expressing the trade-offs between different simultaneous constraints. Today, research is focussing above all on **constraint-driven** approaches, which try to enforce constraints directly, using a simplified quantitative model of the functional relation between performance specifications and parasitics to determine a pattern for the router's behavior.

5.1.1 ROUTERS BASED ON NET CLASSIFICATION

The approach developed at AT&T by Kimble et al. [22] is one of the first layout systems where segregation of analog and digital nets and decoupling of noisy and sensitive signals were pursued. In this approach, both analog and digital parts of a mixed circuit are realized with standard cells. The digital part is implemented separately from the analog part, as shown in Figure 5.1. The layouts of the analog cells are designed in such a way that the pins associated with sensitive nodes, such as

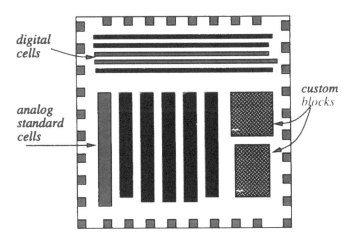

Figure 5.1 AT&T standard-cell system for mixed circuits
(The digital part is implemented separately from the analog standard cells.)

virtual ground of an op-amp, are on the top edges of the cells, and the pins associated with large-swing or insensitive nets (e.g., the output of an op-amp) are on the bottom edges of the cells. The cells are oriented upside-down in alternate rows of the standard-cell layout, as shown in Figure 5.2. As a result, the sensitive and the large-swing/insensitive

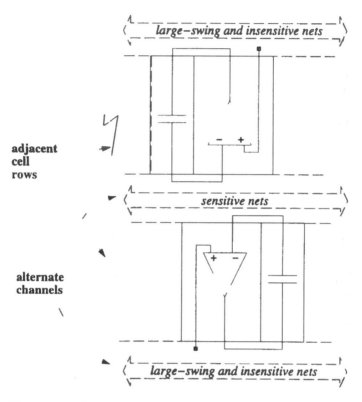

Figure 5.2 Cells are oriented upside-down in alternate rows of the standard-cell layout; as a result, the sensitive and the large-swing/insensitive nets are routed in different channels.

nets are routed in different channels. Within the insensitive/large-swing channels, bias and digital supply nets are routed first along the edge of the cells, followed by analog large-swing nets, analog ground, and digital control signals. Although this technique is not able to account for most of the complex parasitic interaction within an analog or mixed circuit, yet it is effective in decoupling nets by allocating them to separate channels. As a result, after the introduction of this design methodology AT&T was able to report an order of magnitude increase in the productivity and functional "first silicon" for several mixed-signal circuits of various complexity, up to almost 12,000 transistors.

After the AT&T work, several other approaches have been reported which provide more flexibility to the designer during layout design. For example, the idea of keeping sensitive and noisy nets separated by allocating them to different channels has been applied to the floor-plan of most mixed module generators, and in particular switched-capacitor generators like ADORE [39]. The floor-plan for a switched-capacitor filter generated by ADORE is shown in Figure 5.3. All clock lines are above the switch array, and therefore they are separated from the analog lines, routed between switches and capacitors, and between capacitors and the op-amps.

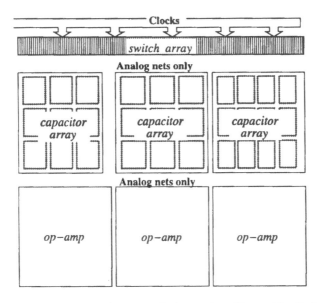

Figure 5.3 Floor-plan for a switched-capacitor filter with all clocks separated from the analog signals

The concept of net classification has been expanded and used in several systems. In [19], four classes are identified: (a) high impedance analog (most sensitive); (b) low impedance analog (less sensitive); (c) digital control signals affecting the analog cells (such as clocks); (d) other digital cells (most insensitive). Nets are routed in order of sensitivity, from class (a) to class (d). Digital and analog nets are routed in different areas. Critical stray resistances are avoided by doubling the width of the nets which are sensitive to voltage drops. Shields are added at the end by a post-processor. In SLAM [7], classification is performed automatically, on the ground of a set of rules and a circuit analysis performed using a knowledge base. Four classes are automatically identified: (a) power; (b) sensitive; (c) noisy; (d) inactive nets. This classification is used to derive constraints on net lengths and minimum separations. In [35] the number of categories is not pre-defined, nets are classified based on their type (bias, supply, noisy, high-sensitivity etc.) as well as on the electrical constraints specified on the parasitics associated with them. Routing is carried out in two phases: first the relative positions of the wires are determined (symbolic routing), then their physical positions are determined in detail from technology information and electrical specifications. ILAC [36] is a tool which automatically generates the layout of analog CMOS cells from netlist information and user specifications. It is targeted to work with the circuits designed by the IDAC [15] design system, developed at CSEM (Centre Suisse d'Electronique et Microtechnique, Neuchatel, CH). Each circuit is partitioned into a number of functional blocks of various complexity, characterized by different alternative realizations called "variants". Once a variant for each block is chosen, a suitable floor-plan and placement are created based on a slicing structure and simulated annealing optimization. Global routing is followed by symbolic detailed routing, based on a scan-line-based incremental channel router, which routes all nets one at a time according to a pre-defined order. Supply nets are routed first, using one layer. Sensitive nets are routed next, trying to keep their lengths short and their parasitics as small as possible. Following are the insensitive nets, and finally the noisy nets. During routing, ILAC also tries to minimize a cost function accounting for undesired coupling, the number of vias, and area.

5.1.2 WEIGHT-DRIVEN ROUTERS

Weight-driven routers try to minimize a cost function given by a weighted sum of items proportional to all parasitics, to wire length, and to the number of bends and vias.

ANAGRAM is an analog router developed at the Carnegie Mellon University (CMU) for analog and mixed circuits in macro-cell design style. In its early implementation [18], it was a channel router based on net classification. It was later modified into a weight-driven router and renamed ANAGRAM-II [5]. It uses a line-expansion algorithm, based on a tile-plane representation of the layout. The cost function takes into account all parasitics associated with every path, although for cross-coupling it uses a simplified model. From its original implementation, ANAGRAM-II has preserved a net classification into "sensitive" and "noisy" nets. The cost function contains a term proportional to the crosstalk between sensitive and noisy nets. ANAGRAM-II can perform symmetric routing of differential nets, even if the placement is not perfectly symmetric, using an algorithm described in detail in Section 5.6.3.

ROAD [27] is an area router for analog layouts, developed at the University of California at Berkeley (UCB). ROAD uses the A* algorithm [34], a modified version of Lee's maze routing [26]. It can perform symmetrical routing for differential architectures, and build shields to improve capacitive decoupling. The cost function is a weighted sum including parasitic resistances, capacitances to ground and individual cross-coupling capacitances. As in other weight-driven tools, the performance degradation of the circuit is only indirectly controlled by the set of weights. In [31], a routing methodology was presented, where weights are defined automatically based on sensitivity analysis performed on the circuit. With this methodology, bounds on performance degradation are not directly enforced, but in case of a constraint violation routing is iterated with a different set of bounds, until the performance specifications are eventually met.

WREN [30] is an optimization-based approach to global and detailed routing, targeted toward analog and mixed circuits at the system level. An aggressive optimization technique, based on simulated annealing, is used to select paths and optimize signal integrity (i.e., robustness to noise in the environment). In practice, randomly selected, alternative paths are chosen for all nets, and each configuration is analyzed using simplified geometric models, which increases the CPU efficiency. The use of simulated annealing increases the computational cost of this approach by orders of magnitude compared with other direct (greedy) approaches. However, a large set of parasitic bounds can be accommodated by this very flexible technique. Details about the algorithms used in WREN are provided in Section 5.4.

5.1.3 CONSTRAINT-DRIVEN ROUTERS

In [17], an algorithm for two-layer gridless channel routing driven directly by analog constraints was reported. The channel-routing problem is represented by a vertical-constraint graph (VCG). Net-to-net crosstalk is controlled by directing undirected edges in the VCG, that is, by enforcing appropriate arrangement of horizontal segments. Coupling between adjacent parallel segments is controlled by keeping a separation between them, and this is achieved by adding weights to the edges of the VCG. The algorithm also attempts to perform single-layer routing for supply nets or signals which are sensitive to voltage drops due to the stray resistances. In [17], the bounds on parasitics are user-defined, and no strategy for their automatic generation is considered.

ART [9] is a constraint-driven channel router, also based on the VCG. ART is able to maintain within their bounds line resistances, capacitances to the substrate, and line-to-line capacitances. The bounds for all parasitics are automatically determined from performance constraints using sensitivity analysis. The parasitic constraints are enforced by mapping the bounds on maximum allowed crossover and adjacency into con-

straints for the VCG. This is achieved with several techniques, described
in detail in Section 5.4. ART is able to introduce shields into the channel
to achieve better decoupling.

Both ANAGRAM-II and ROAD have been modified recently, to accommodate a strategy driven directly by bounds on parasitics. The first
description of such a bound-driven strategy for area routing appeared in
[3], where the modified version of ANAGRAM-II is described. The combined system including a modified version of the placement tool KOAN
and the new constraint-driven router has been called KA-III. The router
is driven directly by bounds on crosstalk, but it still uses weights for all
other parasitics. An extension to constraint-driven control over all parasitics using a set of dynamically adjusted bounds, implemented in ROAD,
has been described in [32]. No artificial trade-offs are made between wire
length and crosstalk. Instead, the cheapest path which violates no parasitic bounds is found. Details about constraint-driven area routing are
reported in Section 5.3.

5.2 ANALOG CONSTRAINTS FOR ROUTING

Analog and mixed circuits are subject to a wide variety of performance specifications on power consumption, noise, frequency response
characteristics, offset, I/O impedance, yield etc. No router (or any other
layout tool) is able to understand, let alone enforce, all these constraints
simultaneously. During routing a large number of parasitics are introduced, which ultimately have a different impact on those specifications.
As shown in Figure 5.4, three types of parasitics are considered:
(1) Stray resistances (R_s),
(2) Stray capacitances to the substrate (C_s), and
(3) Cross-coupling capacitances between interconnections (C_c, C_p).

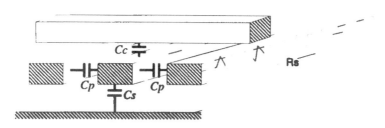

Figure 5.4 Interconnection parasitics

Inductances are relevant in very high-speed circuits, but routers don't
take them into consideration usually, due to the difficulty of modeling. It
is necessary to distinguish between the capacitances of the 2nd and 3rd
group, because their impact on performance is different. With digital

nets, the capacitance toward the substrate and all capacitances between adjacent lines affect (negatively) the speed of the signals. Capacitances between adjacent lines, in particular if digital signals are correlated with opposing closely timed transitions, can also be responsible for glitches and logic faults due to crosstalk noise. With analog signals, the situation is much more complex, because each cross-coupling capacitance between any arbitrary pair of nets has a different impact on the performance functions. For instance, capacitances between input and output nets in an op-amp introduce zeroes in the transfer function, while cross-coupling between the output node and the supply lines degrades the PSRR.

Parasitic control requires the simultaneous control over a large number of parameters whose values vary over different ranges, depend on the lengths, widths and relative positions of several nets, and affect performance in different ways. As an example, consider a circuit containing 100 nets, each with 3 pins. The number of parasitics is: 100 capacitances to the substrate, 300 pin-to-pin resistances, and 4950 cross-coupling capacitances.

During routing, it is much easier to take into account capacitances toward the substrate, than cross-coupling capacitances. In fact, the former are only function of the wire length and width. The latter are also function of the distance and shape of all the surrounding nets, and their computation is much more costly.

As shown in Section 5.1.1, a popular approach used in the past to model the dependency of performance functions on parasitics is net classification. Nets are classified according to the type of signal they carry (stable, large swing, sensitive to noise, noisy etc.) and in particular on their being digital or analog. Classifications introduce considerable simplification in the task of the router, which doesn't need to manage simultaneously hordes of cross-coupling constraints, but only keeps nets belonging to different classes segregated from each other. The efficiency of this approach is considerable, but its results are often unsatisfactory, for the following reasons:

- Net A can be noisy with respect to net B, and sensitive to the noise injected by net C. Noisiness and sensitivity are a relation *between* nets, not an absolute property of a single net. When coupled with other nets carrying signals of comparable power, cross-coupling must be regarded as a bi-directional exchange of power, rather than a situation where one noisy net transfers power to one or more sensitive nets. Several degrees of noisiness and sensitivity can be used, and the nets in a class are required not to be matched with the nets of any other class. However, this ends up over-constraining the router, because the number of restrictions on routing (e.g., no-crossing constraints) grows with the square of the number of classes.

- Cumulative effects are not accounted for. For instance, a bus of several "sensitive" nets running parallel to each other with correlated signals might inject a considerable noise into a single net that inadvertently happens to cross it, and the net becomes sensitive to the bus, although it might be insensitive to any of its parts.

- Net classification requires a knowledge of the circuit that prevents a correct use of the automatic tool by inexperienced designers. As a consequence its usefulness to increase productivity is limited. Moreover, reproducibility of results is not guaranteed, because different users use different classifications (and get different results) depending on their level of expertise and knowledge of the specific circuit.

- Sensitivity of a net to noise from the rest of the circuit might have different impact on different performance specifications. Therefore, different net classifications should be provided for the same circuit, when subject to different sets of performance specifications.

With mixed circuits, net classification is still useful to segregate wide-swing digital signals from sensitive analog nets. However, the evaluation of parasitic criticality should be derived as a reproducible and quantitative function of the set of performance specifications.

In the manual approach to analog layout, the circuit designer provides a list of directions, including a list of the most critical parasitics, to the layout designer. The latter follows a set of rules drawn from his/her experience, trying to keep those parasitics within the specified bounds. This methodology has been encoded in some knowledge-based approaches to layout, such as LADIES [29], SLAM [7] and STAT [28]. However, a more robust and systematic approach is to analyze the relationship between electrical performance and parasitics, and to derive quantitative dependencies between them. The problem of analog routing is divided in two sub-problems of different complexity:

(1) Transform the performance specifications into constraints on parasitics (*constraint translation* or *constraint generation*)

(2) Actually enforce parasitic constraints during routing (*constraint-driven routing*).

This technique is at the basis of all constraint-driven design strategies. Its main advantages are the following:

- If both steps can be performed with sufficient accuracy, the final layout is **guaranteed** to meet all performance specifications.

- It is a quantitative, rather than qualitative approach.

- It is reproducible and objective, independent of the user's experience.

- By virtue of the previous point, not only routing becomes accessible to unspecialized technical personnel, but also it becomes an interesting *learning tool* to understand the parasitic criticality in not-well-known circuits.

Its main drawbacks lay in the difficulty to perform the two steps. If the performance specifications are sufficiently regular and continuous near their nominal value, *sensitivity analysis* can be used for the first step. This approach is described in Section 5.2.1.

5.2.1 SENSITIVITY ANALYSIS FOR CONSTRAINT GENERATION

In what follows, N_p is the number of parasitics, and $\mathbf{p} = [p_1 \ \ldots \ p_{N_p}]^T$ is their array. Each parasitic p_j has a *nominal value* $p_j^{(0)}$, and we denote by $\mathbf{p}^{(0)} = [p_1^{(0)} \ \ldots \ p_{N_p}^{(0)}]^T$ the array of the nominal values. Each performance K_i is a non-linear function of all parasitics $K_i = K_i(\mathbf{p})$ and the array of the N_k performance functions is indicated as $\mathbf{K}(\mathbf{p}) = [K_1(\mathbf{p}) \ \ldots \ K_{N_k}(\mathbf{p})]^T$. All parasitics are subject to variations with respect to their nominal values. Let $\Delta\mathbf{K}(\mathbf{p}) = \mathbf{K}(\mathbf{p}) - \mathbf{K}(\mathbf{p}^{(0)})$ be the array of the corresponding degradations of $\mathbf{K}$ due to such variations, and let $\overline{\Delta\mathbf{K}}(\mathbf{p})$ be the array of the (user-provided) constraints on degradation:

$$\Delta\mathbf{K}(\mathbf{p}) \leq \overline{\Delta\mathbf{K}}(\mathbf{p}) \tag{5.1}$$

The (non-normalized) sensitivity of K_i with respect to p_j is defined as

$$S_{i,j} = \left.\frac{\partial K_i(\mathbf{p})}{\partial p_j}\right|_{\mathbf{p}^{(0)}} \tag{5.2}$$

All sensitivities are arranged in a matrix

$$\mathbf{S} = \begin{bmatrix} S_{1,1} & \cdots & S_{1,N_p} \\ \cdots & \cdots & \cdots \\ S_{N_k,1} & \cdots & S_{N_k,N_p} \end{bmatrix}$$

Sensitivities are computed for each performance function, with respect to all the parasitics that may be introduced or modified by routing. This computation is moderately time consuming, because with the *adjoint technique* of sensitivity analysis [14] it can be performed with only two circuit simulations per performance function, with some limitation for functions in the time domain. Performance degradations are approximated by linearized expressions using sensitivities [11]:

$$\Delta\mathbf{K}(\mathbf{p}) = \mathbf{S}\left[\mathbf{p} - \mathbf{p}^{(0)}\right] \tag{5.3}$$

This linearization is acceptable as long as we assume that degradations are small compared to the nominal values. For matched parasitics, the sensitivity is considered with respect to the mismatch rather than to absolute parameter values. In the general problem formulation, performance constraints are modeled by the following inequalities: [1]

$$\mathbf{S}\left[\mathbf{p} - \mathbf{p}^{(0)}\right] - \overline{\Delta \mathbf{K}} \leq 0 \qquad (5.4)$$

The problem of constraint generation is reduced to the problem of determining an array of *bounds* $\mathbf{p}^{(b)} = [p_1^{(b)} \ \ldots \ p_{N_p}^{(b)}]^T$ for all parasitics, such that inequality (5.4) holds as long as each parasitic remains below its bound, i.e.,

$$\mathbf{S}\left[\mathbf{p}^{(b)} - \mathbf{p}^{(0)}\right] - \overline{\Delta \mathbf{K}} = 0 \qquad (5.5)$$

All bounds must be *feasible* and *meaningful*, i.e., they must be within the range of values $[p_j^{(min)}, p_j^{(max)}]$ that each parasitic p_j can assume in practice. The array of bounds $\mathbf{p}^{(b)}$ must satisfy the following inequalities:

$$\begin{cases} \mathbf{p}^{(b)} - \mathbf{p}^{(min)} \geq 0 \\ \mathbf{p}^{(b)} - \mathbf{p}^{(max)} \leq 0 \end{cases} \qquad (5.6)$$

where $\mathbf{p}^{(min)} = [p_1^{(min)} \ \ldots \ p_{N_p}^{(min)}]^T$ and $\mathbf{p}^{(max)} = [p_1^{(max)} \ \ldots \ p_{N_p}^{(max)}]^T$. In general the constraint generation problem, formulated as (5.5), subject to the feasibility constraints (5.6), has an infinite number of solutions. Choudhury and Sangiovanni-Vincentelli [10] proposed a methodology to find a solution to the constraint-generation problem under particular assumptions. The methodology was implemented in a tool called PARCAR. Although the original formulation of the problem was limited to the particular case of channel routing, PARCAR has been successively extended first to usage with area routing [31] and then to any general layout phase based on tools able to take advantage of bounds on circuit parameters or parasitics [8]. PARCAR chooses, among all possible solutions, the one maximizing the layout tool *flexibility*, which is a measure of how easily the tool is able to meet the constraints. To explain this concept, suppose that the bound for a given parasitic p_j is close to its lower limit $p_j^{(min)}$, and far from its upper limit $p_j^{(max)}$. Then the router is required to maintain p_j within a bound which imposes a tight limit to its variation. If, on the contrary, the bound is close to $p_j^{(max)}$, the effort required is lower, and the constraint is easier to meet. PARCAR uses the

---

[1] The notation $\mathbf{A} \leq 0$ means that every element of array $\mathbf{A}$ is a real number not greater than 0. Similarly, $\mathbf{A} \geq 0$ indicates that every element of $\mathbf{A}$ is a non-negative real number.

following definition for flexibility:

$$\text{flexibility} = 1 - \sum_{j=1}^{N_p} \frac{\left| p_j^{(max)} - p_j^{(b)} \right|^2}{\left| p_j^{(max)} - p_j^{(min)} \right|^2} \tag{5.7}$$

The constraint-generation problem is solved by minimizing a quadratic function subject to linear constraints (5.5) and (5.6), using a standard quadratic programming (QP) package. The quality of the result depends on the estimates of parasitic limits $\mathbf{p}^{(\mathbf{min})}$ and $\mathbf{p}^{(\mathbf{max})}$, which become more and more accurate as layout details are defined during the design. The values of $\mathbf{p}^{(\mathbf{min})}$ and $\mathbf{p}^{(\mathbf{max})}$ are generally not known a priori, and their estimates depend on the layout algorithm used. For example, the minimum value of the cross-coupling capacitance between unrouted nets can be set either to 0, or to the crossover capacitance due to unavoidable crossings. The latter estimate, however, is possible only if the router is able to detect unavoidable net crossings. This is the case for a channel router, where wire paths have been predefined in the global routing phase. With maze routing, on the contrary, the best estimate of the minimum value is 0.

A substantial speed-up of the QP solver is achieved by removing from the problem those parasitics whose cumulative contribution to performance degradation is negligible. A small threshold value $\alpha < 1$ is defined (typically $\alpha = 0.01$). For each performance function K_i, all parasitics are sorted by increasing value. The first n_i parasitics in the sorted list such that:

$$\sum_{j=1}^{n_i} S_j^i \left(p_j^{max} - p_j^{(0)} \right) \le \alpha \overline{\Delta K_i} \tag{5.8}$$

are considered non-critical with respect to the threshold α. This procedure detects the parasitics whose cumulative contribution to performance degradation is small. Notice that the sorting order may be different for different performance functions. Let P_i denote the set of n_i critical parasitics sorted according to performance K_i. When all performance functions are considered simultaneously, the set of non-critical parasitics $P = \bigcap_{i=1}^{N_k} P_i$ is eliminated from further analysis. Obviously, different sorted lists are maintained for each kind of parasitics (capacitances, resistances and inductances) and elimination is carried out separately.

Example: Figure 5.5 shows the schematic of a clocked comparator. The following notation is used: $R_{a,b}$ represents the mismatch between the source resistances of transistors M_a and M_b, and C_n denotes the total capacitance on net n. Two are the performance specifications: offset, denoted by symbol V_{off}, and switching delay, denoted by symbol τ_D. The performance constraints are

$$
\begin{cases}
\tau_D \leq 7ns \\
|V_{off}| \leq 1mV
\end{cases}
\tag{5.9}
$$

We assume that the nominal value of all parasitics is 0, i.e., $\mathbf{p}^{(0)} = [0 \ldots 0]^T$. Simulation with null parasitics yields a nominal value of the switching delay $\tau_D^{(0)} = 4ns$ and null offset. Therefore

$$
\mathbf{K} = \begin{bmatrix} \tau_D \\ V_{off} \\ -V_{off} \end{bmatrix} \qquad
\mathbf{K}(\mathbf{p}^{(0)}) = \begin{bmatrix} 4.0ns \\ 0.0 \\ 0.0 \end{bmatrix} \qquad
\overline{\Delta \mathbf{K}} = \begin{bmatrix} 3.0ns \\ 1mV \\ 1mV \end{bmatrix}
$$

Sensitivity analysis shows that delay is sensitive to stray capacitances, while resistances and mismatch affect only the offset:

$$
\mathbf{p} = \begin{bmatrix} C_{15} \\ C_{16} \\ C_{55} \\ C_{56} \\ R_{1,2} \\ R_{20,23} \\ R_{21,23} \\ R_{20,22} \\ R_{21,22} \\ R_{6,7} \\ R_{3,4} \end{bmatrix} \qquad
\mathbf{S} = \begin{bmatrix}
36ps/fF & 0.0 & 0.0 \\
36ps/fF & 0.0 & 0.0 \\
47ps/fF & 0.0 & 0.0 \\
47ps/fF & 0.0 & 0.0 \\
0.0 & 0.056mV/\Omega & 0.056mV/\Omega \\
0.0 & 0.016mV/\Omega & 0.016mV/\Omega \\
0.0 & 0.016mV/\Omega & 0.016mV/\Omega \\
0.0 & 0.016mV/\Omega & 0.016mV/\Omega \\
0.0 & 0.016mV/\Omega & 0.016mV/\Omega \\
0.0 & 0.011mV/\Omega & 0.011mV/\Omega \\
0.0 & 0.201\mu V/\Omega & 0.201\mu V/\Omega
\end{bmatrix}^T
$$

Notice that because of symmetries, and since the nominal value of mismatch is 0, offset sensitivities in the positive and negative direction are equal. We use the following conservative minimum and maximum parasitic estimates:

$$
\begin{aligned}
C^{(min)} &= 0 \\
C^{(max)} &= 100fF \\
R^{(min)} &= 0 \\
R^{(max)} &= 50\Omega
\end{aligned}
$$

With these estimates, PARCAR yields the following set of parasitic bounds:

$$\mathbf{p}^{(\mathbf{b})} = \begin{bmatrix} 71.96fF \\ 71.96fF \\ 78.52fF \\ 78.52fF \\ 1.0\Omega \\ 7.4\Omega \\ 7.4\Omega \\ 7.4\Omega \\ 7.5\Omega \\ 19.9\Omega \\ 49.5\Omega \end{bmatrix}$$

Here the relation between sensitivity and tightness of bounds is evident. Only a few parameters affect critically the performance of this circuit and therefore need to be bounded tightly. In practice, only the mismatch between the source resistances in the differential pair and the mismatch between the two current mirrors (M_{20}, M_{23}) and (M_{21}, M_{22}) are responsible for offset degradation. ◇

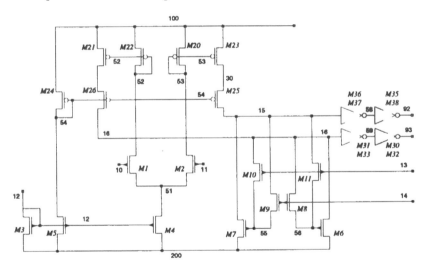

Figure 5.5 Clocked comparator

5.2.2 LIMITS OF
THE CONSTRAINT-GENERATION SOLUTION

It must be pointed out that the set of parasitic bounds computed by optimizing the flexibility function (5.7) is not necessarily the best. For instance, consider the problem shown in Figure 5.6, with $N_p = 2$.

Bounds $p_1^{(b)}$ and $p_2^{(b)}$, together with the nominal values $p_1^{(0)}$ and $p_2^{(0)}$, define a rectangular subset of the feasibility region for the circuit, namely the region of the space (p_1, p_2) where the circuit's performance is acceptable. The points outside the rectangle, but inside the feasibility region, cannot be reached unless a violation of one of the bounds is allowed.

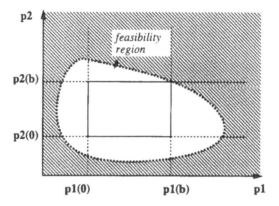

Figure 5.6 Feasibility region and parasitic bounds in a 2-parameter space

This shows that the constraints on parasitics must not be considered as hard bounds, but should be allowed some flexibility. The important thing is to keep the performance degradation within its bounds, i.e., to satisfy (5.4). If during routing a parasitic exceeds its bound, the router can continue, as long as inequality (5.4) holds true. If necessary, the bounds of the parasitics corresponding to the yet unrouted nets can be modified, but this requires the solution of an optimization problem, and therefore it cannot be performed in run-time during routing. However, it is possible to use approximations, for instance, to reduce all bounds by the same relative amount. Assume, without loss of generality, that all parasitics are sorted in such a way that all p_j $(j \leq l)$ are known, that is they depend only on already routed nets, while p_j $(j > l)$ are still undetermined. If for every performance function K_i

$$\sum_{j=1}^{l} S_{i,j}(p_j - p_j^{(0)}) + \sum_{j=l+1}^{N_p} S_{i,j}(p_j^{(b)} - p_j^{(0)}) \leq \overline{\Delta K_i}$$

then any parasitic violation has had no effect (so far) on performance violation. Otherwise, it is necessary to tighten the bounds for the remaining parasitics. This can be done using the following criterion: the new (modified) bounds $p_j^{(b')}$ are given by:

$$p_j^{(b)'} = p_j^{(b)} + \rho(p_j^{(b)} - p_j^{(0)})$$

where ρ is defined as follows:

$$\rho = \frac{\displaystyle\sum_{j=1}^{l} S_{i,j}(p_j^{(b)} - p_j)}{\displaystyle\sum_{j=l+1}^{N_p} S_{i,j}(p_j^{(b)} - p_j^{(0)})}$$

If $N_k > 1$, it is necessary to repeat this computation for each performance function K_i, and then take the largest value: $\rho = \max_i \{\rho(K_i)\}$.

5.2.3 EXTENSIONS TO THE DIGITAL CIRCUITS

Constraint-driven techniques are used for the interconnection of digital circuits too. In particular, the bound generation based on sensitivity analysis can be extended to digital routing to control crosstalk noise and speed constraints. In a recent work by Kirkpatrick and Sangiovanni-Vincentelli [24], definition (5.2) has been extended to sensitivity of performance to the parasitics on digital interconnections. The sensitivity of a digital net n to the noise coming from the surrounding nets is given by the expression

$$S(n) = A(n)D(n)$$

where $A(n)$ is the sensitivity of the voltage on net n to the voltage fluctuations in the adjacent nets, and $D(n)$ is a function whose value is either 1 or 0, depending on (a) the value of the nominal voltage on n; (b) the sign of the noise injected in n; (c) the logic gates driven by net n. As an example, consider the net shown in Figure 5.7. Adjacent switching

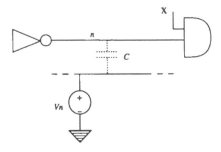

Figure 5.7 Noise is induced on digital net n by adjacent switching nets.

nets are represented by a noise voltage generator V_n capacitively coupled with net n. If the logic value of signal X is "low", $D(n) = 0$, namely no disturbance on net n can have effect on the circuit. However, if signal X has logic value "high", a sufficiently large positive voltage fluctuation of V_n can determine a glitch in the circuit, if the nominal value of the signal on net n is "low". Similarly an error can occur if a negative glitch is induced on a nominally "high" signal on net n.

With this type of extension, and with the growing importance of parasitic effects in modern high-speed technologies, it is apparent that some of the techniques developed for the design of analog circuits are becoming more and more interesting for applications to the digital world.

5.3 ANALOG AREA ROUTING

Maze routing [33] is particularly suitable for analog routing, because of its incremental nature. At every step of the propagation phase each wave front element is the end of a candidate path starting from the source pin. By propagating the wave front element by one step along a grid edge, the corresponding candidate path grows by a segment equal to the length of the grid edge. Hence, the cost of a path can be computed incrementally during propagation.

Weight-driven tools, like ANAGRAM-II and ROAD in their original implementation, are area routers controlled by a cost function. In what follows, we will consider a routing step where the path is increased by one edge e. This can be a grid edge for a maze router, or a new rectilinear path if line search or line expansion is used. The cost of e is a weighted sum of all the incremental contribution to the parasitics of the net, which would derive to the wire by adding to it a segment laying on e. The cost function is given by the following expression:

$$F(e) = L(e) + F_v + F_b + \sum_{p_j \in \mathcal{P}} w_{p_j} p_j \qquad (5.10)$$

where
- $L(e)$ is the length of edge e;
- F_v, F_b are cost contributions in case a via or a bend, respectively, are needed to add edge e to the wire;
- $\mathcal{P}$ is the set of all parasitics associated with this edge: the resistance of a minimum-width wire segment on e, the capacitance to bulk of a minimum-width wire segment on e, the capacitance between a wire segment located on e and every existing wire; and
- w_{p_j} is a weight.

Without constraints, all weights are 0, and the "best" wire path is the one with minimum length, and minimum number of vias and bends. With constraints, however, a trade-off is realized between path length and other undesirable characteristics, such as the resistivity of the routing materials used and the noise from cross-coupled neighbor nets. Each weight expresses the relative importance of the corresponding item in the trade-off. For instance, if resistance is an important constraint, its weight is large, and the grid edges corresponding to high-resistance layers such as polysilicon are very costly. Hence, a longer path using mostly metal will be preferred to shorter ones using polysilicon. Similarly, crosstalk penalty increases the cost near wires implementing critically coupled nets. In Figure 5.8 an example is shown, where the cost function assumes large values near a wire whose cross-coupling must be penalized. The resulting "hill" reduces the propagation speed, and instead of the shortest path an alternative longer path is selected.

In constraint-driven mode, parasitics are eliminated from the cost function, which therefore is reduced to the mere path length, plus the via and bend terms. Parasitics are still measured incrementally on the evolving path, but they are not used to trade off wire length vs. electrical behavior. Instead, each time a new edge is evaluated for propagation, all the parasitics in the candidate path are checked against their bounds. If one of them violates its bound, further expansion along that path is terminated. The bounds used can be dynamically adjusted during routing, as described in Section 5.2.2. With constraint-driven routing, each element of a wave front must store information about the parasitics of the path behind it, namely the resistance to each of the source pins, the total capacitance to ground, and the coupling capacitance with every other existing wire, or at least with those wires whose coupling is critical.

Consider the example of Figure 5.8. With direct bound control, either of the 2 situations shown in Figure 5.9 would occur. If the crossing between the two wires does not introduce excessive cross-coupling, namely if it doesn't violate its bound, then the shortest path, shown in Figure 5.9a, is found. Otherwise, the longer path shown in Figure 5.9b, is found.

The constraint-driven mode offers two advantages with respect to the weight-driven mode. The first is computational speed. Consider the example shown in Figure 5.10, where a pin s is completely surrounded by a wire. Assume first that the crossing between the wires is critical, but acceptable, that is the coupling introduced by a single crossing is within its bound. The portion of grid explored with the weight-driven approach is larger, because the propagation is slower near the intersection with the critically coupled wire. As a consequence, several alternative paths adjacent to the shortest path are explored, even though the shortest path will be the winner at the end. With constraint-driven propagation, on the contrary, the shortest path is found immediately and a smaller portion of the grid needs to be explored. On the other hand, consider the case in which the coupling exceeds the bound for the crosstalk. In this case the constraint-driven propagation is not able to find an interconnection for the second net, but the weight-driven propagation finds a path, although with a bound violation.

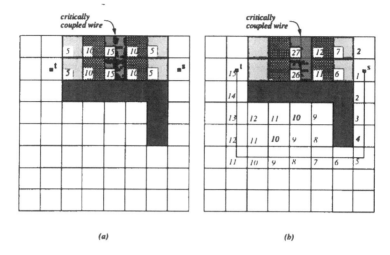

Figure 5.8 Crosstalk penalty
(a) A cost function "hill" surrounds a wire
whose critical coupling must be avoided.
(b) The shortest path is penalized, and a longer path is selected.
(The numbers indicate the integral of the cost
between source and wave front.)

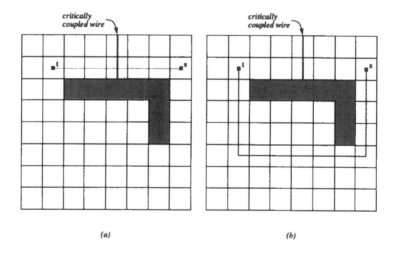

Figure 5.9 Constraint-driven routing
(a) The crossing between the two wires does not violate the coupling bound.
(b) The crossing violates the bound, and a longer path is selected.

The second advantage of constraint-driven routing is that it introduces fewer jo gs and vias. As illustrated in Figure 5.11 with weight-
driven routing nets require often many jogs and vias, in order to turn

around cost function "hills" due to crosstalk constraints, and their paths become tortuous and irregular. With constraint-driven propagation, on the contrary, there are no "hills" to avoid, and the shortest path is always followed unless it introduces a violation.

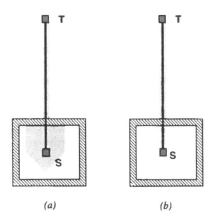

(a)   (b)

Figure 5.10 Critical net crossing
(a) Path found with weight-driven propagation
(b) Path found with constraint-driven propagation

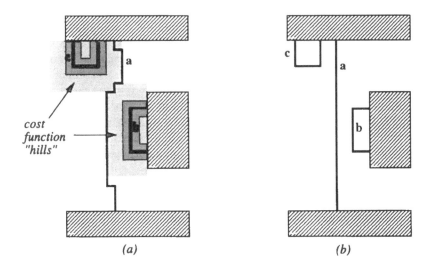

(a)   (b)

Figure 5.11 Crosstalk between net *a* and nets *b* and
c is controlled, but is not very critical.
(a) Path found for net *a* with weight-driven propagation
(b) Path found with constraint-driven propagation, without violations

5.4 ANALOG CHANNEL ROUTING

Most of the channel routers used for analog interconnections use some variant of the vertical constraint graph (VCG) or total constraint graph (TCG) algorithms [6]. The reason is that constraint-graph algorithms allow variable spacing between tracks, and it is possible to control efficiently the proximity between nets. Crosstalk noise can be reduced by enforcing a different order of track assignment, and this is achieved easily with a constraint graph, by directing the edges in a suitable way.

With constraint-graph algorithms, the channel is represented by a graph, whose nodes are the horizontal wire segments, each of which is assigned to a portion of a track. The "weight" of an edge is the minimum distance between center lines of adjacent horizontal segments allowed by design rules, or by parasitic constraints. A directed edge from node A to node B means that the wire segment corresponding to node A must be assigned to a track above that of the wire segment corresponding to node B. The channel routing problem is formulated as finding a set of directed edges between nodes to minimize the longest directed path between the source and sink nodes, which correspond to the upper and lower edges of the channel respectively.

ART uses the VCG model, i.e., each net is assigned to one track. The bounds on parasitics are automatically obtained from performance constraints using the constraint generator PARCAR described in Section 5.2.1, or they can be defined by the user. The constraint on maximum capacitive coupling is first divided into two constraints, one on the maximum *crossover*, and one on the maximum *adjacency* coupling, due to parallel wire segments. These constraints are then mapped into constraints for the VCG. The problem of determining the crossover and adjacency constraints is automatically solved by PARCAR, provided the tool is given good estimates of the minimum and maximum values that these capacitances can physically assume. For each pair of nets A and B, first the "unavoidable crossings" are detected, based on the relative positions of the horizontal segments of the two nets and their pins. If the two nets have no common horizontal span, their crossover capacitance is 0. If instead they have a common horizontal span, if the track of net A is above the track of net B, as shown in Figure 5.12a, the number of crossovers is the number of pins of net B on the upper edge of the channel within the horizontal span of net A, plus the number of pins of net A on the lower edge, within the horizontal span of net B. Vice versa, if the track of net B is above the track of net A (Figure 5.12b), the number of crossovers is the number of pins of net A on the upper edge within the horizontal span of net B, plus the number of pins of net B on the lower edge, within the horizontal span of net A. The minimum and maximum coupling due to crossover between nets A and B, denoted as $C_{A,B}^{(min)}$ and $C_{A,B}^{(max)}$, respectively, are computed considering the cases with the track of net A above or underneath the track of net B, and taking the minimum and the maximum of the two couplings, respectively. If shielding is available (see Section 5.6.4), $C_{A,B}^{(min)} = 0$. Adjacency coupling is due to two contributions, one C_v due to vertical parallel segments, and one C_h due to horizontal segments. The minimum value of both is $C_v^{(min)} = C_h^{(min)} = 0$. The maximum value is obtained by considering the length of the unshielded common span and a separation equal to the minimum allowed by design rules. The unshielded common span for vertical segments is equal to the channel height, and it can be estimated by the channel density.

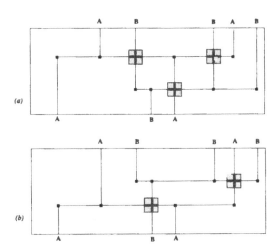

Figure 5.12 Detection of unavoidable crossings
(highlighted by shaded squares)
(a) Net A occupies track above that of net B
(b) Net B occupies track above that of net A

Once the bounds $C_{A,B}^{(b)}, C_v^{(b)}, C_h^{(b)}$ have been computed by PARCAR, ART is able to enforce them by directing some undirected edges of the VCG, or by adding some directed edges, of by increasing the weights of some edges. For instance, if $C_{A,B}^{(b)} < C_{A,B}^{(max)}$, a directed edge is added between the net tracks of A and B, to force them in the relative position which corresponds to the minimum number of crossings, so that $C_{A,B} = C_{A,B}^{(min)}$. The bound on C_h is enforced by requiring either the presence of a shielding net between the tracks of nets A and B, or a minimum separation (weight on the edge) between their tracks. No direct control is kept over C_v; therefore, in case a bound violation occurs it is necessary to add shielding, if possible, between adjacent vertical parallel edges.

In WREN [30], simulated annealing is used to solve the constraint graph, which in this case is a TCG. Each annealing move randomly directs some undirected edges in the TCG, and is followed by a longest-path search of the graph, assigning nets to the tracks. If there are cycles in the TCG, a palette of dogleg topologies is created for each net, and exhaustive search is performed to select the best candidate, among all nets, for inserting a dogleg. The criterion used for this selection is a cost function measuring the criticality of all net crossings as well as proximity between parallel wires. The use of simulated annealing for global routing as well as for detailed routing, and the use of exhaustive search for dogleg insertion, obviously make WREN very slow, but the optimization is performed on the entire configuration space, and therefore it can achieve layouts of remarkable quality.

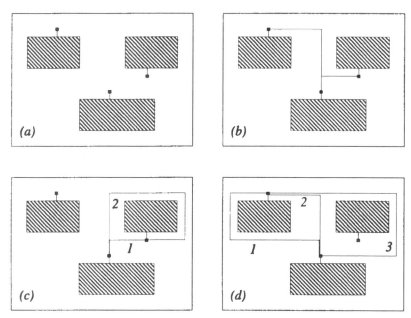

Figure 5.13 Path generation for the segments of a net
(a) Net pins before routing
(b) Minimum spanning tree
(c) Two alternative paths for the 1st segment
(d) Three alternative paths for the 2nd segment

The global router must avoid crossings between critically coupled signals, and detect such crossings when unavoidable. However, the crosstalk caused by mere proximity of parallel wires cannot be detected by a global router. The best it can do is to keep such net pairs in separate channels if possible. One such approach is the one developed in AT&T and described in Section 5.1.1. A different approach, shown in Figure 5.13, has been used in WREN [30]. First, each multi-point net is decomposed into a minimum spanning tree, whose connections are called "segments". A set of alternate topologies is generated, using Lawler's algorithm [25], for each two-point segment of the spanning tree. The annealing process takes care of selecting one of the alternative paths for each segment, in such a way that interactions between critically coupled wires are minimized within the same channel. Notice that the number of alternative configurations is large: N nets, with an average number P of pins per net, correspond to $(P-1) \cdot N$ segments. If each segment has K alternative paths in average, the system has $K^{(P-1) \cdot N}$ different configurations. However, simulated annealing is able to optimize problems characterized by such vast search space, its main drawback being it requires long CPU time to achieve remarkable results.

5.5 PARASITIC MODELS FOR INTERCONNECTIONS

With weight-driven routing, as well as with the constraint-driven approach, parasitics need to be calculated often and accurately. In this section a few commonly used models [12] are reported, in a technology-independent form to be fitted by interpolation with experimental measurements or three-dimensional simulation.

The capacitance between a unit-length wire segment of width w and the substrate is given by:

$$C_u = k_0 + k_1 w$$

where k_0 and k_1 are fitting parameters.

Given two wire segments of widths w_1 and w_2, respectively, running parallel to each other, at a distance d, the unit-length capacitance between them is given by

$$C_u' = k_0' + k_1' w_1 + k_2' w_2 + \frac{k_3'}{d} + \frac{k_4'}{d^2} \qquad (5.11)$$

where k_1', k_2', k_3', k_4' are fitting parameters. If they are on the same layer, parameters k_1' and k_2' are equal, while they can differ if the wire segments are on different layers. Notice that with a parallel-plate approximation k_1', k_2', k_4' would be null. With a $1\mu m$ CMOS process, this would introduce typically a 300 to 400% error in the capacitance calculation.

Crossing wires are represented by the following expression:

$$C_{cr} = k_0'' + k_1'' w_1 + k_2'' w_2 + k_3'' w_1 w_2 \qquad (5.12)$$

where w_1 and w_2 are the two wire widths, and $k_0'', \ldots, k_3''$ are fitting parameters. Notice that this model is not symmetric, that is, in general

$k_1" \neq k_2"$. The model (5.12) for crossing is depicted in Figure 5.14. The fringe effect of wire 1 on the (almost) flat upper surface of wire 2 is $C_{f1} = k_2"w_2$. Analogously, the fringe effect of wire 2 on the lower flat surface of wire i is $C_{f2} = k_1"w_1$. The area (parallel-plate) capacitance due to area overlaps only is $C_p = k_3"w_1w_2$. For CMOS processes of about $1\mu m$ channel length, the two fringe items in (5.12) can be 3 or 4 times as large as the area contribution.

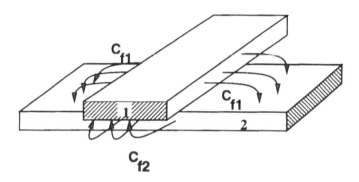

Figure 5.14 Model for the capacitance between crossing wires

5.6 SPECIAL FEATURES

In order to deal with particularly tight constraints, analog routers often provide special utilities, usually not available with routers targeted toward digital design. In this section two of these features are described: algorithms to build symmetric routing and the procedures used to build electrostatic shields to improve capacitive decoupling between interconnections.

5.6.1 WIRE WIDTH CONTROL

Wire width determines the value of stray resistances, stray capacitances toward the substrate, and coupling capacitances of crossing and parallel wires. Moreover, it has a direct impact on reliability, because electromigration effects can take place if the wire width is not dimensioned properly with high current density.

Digital routers don't consider the parasitic effects, but provide wire width control to prevent electromigration, usually with a post-routing optimizer [4, 13]. Although this approach can be applied to all nets in the circuit, it is usually focused on power and ground networks, which are at the same time the longest nets and the most sensitive to electromigration problems due to the high currents they transport. For this reason, the wire-width control problem is often called the Power/Ground

(P/G) problem. With analog constraints, the P/G problem becomes more interesting, because wire width is directly related to resistive and capacitive constraints. Moreover, strong currents are often involved in power stages and fast buffers. In mixed circuits, power and ground derive an additional constraint that they be kept separate to avoid noise transfer from the digital power lines to the analog signals through the substrate capacitive couplings. RAIL [37] is a P/G router for mixed circuits, incorporating substrate coupling effects. Based on simulated annealing, RAIL simultaneously optimizes the topology of the power distribution network, the sizing of each wire segment, and the choice of I/O pad number and location. Constraints are taken into account on DC, AC, and transient electrical performance functions. Starting from an initial configuration, RAIL provides the annealing with a set of "moves" which change the path between pins, altering either the topology of the path, or the width of the wire along the path, or both. A move can also change the set of I/O pads. The annealer tries to minimize a cost function given by the following expression:

$$\text{Cost} = w_1 \cdot A + w_2 \cdot DC + w_3 \cdot AC + w_4 \cdot TR$$

where A is the circuit area, DC, AC, TR are terms representing the violation of electrical constraints, and w_1, w_2, w_3, w_4 are user-defined weights, computed empirically on the basis of the user's experience.

5.6.2 MATCHING

Matching between interconnection parasitics can be enforced in different ways, depending on the type of parasitics involved, and on the maximum mismatch allowed between parasitics.

A technique used sometimes to enforce matching is bounded-length routing [38], defined as follows: we want to route two nets n_1 and n_2, using paths whose lengths l_1 and l_2 differ from one another by less than a certain amount Δl. Let s_1, t_1 be two pins for net n_1, and s_2, t_2 be two pins for net n_2. The two propagations are carried out in parallel, with four waves propagating simultaneously from s_1, t_1, s_2, t_2. Every time the waves from s_1 and t_1 meet on a node n', the length of the path (s_1, n', t_1) is compared with the lengths of all the paths between s_2 and t_2 found so far. Similarly, each path found crossing the waves propagating from s_2 and t_2 is checked against the length of every path between s_1 and t_1. If a match is found, the two paths are back-traced and implemented, otherwise the propagation is carried on. Notice that since two paths are implemented at the same time, caution must be taken to avoid short-circuits between them. For instance, their propagating waves might be restricted to separate portions of the layout. This precaution is not necessary if the matched paths belong to the same net. Figure 5.15

shows bounded length routing for two paths p_1 and p_2 belonging to the same net. While p_1 is of minimum length, p_2 is not minimum. The minimum-length path p_2' cannot be used because there are no paths of such short length connecting s and t_1.

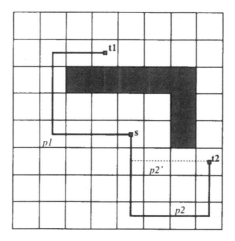

Figure 5.15 Bounded-length routing
for two paths p_1 and p_2 belonging to the same net
(The minimum-length path between s and t_2 is p_2')

Bounded-length routing is often used with microwave circuit routing, and for clock routing, where path length mismatch is directly related to the skew. If the resistivity of all the layers used for the routing is approximately the same, by matching the wire length we achieve matching of stray resistance as well. If these conditions don't hold, resistive matching is not achieved, and it is necessary to enforce tighter constraints on the length of each stretch of every layer used for the interconnections. These constraints cannot be enforced using the bounded-length routing algorithm, but require more complex approaches. Similarly, capacitance (toward the substrate) constraints cannot be enforced with this technique unless all the layers used have approximately the same capacitance per unit length.

A more rigorous approach for matching resistance and capacitance between two paths is the following. At each step of the routing propagation, we compute the total capacitance and the total resistance of the path between the source node and the wave front element at the end of the path. Each time a new edge is added to the path, both resistance and capacitance are updated incrementally by adding the contribution of the new edge to the value computed previously for the partial path. When two wave fronts meet, we stop the propagation only if the values of resistance and capacitance, respectively, match within the allowed tolerance. Otherwise, we ignore the wave crossing and continue the propagation.

When tight matching is required, and above all when matching between cross-coupling capacitances with other nets is important, it is necessary to guarantee that the two paths have the same shape, the same number of bends and vias, etc. In practice, the problem of matching paths is formulated as follows. Paths $p_1 = (s_1, t_1)$ and $p_2 = (s_2, t_2)$ are two matched paths if for every homogeneous (i.e., made of one layer) sub-path (n_1', n_1'') in p_1, there exists a homogeneous sub-path (n_2', n_2'') in p_2, such that the two sub-paths have the same length, the same width, and in the same layer. And vice versa, for each sub-path of p_2 there is one sub-path in p_1 with same length, width and layer.

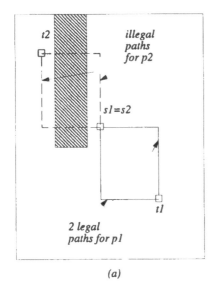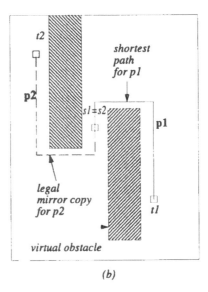

(a)   (b)

Figure 5.16 Algorithm to build matched paths
(a) Find all minimum-length paths between s_1 and t_1.
(b) If p_2 is not feasible because of obstacles,
create "virtual obstacles" and try again.

In an obstacle-free grid, it is easy to build matched paths. Just build one of them, then make a copy of it, eventually rotated and mirrored, and as long as they have no intersections (unless they belong to the same net) they constitute a legal matched pair of paths. With obstacles, on the contrary, it is necessary to use a more complex, iterative algorithm. First build all minimum-length paths between s_1 and t_1, as shown in Figure 5.16a. For each of these paths, that we will denote as p_1^i, try to build a path between s_2 and t_2, equal to p_1^i, except for a translation, rotation and/or mirroring. If one such path p_2^i is found, and it crosses no obstacles, p_1^i and p_2^i constitute a legal matched pair of paths, and the algorithm ends. Otherwise, if no match is found, for each path p_1^i consider each of its copies p_2^i and the obstacle(s) crossed by it. Copy all such obstacles so that their own copies ("virtual" obstacles) are crossed by p_1^i in exactly the same point(s) as they are crossed by p_2^i, as shown in Figure 5.16b. Then remove all paths and search all shortest paths between s_1 and t_1, taking into account the real and the virtual obstacles. At the end of this step, if at least a pair of matched paths has been found, neither of which crosses an obstacle, exit with the cheapest pair found. Otherwise, start all over again with each pair p_1^i, p_2^i and with the obstacles crossed by p_2^i, mirror the obstacles, and find longer pairs of paths. Notice that the iteration can end either successfully with a legal matched pair of path, or unsuccessfully with a set of real and virtual obstacles allowing no legal paths between s_1 and t_1. In this case, no matched pair exists in the layout.

When applicable, a popular technique to enforce tight matching between the parasitics of a pair of nets is to use symmetry, as described in the next section.

5.6.3 SYMMETRIES

Symmetry is a frequent requirement with analog circuits. Analog designers create circuit topologies with symmetric signal paths to exploit the correlations between matched components during the layout phase. During the design of a differential circuit, one assumes implicitly that the layout will exhibit the same symmetries present in the circuit topology. A failure to implement symmetric layout results in most cases in unacceptable performance degradation. Design techniques have evolved, above all for charge-mode MOS circuits, that exhibit sensitivity mainly to differences between circuit parameters, in particular parasitics, and not to their absolute value. For such circuits, the layout must be as close as possible to perfect mirror symmetry.

The main target of symmetric routing is to match parasitics in interconnections. In particular with MOS circuits, which operate often in the charge domain, even small parasitic capacitances can strongly affect circuit performance. When symmetry is required, a symmetry axis is defined in the circuit. Without any loss of generality we can assume that the axis is vertical: it splits the circuit into two halves. Its position in the circuit is determined by the placement, and in what follows it will be indicated with x_s. Symmetry may affect a net in different ways depending on the distribution of its terminals. Let $\mathcal{T}_N$ be the set of all terminals of net N. For each terminal $t \in \mathcal{T}_N$ we can consider the "virtual terminal" t' obtained by mirroring t with respect to the symmetry axis. If (x, y) are the coordinates of terminal t, the coordinates of terminal t' are $(2x_s - x, y)$. If $\forall t \in \mathcal{T}_N$ the virtual terminal t' coincides with a real terminal in $\mathcal{T}_N$, net N is said to be *fully symmetric*. Otherwise, if $\exists t \in \mathcal{T}_N : t' \notin \mathcal{T}_N$, and N is said to be *partially symmetric*. A net is said to be *self-symmetric* if it is symmetric with respect to itself. The wire implementing a self-symmetric net always crosses the symmetry axis. A channel with symmetry constraints belongs to one of the following three categories (see Figure 5.17):

(1) The axis does not pass through the channel, as in channels A and B in Figure 5.17.

(2) The axis crosses the channel in the orthogonal direction, as in channel C.

(3) The axis runs through the channel in the direction of its length, as in channel D.

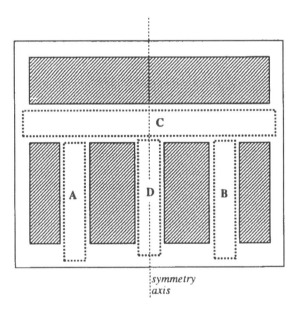

Figure 5.17 Different categories of channels
with respect to the symmetry axis in ART

With channels of type (1), symmetry is achieved in a straightforward way, provided the terminal distribution is perfectly symmetric. One of the two symmetric channels is routed first, and the second one is a mirrored copy. Within the channels of type (2), the approach used depends whether the net crosses the symmetry axis. As shown in Figure 5.18, a self-symmetric net and two symmetric nets which don't cross the axis require no special care, they can be routed on the left half only, and then mirrored to the right side. In ART, symmetric net pairs which intersect the axis require a special connector, shown in Figure 5.19, to allow them to cross. It allows the two interconnections to have the same length and the same number of vias and bends, therefore their match is as good as possible. Although the channels of type (3) are not very dissimilar from the previous case, they are not solved symmetrically in ART.

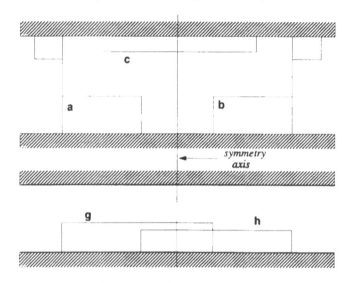

Figure 5.18 Different situations within a symmetric channel
(Nets a,b are symmetric, but don't cross the axis;
net c is self-symmetric; nets g, h are symmetric and cross the axis.)

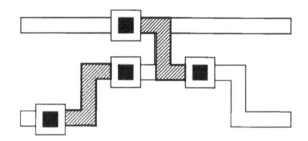

Figure 5.19 Special connector to allow symmetric nets
to cross the symmetry axis

ROAD can find symmetric paths for differential signals only if the placement is perfectly symmetric, although the terminals can be distributed arbitrarily. If the placement is not symmetric, however, non-symmetric blockages can be mirrored before routing, to form "virtual" blockages, as shown in Figure 5.20, so that the layout is symmetrized.

The router then considers the outline determined by the union of real and virtual blockages.

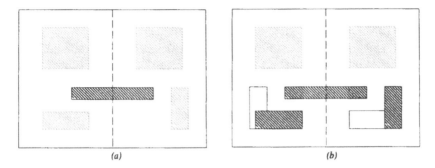

(a)   (b)

Figure 5.20 Generating a virtual symmetric placement
(a) Non-symmetric placement
(b) "Virtual" obstacles are generated to symmetrize the placement.

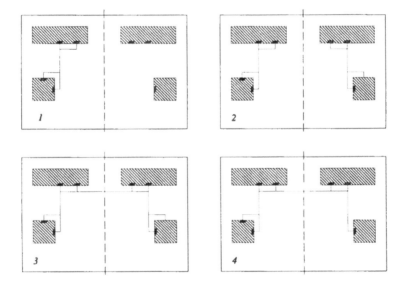

Figure 5.21 The four-step symmetric routing algorithm used in ROAD

The algorithm for symmetric routing in ROAD is as follows (see Figure 5.21:

(1) Every net is built considering only the terminals located on the left side of the symmetry axis and within or on the border of the wiring space that is not contained by any virtual blockage. The segments defined in this way are called *left-hand-side segments*.

(2) Each left-hand-side segment is mirrored with respect to the symmetry axis; its mirrored *right-hand-side segment* is associated to the symmetric net of the left-hand-side segment net. At this point each net corresponds to two interconnections, each connecting a subset of the terminals on a side of the symmetry axis.

(3) The grid is expanded to cover the portions of area occupied by virtual obstacles, but not by real obstacles. Each net is routed spanning the entire set of terminals and the two distributed terminals represented by the paths of the left- and right-hand-side wires.

(4) Segments whose existence is not required for full net connectivity are pruned. Only those branches of non-symmetric nets should be pruned, that don't cross or run close to symmetric nets. In this way, matching between parasitics of symmetric nets is preserved.

Observe that in Step (1) no wires are built for nets having no more than one terminal on the left side of the circuit. Such nets are completely routed only in Step (3).

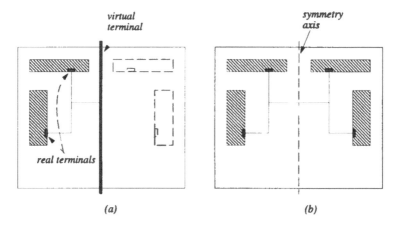

(a)   (b)

Figure 5.22 Symmetric routing of self-symmetric nets
(a) The symmetry axis is seen as a virtual "extended" terminal.
(b) The left half wire is mirrored to the right-hand-side,
and the net is now complete.

In ANAGRAM-II, symmetric nets are routed one by one after routing, instead of being mirrored all together. Although it requires more memory because the whole grid is used all the time, this approach is interesting because it allows routing symmetric and non-symmetric nets without a pre-defined order, and without constraints on symmetric placement. However, it requires perfect symmetry in the distribution of terminals. Partial paths must be checked to make sure that they would cross no blockage or violate design rules, if reflected across the symmetry axis. At each step of the propagation phase, both the new expansion segment, and its symmetric mirror image on the opposite side of the axis are checked. The new segment can be used for the path only if both expansion segments prove legal. ANAGRAM-II is not able to route symmetric nets crossing the symmetry axis, because these would require checking a wire segment against its mirror image. The only case where this is allowed is with self-symmetric nets, because the two halves are connected and no violations are possible. This is achieved by first routing the two halves as mirror images of each other, and then routing the connection between them as a connection to the symmetry axis on either side. The symmetry axis is viewed as an "extended" terminal as shown in Figure 5.22.

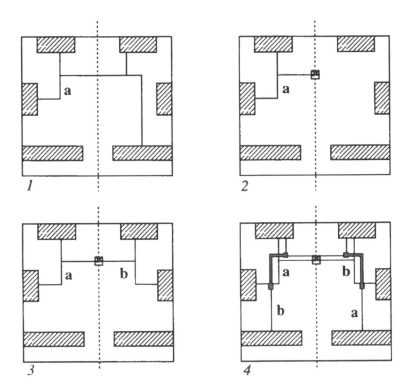

Figure 5.23 Symmetric routing of nets crossing the axis

Non-symmetric nets crossing the symmetry axis have not been solved satisfactorily in any existing area-routing approach. ANAGRAM-II does not allow to route them while ROAD does not guarantee perfect symmetry, because the interconnections between the two halves are routed (in Step 3) without symmetry constraints. A possible solution to this problem is illustrated in Figure 5.23.

(1) One of the two wires is routed with all its pins (on both sides of the axis).

(2) The portion of the wire on the right-hand side of the axis is ripped-up.

(3) The left-hand-side portion of the wire if mirrored to the right. If there is enough space, a special connector like the one shown in Figure 5.19 is added at each crossing point on the axis. Otherwise, the last segment of each wire portion, touching the symmetry axis, must be removed, and the crossing point forbidden from further symmetric routing.

(4) The crossing connector on the axis is connected to all the pins on the right-hand side. The first wire, now complete, is mirrored to obtain the second wire.

5.6.4 SHIELDS

By putting an interconnection, biased with a proper stable voltage (e.g., ground), between wires, the cross-coupling capacitance is reduced considerably. Two types of shields are used, *vertical* and *lateral*. Figure 5.24 shows a vertical shield created during routing to reduce coupling between critical nets. The horizontal wire, realized with layer L_u, runs over a vertical wire segment realized with layer L_l, and the shield is

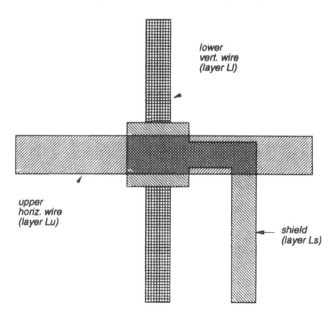

Figure 5.24 Vertical shield

realized with layer L_s. For example, in a 2-metal/1-poly technology, the lower conductor layer L_l could be poly, the upper layer L_u could be metal-2, and the shield layer L_s could be metal-1. Vertical shields require a technology with at least 3 routing layers, and extra area to accommodate the interconnection biasing the shield. They enhance the flexibility of the router, because wire crossing is allowed even with very tight coupling constraints.

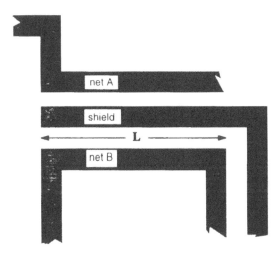

Figure 5.25 Lateral shield

A lateral shield is a wire separating two interconnections on the same layer, as shown in Figure 5.25. In order to determine whether a shield is convenient, it is necessary to compute the area saving it provides, compared with the decoupling technique of keeping the wires far from each other to reduce their capacitance. Let L be the length of the parallel segments of wires A and B, and let $C^{(b)}$ be the bound on maximum capacitance allowed between the two nets. Let d be the minimum distance allowed between the wires without shielding, computed using the model (5.11). We make two assumptions:

(1) The fringe capacitance between the wires is negligible when the shield is present.

(2) The area overhead, due to the fact that the shield must be connected to the proper bias node outside the shielding area, is negligible.

Then the minimum distance between the wires **with** the shield is

$$d^s = 2d_{min} + w_{min}$$

where d_{min} is the minimum distance allowed between wires in that layer (from the design rules), and w_{min} is the minimum wire width (of the shield). Therefore, the shield is convenient if $d' < d$.

Example: Consider the following data:

$$k_0' = 0.00048\, fF/\mu m,$$
$$k_1' = k_2' = -0.023\, fF/\mu m,$$
$$k_3' = 0.306\, fF/\mu m,$$
$$k_4' = -0.032\, fF/\mu m^2,$$
$$w_1 = w_2 = 1\mu m,$$
$$w_{min} = 1\mu m,$$
$$d_{min} = 1\mu m,$$
$$C_{(b)} = 3fF,$$
$$L = 100\mu m.$$

With these numerical values and the capacitive model (5.11), the value of $C^{(b)} = L \cdot C_u{}'$ as a function of the unshielded wire distance d is shown in Figure 5.26. With the given constraint, $d = 4.19\mu m$ and $d' = 3\mu m$. In this case, therefore, shielding is advantageous. The value of $C^{(b)}$ for which the two solutions are equivalent is $C^{(b)} = 5.72fF$.   ◇

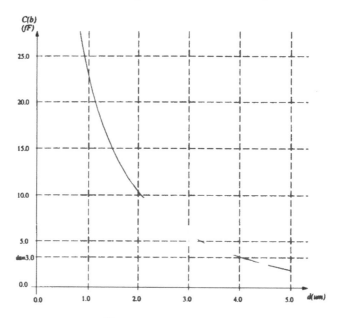

Figure 5.26 Value of $C^{(b)}$ as a function of the unshielded wire distance d

Assumption 1 usually holds with the current technologies, but if the fringe effect is too large, it can be reduced by widening the width w_{min} of the shield. Assumption 2, on the contrary, holds less frequently, because in order to bias the shield it is necessary to build an interconnection to the nearest bias source. However, the impact of this on the total area of the circuit is difficult to evaluate before routing the entire shield.

A shield can be either a wire segment used for interconnection within the circuit, and routed between the wires to be decoupled, or a stub, eventually open on one side, without other function except to serve as a shield. A common problem with shields is that when routing is followed by a compaction phase, the compactor is usually not able to recognize the functionality of the dangling-end shielding stubs, which therefore are removed. This kind of structures should be used only if no compaction follows, unless the compactor is able to recognize and maintain shields, as in [16].

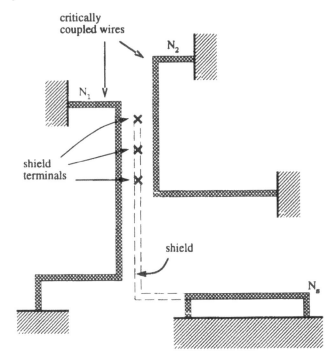

Figure 5.27 Procedure to build lateral shield

5.6.5 BUILDING LATERAL SHIELDS

Lateral shields are built after all wires have been routed. The procedure is illustrated in Figure 5.27. Let N_1 and N_2 be two critically coupled wires to be reciprocally shielded, and let N_s be the net to which the shield has to be connected.

(1) Find a grid node lying between N_1 and N_2. If more than one of such nodes exists, choose the one closest to the median position between the wire segments.

(2) Create a temporary pin on each of the selected nodes, and route net N_s through all of them. Connect net N_s to the substrate, or to the nearest existing pin or wire segment for that net.

In Step 1, the presence of a grid node is required between wire segments. If the router uses a dynamically allocated non-homogeneous grid, it is possible to create new grid nodes when necessary, as long as local area congestion does not prevent this operation. If the local region is over-crowded, shielding is not possible, and extra space should be added in the placement, or partial re-route should be tried.

5.6.6 BUILDING VERTICAL SHIELDS

Vertical shields can be built if at least three layers are available for interconnections, as shown in Figure 5.24. Two wire segments crossing on the upper and lower layers are separated by a rectangular plate in the intermediate layer, connected to ground. The rectangular plate must be large enough to shield the fringe effect between the crossing wire segments. For this structure to be physically possible, it is necessary to route the crossing wire segments on the proper layers, and leave the intermediate layer available. If the lower layer is polysilicon, and the upper layer is metal-2, the net routed underneath must be non-critical with respect to stray resistance and to capacitance toward the substrate, which are both large for polysilicon.

REFERENCES

1. P. E. Allen and E. R. Macaluso, "AIDE2: An Automated Analog IC Design System," in *Proc. IEEE CICC*, 1985, pp. 498–501.

2. R. J. Bowman and D. J. Lane, "A Knowledge-Based System for Analog Integrated Circuit Design," in *Proc. IEEE ICCAD*,1985, pp. 210–212.

3. B. Basaran, R. A. Rutenbar, and L. R. Carley, "Latchup-Aware Placement and Parasitic-Bounded Routing of Custom Analog Cells," in *Proc. IEEE ICCAD*, November 1993, pp. 415–421.

4. S. Chowdhury and M. Breuer, "The Construction of Minimal Area Power and Ground Nets for VLSI Circuits," in *Proc. IEEE/ACM DAC*, 1985, pp. 794–797.

5. J. M. Cohn, D. J. Garrod, R. A. Rutenbar, and L. R. Carley, "KOAN/ANAGRAM II: New Tools for Device-Level Analog Placement and Routing," *IEEE J. Solid State Circuits*, vol. 26(3), 1991, pp. 330–342.

6. H. H. Chen and E. S. Kuh, "Glitter. A Gridless Variable-Width Channel Router," *IEEE Trans. CAD*, vol. CAD-5, n. 4, Oct. 1986, pp. 459–465.

7. D. J. Chen, J. C. Lee, and B. J. Sheu, "SLAM: A Smart Analog Module Generator for Mixed Analog-Digital VLSI Design," in *IEEE ICCD*, 1989, pp. 24-27.

8. E. Charbon, E. Malavasi, and A. Sangiovanni-Vincentelli, "Generalized Constraint Generation for Analog Circuit Design," in *Proc. IEEE IC-CAD*, November 1993, pp. 408-414.

9. U. Choudhury and A. Sangiovanni-Vincentelli, "Constraint-Based Channel Routing for Analog and Mixed-Analog Digital Circuits," in *Proc. IEEE ICCAD*, November 1990, pp. 198-201.

10. U. Choudhury and A. Sangiovanni-Vincentelli, "Constraint Generation for Routing Analog Circuits," in *Proc. IEEE/ACM DAC*, June 1990, pp. 561-566.

11. U. Choudhury and A. Sangiovanni-Vincentelli, "Use of Performance Sensitivities in Routing of Analog Circuits," in *Proc. IEEE Int. Symposium on Circuits and Systems*, May 1990, pp. 348-351.

12. U. Choudhury and A. Sangiovanni-Vincentelli, "An Analytical-Model Generator for Interconnect Capacitances," in *Proc. IEEE CICC*, May 1991, pp. 861-864.

13. R. Dutta and M. Marek-Sadowska, "Automatic Sizing of Power/Ground Networks in VLSI," in *Proc. IEEE/ACM DAC*, 1989, pp. 783-786.

14. S. W. Director and R. A. Rohrer, "The Generalized Adjoint Network and Network Sensitivities," *IEEE Trans. on Circuit Theory*, vol. 16, August 1969, pp. 318-323.

15. M. Degrauwe et al., "IDAC: An Interactive Design Tool for Analog CMOS Circuits," *IEEE Journal of Solid State Circuits*, vol. SC-22, No.6, December 1987, pp. 1106-1116.

16. E. Felt, E. Malavasi, E. Charbon, R. Totaro, and A. Sangiovanni-Vincentelli, "Performance-Driven Compaction for Analog Integrated Circuits," in *Proc. IEEE CICC*, May 1993, pp. 1731-1735.

17. R. S. Gyurcsik and J.-C. Jeen, "A Generalized Approach to Routing Mixed Analog and Digital Signal Nets in a Channel," *IEEE Journal of Solid State Circuits*, vol. 24, n. 2, April 1989, pp. 436-442.

18. D. J. Garrod, R. A. Rutenbar, and L. R. Carley, "Automatic Layout of Custom Analog Cells in ANAGRAM," in *Proc. IEEE ICCAD*, November 1988, pp. 544-547.

19. I. Harada, H. Kitazawa, and T. Kaneko, "A Routing System for Mixed A/D Standard Cell LSI's," in *Proc. IEEE ICCAD*, November 1990, pp. 378–381.

20. R. Harjani, R. A. Rutenbar, and L. R. Carley, "Analog Circuit Synthesis for Performance in OASYS," in *Proc. IEEE ICCAD*, November 1988, pp. 492–495.

21. G. Jusuf, P. R. Gray, and A. Sangiovanni-Vincentelli, "CADICS - Cyclic Analog-To-Digital Converter Synthesis," in *Proc. IEEE ICCAD*, November 1990, pp. 286–289.

22. C. D. Kimble, A. E. Dunlop, G. F. Gross, V. L. Hein, M. Y. Luong, K. J. Stern, and E. J. Swanson, "Autorouted Analog VLSI," in *Proc. IEEE CICC*, 1985, pp. 72–78.

23. H. Y. Koh, C. H. Séquin, and P. R. Gray, "OPASYN: A compiler for CMOS operational amplifiers," *IEEE Trans. on CAD*, vol. 9, n. 2, February 1990, pp. 113–126.

24. D. A. Kirkpatrick and A. L. Sangiovanni-Vincentelli, "Techniques for Crosstalk Avoidance in the Physical Design of High-Performance Digital Systems," in *Proc. IEEE ICCAD*, May 1994.

25. E. Lawler, *Combinatorial Optimization: Networks and Matroids*, Holt, Reinhart and Winston, 1976.

26. C. Lee, "An algorithm for path connections and applications," *IRE Trans. Electron. Computer*, vol. EC-10, September 1961, pp. 346–365.

27. E. Malavasi, M. Chilanti, and R. Guerrieri, "A General Router for Analog Layout," in *Proc. COMPEURO '89, Hamburg*, May 1989, pp. 549–551.

28. S. W. Mehranfar, "STAT: A Schematic to Artwork Translator for Custom Analog Cells," in *Proc. IEEE CICC*, May 1990.

29. M. Mogaki, N. Kato, Y. Chikami, N. Yamada, and Y. Kobayashi, "LADIES: An Automatic Layout system for Analog LSI's," in *Proc. IEEE ICCAD*, November 1989, pp. 450–453.

30. S. Mitra, K. Sudip, K. Nag, R. A. Rutenbar, and L. R. Carley, "System-level Routing of Mixed-Signal ASICs in WREN," in *Proc. IEEE ICCAD*, November 1992, pp. 394–399.

31. E. Malavasi and A. Sangiovanni-Vincentelli, "Area Routing for Analog Layout," *IEEE Trans. on CAD*, vol. 12, n. 8, August 1993, pp. 1186–1197.

32. E. Malavasi and A. Sangiovanni-Vincentelli, "Dynamic Bound Generation for Constraint-Driven Routing," *Submitted to IEEE/CICC-95*, May 1995.

33. T. Ohtsuki, "Maze-Running and Line-Search Algorithms," in *Layout Design and Verification*, ch. 3, T. Ohtsuki Ed., North Holland, 1986, pp. 99–131.

34. J. Pearl, *Heuristics: Intelligent Search Strategies for Computer Problem Solving*, Addison-Wesley, Reading, MA, 1984.

35. S. Piguet, F. Rahali, M. Kayal, E. Zysman, and M. Declercq, "A new Routing Method for Full Custom Analog IC's," in *Proc. IEEE CICC*, May 1990, pp. 2771–2774.

36. J. Rijmenants, J. B. Litsios, T. R. Schwarz, and M. G. R. Degrauwe, "ILAC: An Automated Layout Tool for Analog CMOS Circuits," *IEEE Journal of Solid State Circuits*, vol. 24, n. 2, April 1989, pp. 417–425.

37. B. R. Stanisic, N. K. Verghese, D. J. Allstot, R. A. Rutenbar, and L. R. Carley, "Addressing Substrate Coupling in Mixed-Mode ICs: Simulation and Power Distribution Synthesis," *IEEE Journal of Solid State Circuits*, vol. 29, n. 3, March 1994, pp. 226–237.

38. Y. T. Wong, M. Pecht, and G. Li, "Detailed Routing," in *Placement and Routing for Electronic Modules*, Marcel Dekker Inc., N.Y., 1993, pp. 181–219.

39. H. Yaghutiel, A. Sangiovanni-Vincentelli, and P. R. Gray, "A Methodology for Automated Layout of Switched-Capacitor Filters," in *Proc. IEEE ICCAD*, 1986, pp. 444–447.

Chapter 6

SEGMENTED CHANNEL ROUTING

Kaushik Roy

Field Programmable Gate Arrays (FPGAs) combine the flexibility of mask programmable gate arrays with the convenience of field programmability. These devices can be used for rapid prototyping of designs with a very short turnaround time. Different types of FPGA architectures are presently available from different manufacturers. In this chapter we will consider channeled architecture FPGAs [1,2]. Figure 6.1 shows such an architecture. The logic modules which implement various types of logic functions are placed in predefined rows. Channels are defined in between rows of logic modules for routing of nets. The routing resources are limited as the horizontal routing tracks are already laid out in the channel. However, the tracks are segmented. For example, the topmost track of Figure 6.1 is broken up into two segments a and b, which can be connected together if required by programming a *horizontal* antifuse (*hfuse*). Each input and output of a logic module is connected to a dedicated vertical segment. *Cross* antifuses (*cfuses*) are located at the crossing of each horizontal and vertical segments. Programming these antifuses produces a bi-directional connection between the horizontal and vertical segments. To serve as links between different channels there are vertical feedthroughs passing through the logic modules.

For these types of architectures, channel routing poses a unique problem. The antifuses have to be programmed in such a way so as to efficiently use the scarce routing resources to meet critical path delay requirements. Horizontal antifuse programming adds resistance and capacitance to a net, and depending on the technology used, the resistance and capacitance values can be detrimentally high. And hence, the critical nets should try to use the minimum number of horizontal segments possible. The channel router obtains its inputs from a global placement and routing tool, which specifies the net list between two rows of terminals across a two-layered channel.

It should also be observed that the routability and the performance obtained by a channel router depends to a large extent on the channel architecture. In particular, it is very important to design proper channel segmentation to obtain the required performance from a design while achieving routability for a large class of designs. If a short net is assigned to a large segment, not only a portion of the segment is wasted, but also each unprogrammed antifuse contributes positively toward a larger

207

routing capacitance. Each unprogrammed antifuse is associated with a small capacitance. If a large number of such unprogrammed antifuses are present in the architecture, the routing delays can be significant. Programmed antifuses are associated with the programmed resistance and a large number of such resistance on a net increases the RC time constant of the net. Hence, a strong correlation between the spatial distribution of segments and nets is required to achieve the best performance and routability.

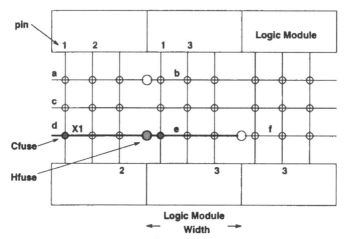

Figure 6.1 FPGA channel architecture

The unsegmented channel routing problem can be efficiently solved using the algorithms given by Yoshimura and Kuh [3]. The vertical and the horizontal constraint graphs completely characterize the channel routing problem. Using the horizontal constraint information, nets are merged on the vertical constraint graph so that the number of tracks is minimized. Rivest and Fiduccia have reported a *greedy* channel router which is quick and simple [4]. The greedy router wires up the channel in a left to right column by column manner, wiring each column completely before starting the next. The segmented channel problem has to be tackled from a different angle as the tracks are already laid out. Greene et al. have used the concept of *frontiers* to solve the general segmented channel routing problem [5]. However, they have no way to improve the quality of the solution. They have also shown that certain special cases of the problem can be solved in polynomial time. Brown et al. have considered the problem of detailed channel routing for a different type of FPGA architecture [6, 7].

A bounded search algorithm can be used to solve the segmented channel routing problem. The quality (or *cost*) of a solution is judged by the total number of segments used by the critical nets and the length of the segments assigned to the nets of the circuit. By investigating cost locally, a greedy search assigns nets to the best available segment(s). The algorithm resorts to backtracking if an initial solution is not obtained. The cost of the greedy solution is used as the the initial cost bound. The solution space is implicitly enumerated to come up with better solutions whose costs are used as the new bound for further search. The search can stop whenever a reasonable solution is obtained. Each net is allowed to use at most one track within any particular channel due to a technology constraint which does not allow programming of antifuses connected in an L-shaped fashion. Programming antifuses connected in such a way can lead to programming two antifuses at the same time which can have impact on reliability. And hence, the choices available at each level of the solution tree is also limited by this constraint. The channel router was investigated using different channel segmentation schemes for channel routability, performance, and routing track requirements. The segmented channel architectures were obtained based on a large set of net length distributions from benchmark designs. The details of channel segmentation methodology is given in Section 6.4.

The preliminaries and some basic definitions for the segmented channel routing problem are given in Section 6.1. Section 6.2 describes the details of the cost function and the bounded search algorithm for channel routing. Section 6.3 briefly describes the inherent fault tolerance issues associated with channeled architecture FPGAs. Section 6.4 considers channel segmentation schemes and its effect on performance and routability. Section 6.5 considers the implementation issues and discusses the results of our analysis. Finally, the conclusions are given in Section 6.6.

6.1 PRELIMINARIES AND DEFINITIONS

The following definitions will be useful in formulating the segmented channel routing problem.

Definition 6.1 A segment j is said to *belong* to track T, if segment j lies on track T. A track consists of a set of segments. For example, track 1 of Figure 6.1 consists of segments a and b.

Definition 6.2 A segment j is said to be *unassigned* if no net has been assigned to that segment.

Each segment j is characterized by the track to which it belongs to and its leftmost and rightmost x-coordinates. The leftmost x-coordinate of j is given by $left_j$ and the rightmost x-coordinate by $right_j$. The length of j is denoted by $L_j = (right_j - left_j)$. Similar definitions hold for the nets.

Consider a rectangular segmented channel with two rows of terminals along its top and bottom sides. A number between 0 and P is assigned to each terminal. Terminals with the same number x, $x \leq P$, must be connected by the same net x, while those with number 0 are unconnected terminals. Figure 6.2 shows a rectangular segmented channel where net 1 can be assigned to segments a and b, or to segments f and g, or to segment d. Once net 1 is assigned to some segment(s), then that segment(s) cannot be used by any other net. It should be observed that the first two choices for net 1 require *hfuse* programming, whereas the last choice does not. However, the same number of *cfuse* programming is required for all the above choices.

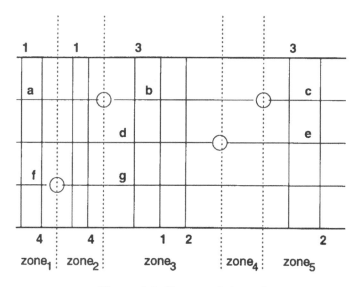

Figure 6.2 Segmented channel

The segmented channel routing problem can be stated as follows: Given a channel configuration with two rows of terminals along the top and bottom sides of the channel, find an assignment of nets to the segments such that each net x, $x \leq P$, is assigned to a set of previously unassigned segments S belonging to track T such that the segments cover the span of net x and a cost function is minimized. The details of the cost function will be described in Section 6.2.1.

The track segments and the nets (refer to Figure 6.2) form two sets of nodes (S_s and S_n respectively) of a bipartite graph G, shown in Figure 6.3. An edge exists between nodes $x \in S_n$ and $j \in S_s$ if and only if net x can be assigned to segment j. The set of segments are grouped together. Segments $j_1, j_2, ..., j_m$ are in a group, $group_t$, if the segments belong to the same track, t.

Definition 6.3 Net x is said to match a set of adjacent segments $j_q, j_{q+1}, ..., j_{q+l}, \in group_t$ if there exist edges emanating from node x to nodes

$j_q, j_{q+1}, ...j_{q+l}$, and there exists no $j_p \in group_t$ such that there is an edge between x and j_p. For example, net 1 of Figure 6.3 matches segments $a, b \in group_1$. If net x is matched to a set of segments S, then the set S and node j are removed from graph G.

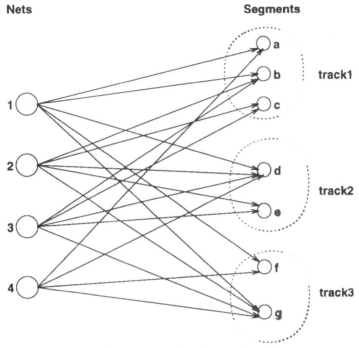

Figure 6.3 Bipartite graph, G

The segmented channel routing problem can be transformed into a matrix row matching problem for the ease of solving it using a digital computer. A given channel is divided into a number of *zones* according to the leftmost and rightmost coordinates of each segment. A vertical line is drawn at each $left_j$ and $right_j$ of segment j, $\{\forall j \mid j \in S_s\}$. Let us consider the channel of Figure 6.2. There are a total of five zones stretching across the channel. The mth zone is given by $zone_m$. A segment can stretch across different zones and a zone includes different segments belonging to different tracks. If the x-coordinate interval in which $zone_m$ exists overlaps with the x-coordinate interval in which net x exists, then net x is said to belong to $zone_m$. For example, net 1 of Figure 6.2 belongs to zones $zone_1$, $zone_2$, and $zone_3$; and net 4 belongs to $zone_1$ and $zone_2$.

We define two matrices - the net matrix, M_n and the segment matrix, M_s. M_n is a $U \times V$ and M_s is a $W \times V$ matrix, where U, V, and

W represent the total number of nets, the total number of zones and the total number of segments present in a channel, respectively. The (x, k)th element of M_n, given by $M_n(x, k)$, is 1 if net x is present in $zone_k$; otherwise the entry is 0. Similarly, $M_s(j, k)$ is 1 if segment j belongs to $zone_k$; otherwise $M_s(j, k)$ is 0. Figure 6.4 shows the M_n and M_s matrices for the channel of Figure 6.2.

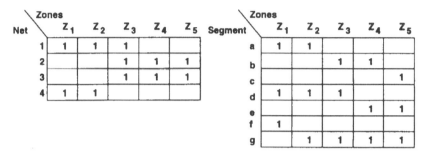

Net \ Zones	Z_1	Z_2	Z_3	Z_4	Z_5
1	1	1	1		
2			1	1	1
3			1	1	1
4	1	1			

Segment \ Zones	Z_1	Z_2	Z_3	Z_4	Z_5
a	1	1			
b			1	1	
c					1
d	1	1	1		
e				1	1
f	1				
g		1	1	1	1

Figure 6.4 Matrices M_n and M_s

Matrices M_n and M_s have only $(0,1)$ entries. Hence, we can define logical OR operation on the rows of these matrices. ORing of two rows R_i and R_j (given by $OR(R_i, R_j)$) is equivalent to column by column ORing of the two rows.

Definition 6.4 A row R_x of matrix M_n is said to match with row R_j of matrix M_s if R_j has 1 in every column location that R_x has a 1. Row 4 of matrix M_n of Figure 6.4 can be matched to row d of M_s. In fact, a row R_x of matrix M_n can be matched with a set of rows $S_x = R_j R_{j+1} \ldots R_k$ of M_s, if R_x matches with $OR(R_j, R_{j+1}, ..., R_k)$, and the set S_x belongs to $group_t$. In general, we are interested in minimal matching of row R_x with the set of rows from $group_t$ such that the cardinality of S_x is minimum. The minimum cardinality of S_x is given by N_x. For example, row 2 can be matched to row $OR(b, c)$ of Figure 6.4.

Definition 6.5 A k-segment matching problem is one in which each row R_x of M_n can be matched with at most k rows of M_s. 1-segment matching is the simplest one where each net has to be matched to only one track segment.

The channel routing problem in terms of matrices M_n and M_s can be be formulated as a matrix row matching problem. Each and every row of M_n has to be minimally matched with rows of M_s so as to minimize a cost. The simplest cost can be $\sum_{R_s \, of \, M_n} N_x$.

If the total number of 1s in any column k of a $(0,1)$ matrix A is given by $NC_k{}^A$, then a *necessary condition* for a channel to be routable can be expressed as:

$$\forall k; \quad NC_k{}^{M_n} \leq NC_k{}^{M_s}$$

The number of segments in any particular zone has to be more than or equal to the number of nets crossing that zone.

Green et al. have shown that K-segment $(K > 1)$ channel routing problem is equivalent to the problem of Numerical Matching with Target Sums [8] and hence, is NP-complete [5].

6.2 BOUNDED SEARCH ALGORITHM

In order to efficiently solve the matrix row matching problem of Section 6.1, we resort to a bounded search approach. This section also discusses cost functions and net ordering schemes for an algorithm for this approach.

6.2.1 COST FUNCTIONS

The cost of a solution is determined by the number of segments used by critical nets of the channel, and the length of the segments used by different nets. The number of *hfuses* to be programmed for net x is $N_x - 1$, and is one less than the number of segments used by the net. Programming these antifuses can increase the critical path delays of nets. In terms of the matrix matching problem, the number of rows of M_s used to match each row R_x of M_n gives the total number of segments used by net x. If net x of length $L_x = (right_x - left_x)$ is matched to p segments, each of length L_j $(j = m, ..., m+p)$, starting from segment m, then $(\sum_{j=m}^{m+p} L_j - L_x)$ gives a measure of unused segment(s) length. The cost C of matching a net x with one or more segments (rows of M_s) for K-segment (determined by the user depending on programmed antifuse resistance) routing is

$$C = w_1.\alpha + w_2.\beta + w_3.\gamma \qquad (6.1)$$

where

$$\alpha = \frac{\sum_{j=m}^{m+p} L_j - L_x}{\sum_{j=m}^{m+p} L_j}$$

$$\beta = \frac{N_x}{K+1}$$

and

$$\gamma \in \{0, \frac{1}{3}, \frac{2}{3}\}$$

where α and β are penalties for segment length wastage and *hfuse* usage, respectively, and are both positive and less than 1. γ is 0 if the adjacent segments of L_j have already been assigned to other nets, else γ is either $\frac{1}{3}$ or $\frac{2}{3}$ if either one or both adjacent segments are unassigned, respectively.

The values are normalized between 0 and 1 and are selected experimentally. It can be noted that γ is geared towards compacting channels and obtaining an initial solution, and is not a performance measure. Hence, the penalty for assigning nets to segments whose adjacent segments are unassigned is ignored during bounded search to find a minimum cost solution. The weights w_1, w_2 assigned to the wastage factor, and the antifuse usage factor respectively, are determined by the technology under consideration. For example, the metal-metal horizontal antifuse has a much lower programmed resistance than a programmed poly-metal antifuse, and hence, w_2 for the latter technology should be higher than the metal-metal antifuse technology. Usually, w_2 is greater than w_1, and w_3 is much smaller than w_1.

6.2.2 THE ALGORITHM

An algorithm using greedy heuristics can be used to find a bound on the cost of the solution. The algorithm appeared to produce a good cost bound for most of the problems that we worked on. Figure 6.5 shows the greedy algorithm. The row R_x of matrix M_n with the longest

Procedure Greedy:

While there are some unassigned R_x of M_n {
    Best_match(R_x);
    If there is no match
      Backtrack;
    Else {
      Remove matched row from M_s;
      Activity = Activity + M_{ij};
      Update cost of partial solution.
  } }

Figure 6.5 Greedy algorithm

net is selected first. Ties can be broken arbitrarily. The routine *Best-match(R_x)* finds the choices available for matching R_x with rows of M_s. The choices are ordered in terms of the cost. The one with the lowest cost is selected for matching R_x. It should be noted that the greedy heuristic selects the local best choice for each row, which may not produce an overall optimum solution. Once R_x is matched with row(s) of matrix M_s, the matched and the matching rows are removed from the matrices, thereby reducing the size of the problem for the next iteration. The algorithm backtracks to the previous choice point if *Best-match(R_x)* does not find a choice for net x. No solution exists if the complete search fails to generate a solution.

The complexity of the 1-segment greedy algorithm is $O(UW)$, where U and W are the number of rows for M_n and M_s, respectively. The k-segment matching problem can lead to several choices for net x. It includes the choices available for 2, 3,..., $k-1$ segment problems. For a 2-segment problem, in the worst case, there can be $\binom{n}{2}$ choices, where n is the number of unassigned segments. However, the number of choices is drastically reduced due to the fact that net i can only be matched to a set of unassigned segments S_i, such that S_i belongs to one particular *group*.

If the user is satisfied with the cost C of the solution obtained using the greedy heuristic, the program can stop. Otherwise, the solution space is implicitly enumerated using C as the bound on cost. The bounded search algorithm can be better understood by considering a directed tree structure of Figure 6.6. The root node of the tree represents the problem to be solved. All edges are directed away from the root. Any intermediate node represents a partial solution to the problem. The leaf nodes represent different solutions. The intermediate nodes have a partial solution cost attached to them. Each node stores the choices available for matching the topmost row of M_n with the rows of M_s and the respective costs. The choices are ordered in terms of increasing cost. It can be observed that the row size of the matrices M_n and M_s decreases as we go down the tree from its root.

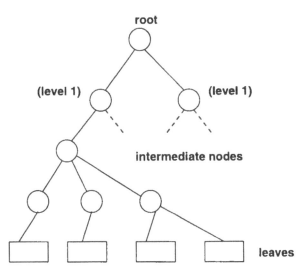

Figure 6.6 The solution tree

The leftmost edges of the tree starting from the root node represent the greedy solution with cost C. The algorithm backtracks from the

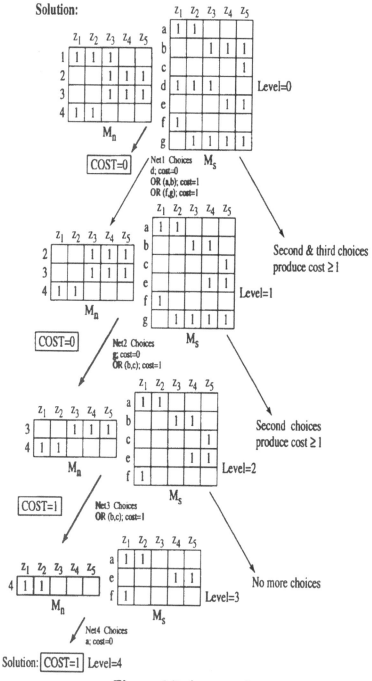

Figure 6.7 An example

solution node, and goes up the tree. The immediate parent node p of the leaf node l is visited first. The best unused choice at node p is selected to match the top row of M_n at that node. If that choice produces an intermediate cost which is more than or equal to the cost C, the algorithm goes up one more level (parent of node p). Else, the tree is expanded from node p. Anytime a solution with cost $C_m < C$ is obtained, C_m becomes the new cost bound. If a solution is infeasible at any point, the algorithm backtracks to the previous choice point and enumerates other choices, if available, the details of which are given in Section 6.2.3.

Figure 6.7 shows the application of the algorithm on a simple example. Matrices M_n and M_s represents the channel of Figure 6.2. The objective function is to minimize a simple cost function, $\sum_{x \in all\ nets} cost_x$, where $cost_x = N_x - 1$. Level 0 of the tree shows the initial matrices. Row 1 (net 1) of M_n is selected for matching at level 1 of the tree. The available choices for matching row 1 are given by rows: d, $OR(a, b)$ and $OR(f, g)$. The first choice has a cost of 0 unit, whereas the other two choices require programming one *hfuse*. Hence the cost for the later two choices are 1 unit each. The algorithm selects the local best choice, which is segment d in this example. The rows 1 and d are removed from matrices M_n and M_s, respectively, and choice d is marked *used*. It can be observed that $OR(d, e)$ is not a choice for net 1, because segment d has already been considered as a choice for that net. The greedy search goes down the left edges of the tree until a solution (leaf node) is obtained. If a solution cannot be obtained, the algorithm backtracks to select other choices. In this example, the greedy search produces a solution with cost of 1 unit. The algorithm now backtraces up the tree levels from the solution node. At level 3 there are no available *unused* choices. However, there are choices available at level 2 and level 1. Those choices produce intermediate costs which are equal to the cost of the current best solution obtained. Hence, the solution space is not enumerated. The greedy search has provided the minimum cost bound for this problem.

6.2.3 NET ORDERING SCHEMES

The order in which nets are selected for matching can affect the amount of search required to arrive at a solution. This can be best understood by considering the example of Figure 6.8. If the nets are ordered as shown in the M_n matrix, the greedy search will assign net 1 to segment a, net 2 to segment d, and finally net 3 to segments e, f, g, h, i, and j, producing a solution which requires five-*hfuse* programming. Minimum cost solution requiring one-*hfuse* programming can then be obtained using the bounded search on this cost. The solution assigns net 1 to segments b and c, net 2 to segment d, and net 3 to segment a. However, the same solution can be quickly obtained using greedy search with a different net ordering scheme such as nets 3, 2, and 1.

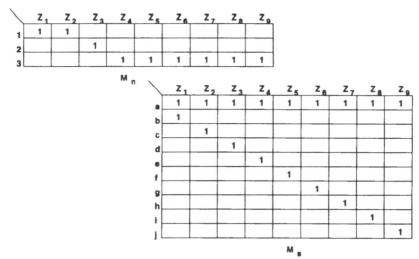

Figure 6.8 Net ordering

Three net ordering schemes have been considered:

(1) Ordering nets according to length - longest first (LF).

(2) Ordering according to the number of 1-segment choices available for each net. Two schemes have been used - static ordering (CS) and dynamic ordering (CD).

(3) Ordering according to left edges $(left_x)$ of nets (LE).

CS ordering is similar to LF ordering, because the longer nets usually have less number of 1-segment choices available for matching. The dynamic net ordering scheme (CD) is similar to CS; however, the nets are reordered after a net has been matched to segment(s). This is required because the number of 1-segment choices of the unassigned nets changes after each assignment of a net to segment(s). Left edge (LE) ordering of nets can be used to make backtracking (to find an initial solution) more efficient. Nets x and $x-1$ will usually have considerable x-coordinate overlap due to this ordering scheme, and hence, the available segment choices for matching these nets will also overlap. Nets x, and $x-1$ are dependent on each other for their segment choices. If the algorithm does not find a valid match for net x, it will backtrack to the previous level to undo the assignment of net $x-1$ to segment j. Net $x-1$ will be assigned to the next best available choice, if any, and proceed. If such a choice is unavailable, the algorithm backtracks to the previous level and try undoing the match for net $x-2$ and so on until a solution is obtained. If no solution is obtained even after exhausting all choices implicitly, then a solution does not exist with such a segmentation scheme. This emulates dependency directed backtracking [9].

The critical nets determined by a timing verifier are routed first so that they can be assigned the best possible segments.

6.3 FAULT TOLERANCE
DUE TO INHERENT REDUNDANCY

The FPGA architecture of Figure 6.1 lends to limited fault tolerance. The cross antifuses (*cfuses*) are located at the crossing of each vertical line, connecting the pin of a logic module, and the horizontal routing tracks. Typically, the router connects each pin to one horizontal track. The rest of the *cfuses* on that vertical line are left unprogrammed. Similarly, most of the *hfuses* also remain in the unprogrammed state because only few of the nets require more than one segment for routing. So it may be possible to reconfigure routing in such a way that the antifuse faults do not cause an error to occur during normal operation.

Two types of faults in the antifuses are considered. The first type of faults can be diagnosed *a priori*, i.e., before programmation and are called *type 1* faults. After detecting the type 1 faults, routing can be reconfigured around the faulty *cfuse* or *hfuse* if possible. A shorted (or already programmed) *cfuse* or *hfuse* is an example of type 1 fault. Such faults can be easily diagnosed by checking the continuity of the tracks.

Type 2 faults cannot be diagnosed *a priori*. Let us consider an antifuse which shows a normal behavior (open) in the unprogrammed state but after programmation does not produce a low resistance connection. Such a fault cannot be diagnosed before the FPGA is programmed, and hence, routing reconfiguration is not possible.

Type 1 faults for horizontal antifuses: Due to this fault the two segments (a and b) adjacent to a faulty *hfuse* get electrically connected. The fault produces a modified (faulty) segment matrix M_s. The rows corresponding to segments a and b are ORed together to form a new row. Fault tolerance during layout is achieved if routing ensures any one of the following criteria:

(1) Some net x is assigned to segments a and b.
(2) Segment a is assigned to some net x, and segment b is unused.
(3) Segment b is assigned to some net x, and segment a is unused.
(4) Segments a and b are unassigned.

Type 1 faults for cross antifuses: Let us consider a faulty *cfuse* at the crossing of a vertical line connecting a logic module pin to net x and the horizontal segment a. This fault causes the pin (or net x) to get electrically connected to segment a. The effect of this fault can be simulated by removing the row corresponding to segment a from matrix M_s and assigning the segment to net x. It should be noted that net x is removed from matrix M_n only after it has been assigned to segment a or to any other segment(s). Fault tolerance can be achieved only if the router satisfies any one of the following criteria:

(1) Segment a is logically assigned to net x.

(2) Segment a is not logically assigned to any net.

From the above discussions it is very clear that the effect of the *type 1* faults can be simulated by a modified M_s matrix and hence, our algorithm easily handles *type 1* faults to achieve fault tolerance during routing.

There are t number of *cfuses* on each vertical line connecting the pins to the different tracks, where t is the number of tracks. Of those *cfuses* only one is programmed to connect the pin to the corresponding track segment. This inherent redundancy in the architecture suggests that it might be possible to depopulate the *cfuses* along the vertical lines. This can potentially decrease the capacitive loading from the antifuses. If metal-metal low programmed resistance amorphous silicon [2] antifuse technology is used, the leakage current from the unprogrammed antifuses can be high and can have adverse effect while programming other fuses. Hence, the depopulation scheme can also reduce the leakage current during programmation for such a technology. The *cfuses* can be depopulated based on the segmentation scheme and should be such that routability is not adversely affected.

6.4 CHANNEL SEGMENTATION

The routing resources are fixed for FPGAs, and hence, the channel segmentation scheme has an enormous effect on performance and routability of a channel. Efficient usage of horizontal antifuses (*hfuses*) is required for routability and for meeting the critical path delay (performance) requirements. Intuitively, a strong correlation between the segment length and the net length distributions within a channel is very desirable . However, the mere existence of a unique segment of acceptable length for every net in a channel does not guarantee 100% routability. This is due to the fact that an additional factor, the *location* of a segment with respect to a net span in a channel is also important in determining whether that segment can be used for routing that net. It is imperative, therefore, to consider the *spatial* distributions for segments and nets. An automated channel architecture synthesis algorithm based on simulated annealing has been developed which attempts to maximize the correlation between the spatial distribution of segments and nets [16]. The set of benchmark net distributions were obtained from Texas Instruments' gate array designs. The optimization procedure can be also used to produce channel architectures for net length distributions which can be represented in terms of equations.

It can be observed that routability can be enhanced by increasing the number of tracks and/or using a more granular segmentation scheme. If the number of tracks is kept fixed, maximum channel routability is obtained when each track is made up of segments each of minimum length. A single logic module width (Figure 6.1) is the minimum segment length that we have considered. Let us consider the cost function of Section 6.2.1 with $w_3 = 0$. The cost of assigning net x to segment(s) is given by $C_x = w_1 \alpha + w_2 \beta$. With minimum length segmentation scheme a smaller value of α is obtained, signifying better channel utilization or less segment wastage. However, with such a segmentation scheme N_x is large, requiring larger number of *hfuse* programming, and hence, performance may deteriorate depending on the value of w_2. Weight w_2 simulates different electrical characteristics of antifuses, and can influence the routing architecture.

Routability of the FPGAs requires the presence of a large number of *cfuses* and *hfuses* in the channel. The antifuse requires multiple programming pulses to successfully form the electrical connection between logic modules or segments. Depending on the circuit size and device types, 3,000 to 20,000 antifuses are typically programmed [12]. Cross antifuse (*cfuse*) programming introduces certain routing constraints which influence channel architecture synthesis. The following section details the *cfuse* programming methodology, the details of which will be required in proper understanding of channel architecture synthesis.

6.4.1 NET CROSS ANTIFUSE PROGRAMMING AND CHANNEL ROUTING

Each *cfuse* in the channel can be individually addressed for programmation. Let us consider antifuse X_1 of Figure 6.1. Programming X_1 requires charging the vertical line crossing X_1 to a voltage *VPP* and the segment *d*, crossing X_1 horizontally, to *GND* (0V). *VPP* is determined by the antifuse technology under consideration. A voltage stress of *VPP* across the antifuse for a certain period of time creates a low resistance (technology dependent) connection between the vertical line and the horizontal segment. All other vertical and horizontal lines are charged to a voltage of *VPP/2*. It can be noted that the other unprogrammed *cfuses* experience a voltage stress of 0 volts or *VPP/2* volts, and hence remain unprogrammed.

As noted in Section 6.2, the channel routing problem can be formulated as an assignment problem where each net within a channel is assigned to one or more unassigned segments. Each net is allowed to use at most one track due to a technology constraint which does not allow programming of antifuses connected in an L-shaped fashion [1]. The programming of such antifuses can lead to programming two antifuses at the same time which can degrade the performance of the programmed antifuses. Let us consider net 3 of Figure 6.9. If the net were routed by programming *cfuses* Y_1, Y_2, Y_3, and Y_4 (using 2 tracks) then Y_1, Y_2, and Y_3 form an "L". After programming any two of the antifuses forming the "L", and while trying to program the third one using the technique described above, antifuse Z also get programmed at the same time. The electrical connection due to the programmation of antifuse Z does not logically affect routing. However, whenever two or more antifuses gets programmed simultaneously using the same programming pulse, the reliability of the connections thus produced may not be very predictable.

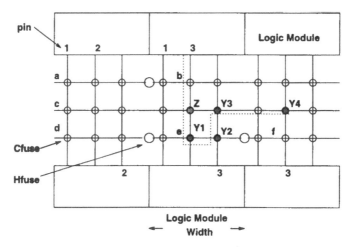

Figure 6.9 Programming *cfuses*

For the purposes of architecture synthesis, we can use a fast, greedy 1-segment routing scheme which can be efficiently used to optimize the channel architecture. The nets are ordered in decreasing order of length while the segments are ordered in increasing order of length. In the greedy scheme, the nets are assigned to the first available segment in that order. Once a net is assigned to a segment, the corresponding segment is made unavailable for further routing of any other net. It can be observed that the first segment assigned to the net is the best match in terms of segment length wastage. If there are m nets in a channel consisting of n track segments, then the worst case routing complexity is $O(mn)$ with the above ordering scheme. It takes $O(m \log(m))$ and $O(n \log(n))$ time to sort the nets and the segments, respectively.

The cost of routing is determined by the number of segments used by the critical nets in a channel, and the length of the segments assigned to different nets. The number (H_x) of horizontal antifuses to be programmed is equal to the number of segments assigned to the corresponding net (x) minus one. Note that the number of *cfuses* to be programmed for a net remain constant irrespective of the number of segments assigned to net x, and hence, the delays incurred due to programming the *cfuses* is not considered during routing. Depending on the technology, the resistance and the capacitance associated with programmed *hfuses* can be detrimentally high. If a net x of length L_x is routed with p ($1 \leq p \leq K$, maximum of K segments allowed) segments, each of length L_j ($j = 1, ..., p$), then $\{(\sum_{j=1}^{p} L_j) - L_x\}$ gives a measure of the wasted (or excess) segment(s) length. For K-segment routing, in which a maximum of K segments ($K - 1$ *hfuses*) can be used by each net, we define the cost C of routing a net x as

$$C_x = w_1.\alpha + w_2.\beta \qquad (6.2)$$

where

$$\alpha = \frac{\sum_{j=1}^{p} L_j - L_x}{\sum_{j=1}^{p} L_j}$$

and

$$\beta = \frac{H_x}{K}$$

where α and β are penalties for segment length wastage and horizontal antifuse usage, respectively, and are both positive and less than 1. The weights w_1, w_2 assigned to the wastage factor, and the antifuse usage factor respectively, as described in Section 6.2.1. The total cost of routing all the nets in a channel is $\sum_p C_p$, $0 < p \leq V$, where V is the total number of nets in a channel.

6.4.2 CHANNEL ARCHITECTURE SYNTHESIS

Given a channel architecture, the *routability* and the *performance* are solely dependent on the net length distribution and the location of the nets across the channels. Hence, to come up with a good segmentation scheme we consider the net length and their spatial distributions across the channels for a wide range of examples. The distributions were obtained from a set of gate-array benchmark designs with a given channel length.

6.4.2.1 ROUTABILITY AND ROUTING DELAY

It can be observed from Figure 6.1 that if the routing tracks are finely segmented, routability is enhanced. However, for routing the longer or more critical nets, a larger number of horizontal antifuses will probably be required. The resistance and capacitance of the programmed antifuses can be detrimentally high for some of the critical nets. If all the track segments are long, routability decreases because of the decrease in the number of available segments for routing. Hence, a judicious mixture of long and short segments are required to get the best performance and routability from the architecture.

Definition 6.6 The *length* of a net in a channel is given by the maximum and minimum x-coordinate span of the net.

Definition 6.7 The *width* of a channel is the x-coordinate span of the channel.

The net length distributions across a channel is given by the number (n_j) of nets of length (l_j) present in the given channel. Lengths are quantized to the logic module width (shown in Figure 6.1), which is the minimum segment length in our case. Given a set of p net length distributions D_i, $0 < i \leq p$, (each consisting of a set of nets N_i), the total length of nets for each distribution D_i is given by $L_i = \sum_j n_j l_j$. Let the given channel have a set of segments S, where the total length of the segments is given by T.

6.4.2.2 NET-SEGMENT CORRELATION

For best performance and routability, the segment length distribution and the segments' spatial location within the channel should very closely follow the net length and their spatial distributions. We determine the correlation between the net and segment distributions within a channel in terms of 1-segment routing and the segment wastage factor.

To solve the segmentation problem efficiently, the channel is divided into q equal zones by imaginary vertical lines, the minimum zone length being equal to the length of a logic module. Let us consider Figure 6.10 as an example. There are 6 zones, $zone_1$ through $zone_6$ in the figure. A minimum segment length of one logic module width is used in all our discussions. The track segments are ordered in terms of increasing length. Each segment s carries information on its spatial location within the channel — the leftmost (s_l) and the rightmost (s_r) zones that the segment spans across. For example, for segment c in Figure 6.10, $c_l = 1$ and $c_r = 3$, whereas for segment f, $f_l = 1$ and $f_r = 1$. A channel is divided into M zones, where M is the number of logic modules per row of channel. A net $n \in N_i$ whose left most and right most zones are given by n_l and n_r is 1-segment routable if there exists an unassigned segment $g \in S_a$ (the set of available segments) such that

$$g_l \leq n_l \quad \text{and} \quad g_r \geq n_r$$

The set of unavailable segments (already assigned to some nets) is given by $S_u = (S - S_a)$. Once such an assignment is made, segment g is made unavailable for any other net. If it is possible to find such assignments for all the nets in a channel, the set of nets are said to be *1-segment routable* with respect to that segmentation scheme. Let us consider net 1 of Figure 6.10, where $M = 6$, and $net1_l = 1$ and $net1_r = 3$. To start with, all the segments in the channel belong to set S_a, and set S_u is empty. Condition (1) is satisfied for net 1 with respect to segment c. As $c \in S_a$, net 1 can be assigned to segment c, and hence is 1-segment routable. Segment c is made unavailable for any other net by moving segment c from set S_a to set S_u ($c \ni S_a$, $c \in S_u$). However, there are no segments in set S_a which satisfies condition (1) for net 2 after net 1 is routed with segment c, making net 2 1-segment unroutable. It should be noted that net 1 would have been 1-segment *unroutable* if net 2 were routed first using segment c.

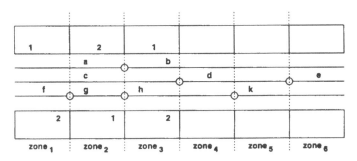

Figure 6.10 Zones in a channel

The nets are ordered in terms of the leftmost coordinates. If net n is assigned to segment g then the segment wastage factor is given by

$$\alpha_n = \frac{(g_r - g_l) - (n_r - n_l)}{(g_r - g_l)}$$

where $(g_r - g_l)$ and $(n_r - n_l)$, respectively, denote segment length and net length in terms of the number of zones they span across.

Routing produces a set of 1-segment routable nets N_{ir} and a set of 1-segment unroutable nets N_{iu}. For *1-segment routable* channels, N_{iu} is empty. However, the given set of nets N_i are possibly K-segment routable even though single segment routing is not possible. In order to comprehend K-segment routing, the set of *1-segment unroutable* nets N_{iu} (after initial 1-segment routing) and the given channel are considered. Obviously, condition (1) cannot be satisfied for any net $n \in N_{iu}$ with any segment $g \in S_a$. The maximum x-coordinate overlap, θ_{ng} of a net $n \in N_{iu}$ with any segment $g \in S_a$ can be used as a heuristic measure of K-segment routability of net n. The larger the value of θ_{ng}, the larger is the measure of K-segment routability. Segment g is moved from set S_a to the set S_u after such an overlap is established. Let us again consider the nets and the channel of Figure 6.10. After routing net 1, net 2 is not

1-segment routable, and hence is a member of set N_{iu}. However, net 2 has a large overlap ($\theta_{2a} = 2$, where 2 zones is the length of the overlap between segment a and net 2) with unassigned segment a. Though condition (1) is not satisfied with respect to segment a, net 2 can be routed using 2 segments – a and b.

It can be observed that exact K-segment routability could have been used as a measure of correlation between S and N_i. However, that would be very CPU time intensive. We obtained good results using *1-segment routability* and *segment wastage* as the measure of correlation.

6.4.2.3 SYNTHESIS TECHNIQUE

Channel segmentation requires the optimization of the horizontal antifuse usage and segment wastage for different net distributions. Simulated annealing [18] was used to efficiently generate such an architecture. We use net distributions from a large set of gate array benchmark designs to optimize our channel segmentation architecture. The routing cost for channel i, having the set of nets N_i, and a set S of already laid out segments is given by

$$C_i = \nu_r \cdot \mid N_{iu} \mid + \nu_w \cdot \sum_{k \in N_{ir}} \alpha_i + \nu_o \cdot \sum_{k \in N_{iu}} \frac{\theta_{kg}}{k_r - k_l} \qquad (6.3)$$

where, $\mid N_{iu} \mid$ represents the cardinality of set N_{iu} and is equal to the number of unroutable nets in channel i. θ_{kg} corresponds to the maximum overlap of net k with an unassigned segment g. The first term in the equation corresponds to the number of *1-segment unroutable* nets, the next term is due to the segment wastage factor, while the last term corresponds to percentage segment overlap for the unroutable nets. Let us again consider Figure 6.10. For the 1-segment unroutable net 2, the segment overlap with an unassigned segment a is $\frac{2}{3}$ ($\theta_{2a} = 2$ and ($net2_r - net2_l$) = 3) while the segment overlap with segment f is $\frac{1}{3}$ ($\theta_{2f} = 1$ and ($net2_r - net2_l$) = 3). ν_r, ν_w, and ν_o are the weights associated with the corresponding factors. The routability weight ν_r is much higher than ν_w or ν_o as it relates to both *routability* and *performance*. The *1-segment routable* nets with very low segment wastage usually have lower interconnect delays than nets requiring two or more segments for routing due to the presence of programmed horizontal antifuse(s). In fact, the exact routing delay depends on the number of unprogrammed *cfuses* on the segment(s), the length of the segment(s), the resistance of any programmed *hfuse(s)*, and *cfuses*. The weight ν_o associated with the overlap factor is a small negative number. For unroutable nets we consider a larger overlap to be better – the net has a higher probability of getting routed using two or more segments. For p different channels

(p sets of nets, N_1, N_2, ..., N_p), the total cost of routing, C, using the same set of segments S is given by $C = \sum_{i=1}^{p} C_i$. The usage of *1-segment routability* and the *overlap factor* represents a novel way of efficiently predicting the K-segment routability of a segmented channel architecture.

Due to the complex nature of the cost function, simulated annealing was used. Annealing starts by assigning an arbitrary segmentation for a channel of given width and a given number of tracks. Two moves are allowed in this specific annealing algorithm — merging of two adjacent segments in a track and breaking of a segment within a track into two segments such that the broken segments add up to the original segment length. The segments are randomly picked for either merging or breaking. Merges or breaks are also determined randomly. It is not possible to break a segment of length equal to the width of a single logic module. After each move the cost C_i is calculated for each of the sets of net distributions, $N_i,....N_p$. If C decreases from its previous value the move is accepted. However, a move with a higher C is accepted with a probability $e^{\frac{-|\delta(C)|}{Temp}}$, where $|\delta(C)|$ is the absolute value of the change in cost C and $Temp$ is the annealing temperature. At a higher temperature a higher cost move is accepted with a higher probability than at a lower temperature. This helps the optimization technique to move out of local minima of the cost function. The cooling schedule used is the same as the one proposed in [19].

Let us consider a simple example to illustrate the architecture synthesis methodology. Figure 6.11a shows an initial channel segmentation. There are three routing tracks having uniform segmentation, each segment having a length of two logic modules. The channel is divided into six equal zones. Two net distributions, N (only three nets) and N' (only two nets) are also shown in the figure. The pin numbers with primes correspond to the distribution belonging to N'. It can be observed that for such a segmentation, net 2 of distribution N can be assigned to segment c (or f, or k having same cost), while net $2'$ of distribution N' can be assigned to segment e (or b, or h having same cost). However, all the other nets are not 1-segment routable. If we assume that $\nu_r = 100$, $\nu_w = 2$, and $\nu_o = 0$, then the cost of routing using Equations 6.2 and 6.3 is given by:

$$C_N = 100 * 2 + 2 * 0 = 200$$
$$C_{N'} = 100 * 1 + 2 * 0 = 100$$

Hence, the total cost after the first iteration is

$$C_1 = C_N + C_{N'} = 300$$

Now the algorithm can randomly pick a segment (say b) and merge it with segment a (Figure 6.11b). For such a move, the change in cost

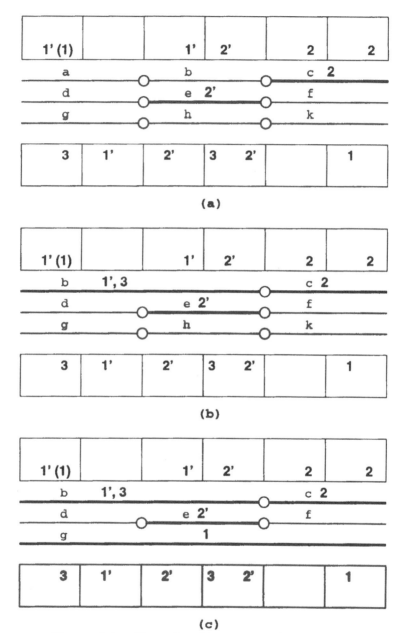

Figure 6.11 Channel architecture synthesis example

is calculated. Net $1'$ can now be assigned to segment b, and net $2'$ to segment e, making distribution N' 1-segment routable. For distribution N, net 2 and 3 are assigned to segments c and b, respectively. Note that net $1'$ is also assigned to segment b. That is allowed because the nets belong to different distributions. The portion of segment wasted when net $1'$ of length 3 zones is assigned to segment b of length 4 zones is $\frac{4-3}{4} = 0.25$. It can be observed that for all the other routed nets, the segment wastage factor is 0 (this of course does not include the wastage due to segment quantization). The nets assigned to the segments are also shown in the figure on the top of each segment. Net 1 of distribution N is 1-segment *unroutable*. Hence, the new cost of routing is given by:

$$C_N = 100 * 1 + 2 * 0 = 100$$

$$C_{N'} = 100 * 0 + 2 * \frac{4-3}{4} = 0.5$$

The total cost after the second iteration is

$$C_2 = C_N + C_{N'} = 100.5$$

The change of cost $\delta(C) = C_1 - C_2 = 199.5$, and hence the move of merging b and a together is accepted. After a few iterations, the synthesis algorithm came up with an architecture shown in Figure 6.11c, which indeed produces the lowest cost of routing:

$$C_N = 100 * 0 + 2 * 0 = 0$$

$$C_{N'} = 100 * 0 + 2 * \frac{4-3}{4} = 0.5$$

and hence,

$$C = C_N + C_{N'} = 0.5$$

6.5 IMPLEMENTATION AND RESULTS

The net matrix M_n and the segment matrix M_s are (0,1) matrices. It is memory inefficient to store the (i,j)th entry of each matrix in one computer word. If the computer word size is m, and the number of columns in each of the matrices is less than or equal to m, then an m-bit word can be used to store an entire row a matrix. It should also be noted that the column by column ORing between two rows a matrix is a slow sequential operation. However, bit by bit ORing of two words can be done in parallel and is much faster. Hence, the CPU time can be reduced by using words to store net and segment information.

The information about the track segments and their *groups* are available from matrix M_s. Adjacent rows of M_s correspond to adjacent segments in a particular channel. For example, the first three rows of M_s of Figure 6.1 represent segments a, b, and c, respectively, belonging to $group_1$, followed by segments d and e of $group_2$ and so on. Logical ORing of all rows belonging to a *group* produces 1s in all columns of M_s, and can be used to identify the different *groups*.

The algorithms for segmented channel routing and channel architecture synthesis have been implemented in C. No benchmarks are presently available for segmented channels. Comparisons with other segmented routers are not possible due to the lack of published results. We used the combinational logic synthesis benchmarks from MCNC to evaluate the algorithms. The designs were logic synthesized using Texas Instruments TPC1010 based library. A simulated annealing based modified TimberWolfSC algorithm was used for placement and global routing to minimize the net congestion in each zone of a channel [11,17].

Channel architecture synthesis requires a set of benchmark net distributions. The set of net distributions in different channels for our experiments were obtained from a set of gate array benchmark designs of Texas Instruments. These benchmark designs pertained to different types of circuits such as disk controllers, microcontrollers, data paths etc. About 300 benchmark channels were used for synthesizing the architecture. The number of nets in each channel varied from 43 to 87. The benchmarks were all logic synthesized using TPC1010 library [12]. Placement and global routing on all these benchmarks were done using a simulated annealing based placement tool [17]. We found that net distributions generated using different kinds of placement tools, like min-cut, were not drastically different. Hence, we decided to use the modified TimberWolf based placement tool to generate all our benchmark distributions. However, it should be noted that our aim was to develop an efficient way to automatically generate channel segmentation. Any other placement tool can be used with equal efficiency to generate the benchmark distributions and then can be used for synthesizing the architecture. In fact, the tool can be used to synthesize architectures even when the net distributions are *a priori* known to follow any given distribution which can be represented by an equation or represented graphically.

Uniform segmentation was used as an initial state for simulated annealing where each segment was of length 2 logic modules. For all our experiments a track length of 44 logic modules have been used. Hence, there were 44 zones. The channel architecture obtained using our synthesis procedure was compared with existing industrial segmentation schemes having 12 or 22 tracks. The FPGA template (only the number of rows) were varied to fit different designs. Each of the channels had identical segmentation. The segmentations were also compared with a uniform channel segmentation scheme with 2 logic module long segments. The latter channel architecture is very good for routability. However, routing requires programming a large number of horizontal antifuses which degrades performance.

A segmented channel router [15], described in Section 6.4.1, was used for evaluating the segmented channel architectures. The channel router uses the cost described in Section 6.4.1 to assign a set of nets to one or more segments (K-segment routing). Weights of $w_1 = 1$ and $w_2 = 20$ were used. Therefore, there was a heavy penalty on antifuse usage. Once the channel routing is done, the complete layout knowledge enables accurate delays to be generated for all interconnections. These interconnect delays are then used by the timing analyzer to determine the critical paths, the details of which are given in [13]. For the FPGA interconnects, the three primary sources of delays are programmed horizontal and cross antifuses and the usual delay due to routing in metal, and the delay due to unprogrammed antifuses. The first step toward generating the interconnect delays is transforming the nets to RC-networks. We represent the RC-network as a graph with the nodes being cell pins. Each net segment and the horizontal and cross antifuses are modeled by lumped RC delays. The block delays used were 1 ns.

Table 6.1 Results for f104667 with 12 tracks

Channel	Nets	Hfuses			Unrouted nets			Seg. wastage (%)		
		SAS	I	U	SAS	I	U	SAS	I	U
1	23	0	0	29	0	0	0	56.1	54.2	33.0
2	31	0	1	57	0	1	0	55.3	-	28.9
3	50	1	0	78	0	2	0	54.6	-	31.9
4	55	0	5	74	0	6	0	55.0	-	31.5
5	66	0	2	75	0	11	0	61.5	-	34.7
6	61	0	1	55	0	9	0	67.8	-	36.4
7	40	0	0	64	0	0	0	55.4	64.6	30.4
8	40	0	0	47	0	0	0	62.0	71.7	34.4
9	19	0	0	24	0	0	0	57.1	56.0	38.1

Table 6.1 shows the details of the results obtained for an industry design example *f104667*. The first two columns show the channel number and the number of nets in each channel. The number of horizontal antifuses programmed, the number of unroutable nets in each channel, and the percentage segment wastage for our Simulated Annealing based Segmentation (SAS) scheme are compared with an existing industrial segmentation scheme (*I*) with 12 tracks. The results were also compared with a uniform segmentation scheme (*U*) having each segment of length two logic modules. Five of the nine channels were not routable with architecture *I*. The SAS scheme completely routed all the channels and

used lesser number of horizontal antifuses. Uniform segmentation used lot of horizontal antifuses due to its granular segmentation architecture. However, as expected, segment wastage was considerably lower for the U scheme than the other two architectures. Percentage segment wastage in the last column was calculated using actual net and segment lengths. A '–' appears in the percentage segment wastage column whenever there were unroutable nets in a channel.

Table 6.2 Results of routing on channels having uniform segmentation, each segment of length 2 logic module width

Example	Num. of channels	Tot. num. nets	Ave. net density	Hfuses	Seg. wastage (%)
f104780	19	1074	12.1	2713	31.8
f104243	14	770	9.3	1452	32.4
cf92382a	19	1174	8.7	1656	39.44
f103918	19	1136	11.3	2540	28.4
duke2	8	496	13.9	1408	26.7
bw3	8	317	6.9	406	27.9

Table 6.3 Results of routing on SAS architecture vs. I architecture

Example	Hfuses		Unrouted nets		Seg. wast. (%)		Timing improv.(%)
	SAS	I	SAS	I	SAS	I	
f104780	44	56	0	6	44.5	45.9	-
f104243	0	1	0	0	44.0	44.8	6.1
cf92382a	0	0	0	0	48.9	52.0	5.2
f103918	15	20	0	0	43.3	43.5	9.3
duke2	11	12	0	0	43.2	44.9	5.5
bw3	36	44	0	13	52.0	45.6	-

Table 6.2 summarizes the statistics of some other examples (not used as benchmark designs in generating the channel architecture) and shows the routing results on a channel with uniform segmentation (U), each segment having a length of 2 logic modules. This table can be used for comparing the results obtained using SAS. Twenty-two routing tracks were used. The total number of channels for each example and the total number of nets in all the channels are shown in columns 2 and 3, respectively. The average *net density* over all the channels is shown next. *Net density* of a channel is defined as the maximum number of nets crossing any imaginary vertical line through the channel and is a measure of the number of tracks required for *unsegmented routing* as in gate arrays. A large number of antifuses were used because of the very granular segmented channel. Percentage segment wastage in the last column represents the average over all the channels for the example under consideration and can be compared to Table 6.3 shown below.

Again due to the fine granularity of the architecture, segment wastage is low. Such an architecture is good for routability, and hence, there were no unroutable nets. It must be noted that the number of *hfuses* used and the percentage segment wastage are both measures for both performance and routability as noted in Section 6.2.

Table 6.3 summarizes the results of routing on the SAS architecture using the industrial and MCNC benchmark designs of Table 6.2. Twenty-two routing tracks were used. The results were compared with architecture *I*. Architecture SAS produced better results in terms of channel routability and percentage segment wastage. For example *bw3*, there were 13 unroutable nets. However, architecture SAS could route all the nets. The unroutable channels were not considered in segment wastage calculation. The lower percentage segment wastage for *I* for example *bw3* may be attributed to the unroutable nets. For smaller examples both architectures produced comparable results. The last column shows the percentage improvement in critical path delays as a result of using the SAS architecture as compared to the industrial architecture. The critical path delays were computed by a post-layout RC-extractor and timing analyzer. For each design there was an improvement in routability (in which case timing improvement is not provided since measuring critical path delay for an unroutable design does not make sense) or an improvement in the critical path delay. It may also be noted that the critical path delay increases with the increase of *hfuse* usage and segment wastage as assumed earlier.

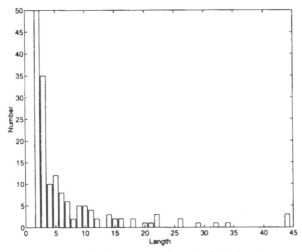

Figure 6.12 Segment length distribution

Figure 6.12 shows the statistics of SAS architecture. The x-axis represents the segment length in units of logic module width and the y-axis represents the number of segments of any particular length present in the channel. The synthesis program to generate this channel architecture with 22 tracks took about 12 hours of CPU time on an HP 425 workstation. It should be noted that architectures are generated once, and hence, a large CPU time can be justified. The segment length distribution closely follows the benchmark net length distributions used to generate the channel. There are a number of small segments and a few large segments distributed across the channel of length 44 logic modules.

6.6 CONCLUSIONS

A bounded search approach to segmented channel routing and the associated channel architectural issues have been presented. The algorithm is complete. In general, depending on the user's choice of cost, solutions with varying quality can be produced. The segmented channel routing problem is NP-complete, but our experience shows that using proper heuristics it is possible to obtain good solutions efficiently using the greedy search. The CD and LE ordering schemes for nets seems to work best for obtaining a greedy solution. The effects of channel segmentation schemes on routability and performance of a channel have also been shown. More granular segmentation produces very routable architectures, but a lot of horizontal antifuses have to be programmed which can degrade performance due to the delays introduced by programmed antifuses. The performance and routability of a channel are heavily dependent on the segmentation scheme and the distribution of nets within the channel. The correlation between the set of nets and the segments is determined using 1-segment routing, percentage of segment wastage, and the overlap factor — a higher correlation corresponds to a lower number of unroutable nets and lower segment wastage. Though simulated annealing is computation intensive, it can certainly be used for architecture development. Architectures have to be optimized only once, and hence, we can afford to spend more computer time.

The inherent nature of segmented channel routing problem lends to limited fault tolerance. If certain segments and antifuses are known to be faulty, it might be possible to get around that problem by assigning nets only to the available 'good' segments. Such faults are easily modeled and handled by the algorithm.

ACKNOWLEDGMENT

The author would like to thank Sudip Nag of Carnegie-Mellon University, Ashwin Shah, Mahesh Mehendale, and Ching-Hao Shaw of Texas Instruments for their valuable comments and suggestions.

REFERENCES

1. A.E. Gammal, J. Greene, J. Reyneri, E. Rogoyski, K. El-Ayat, and A. Mohsen, "An Architecture for Electrically Configurable Gate Array," *IEEE Journal of Solid State Circuits*, Vol. 24, No. 2, April 1989, pp. 394-398.

2. J. Birkner et al., "A Very High Speed Field Programmable Gate Array Using Metal to Metal Antifuse Programming Elements," *IEEE Custom Integrated Circuits Conference*, May 1991, pp. 1.7.1-1.7.6.

3. T. Yoshumura, and E. Kuh, "Efficient Algorithms for Channel Routing," *IEEE Trans. on CAD*, vol. CAD-1, January 1982, pp. 25-35.

4. R. Rivest, and C. Fiduccia, "A Greedy Channel Router," *Design Automation Conference*, 1990, pp. 567-572.

5. J. Green, V. Roychowdhury, S. Kaptanaglu, and A. Gammal, "Segmented Channel Routing," *Design Automation Conference*, 1990, pp. 567-572.

6. S. Brown, J. Rose, and Z. Vranesic, "A Detailed Router for Field Programmable Gate Arrays," *Intl. Conference on CAD*, 1990.

7. H.-C. Hsieh et al., "A Second Generation User-Programmable Gate Array," *IEEE Custom Integrated Circuits Conference*, 1987, pp. 515-521.

8. M. Garey, and D. Johnson, *Computers and Intractability: A Guide to the Theory of NP-Completeness*, Freeman, 1979.

9. R. Kunda, P. Narain, J. Abraham, and B. Rathi, "Speed Up of Test Generation Using High-Level Primitives," *Design Automation Conference*, 1990, pp. 594-599.

10. W. Elmore, "The Transient response of Damped Linear Networks with Particular Regard to Wideband Amplifiers," *Journal of Applied Physics*, June 1948.

11. C. Sechen, and A. Sangiovanni-Vincentelli, "The TimberWolf Placement and Routing Package," *IEEE Custom Integrated Circuits Conference*, 1984.

12. *Field Programmable Gate Array – Application Handbook*, Texas Instruments, 1992.

13. S. Nag and K. Roy, "Iterative Performance and Wirability Improvement for FPGAs," *IEEE Design Automation Conference*, 1993, pp. 321-325.

14. K. Roy and M. Mehendale, "Optimization of Channel Segmentation for Channeled Architecture FPGAs," *IEEE Custom Integrated Circuits Conf.*, 1992, pp. 4.4.1-4.4.4.

15. K. Roy, "A Bounded Search Algorithm for Segmented Channel Routing of Channeled Architecture FPGAs and Associated Channel Architecture Issues," *IEEE Trans. on Computer-Aided Design*, November 1993, pp. 1695-1705.

16. K. Roy and S. Nag, "Automatic Synthesis of FPGA Channel Architecture for Routability and Performance," *IEEE Trans. on VLSI Systems*, December 1994, pp. 508-511.

17. C. Shaw, M. Mehendale, D. Edmondson, K. Roy, M. Raghavander, D. Wilmoth, M. Harward, R. Landers, and A. Shah, "An Architecture Evaluation Framework," *ACM FPGA-92 Workshop*, Berkeley, 1992, pp. 15-20.

18. S. Kirkpatrick, C. Gellat, and M. Vecchi, "Optimization by Simulated Annealing," *Science*, Vol. 220, N. 4598, May 1983, pp. 671-680.

19. M. Huang, F. Romeo, and A. Sangiovanni-Vincentelli, "An Efficient General Cooling Schedule for Simulated Annealing," *Intl. Conf. on Computer-Aided Design*, 1986, pp. 381-384.

Chapter 7

RESTRICTED ROUTING
FOR CMOS GATE ARRAYS

Bin Zhu[1], Aiguo Lu[2] and Xinyu Wu

With the increasing commercial requirements for application-specific integrated circuits (ASICs), gate array technology is playing an important role in VLSI design. In this design methodology, the partially fabricated master slice is preprocessed, on which the devices are placed in a simple arrangement without connections so that various logic circuits can be implemented. Accordingly, only mask patterns for via holes and metal wirings need to be customized, which results in low design cost and fast turnaround time. CMOS technology occupies an important place in gate arrays, accounting for about 80 percent of the total gate array products. This is mainly due to its high speed, high integration level and low power consumption.

7.1 TECHNOLOGICAL FEATURES
OF CMOS GATE ARRAYS

An example of the master slice for CMOS gate arrays is shown in Figure 7.1.

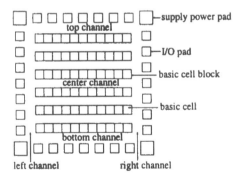

Figure 7.1 A master slice of the CMOS gate arrays

---

[1]The work was performed while visiting Lehrstuhl Prof. Dr. Günter Hotz, Department of Computer Science, University of Saarland, Germany.

[2]Currently with the Department of Electrical and Electronic Engineering, University of Bristol, United Kingdom.

237

Basic cells with equal height are placed adjacent to each other in the cell rows, called *basic cell blocks*. Two kinds of basic cells are shown in Figures 7.2a and 7.2b, the former with six pairs of transistors, and the latter with four pairs of transistors. Generally, two pairs of transistors are defined as a *standard gate*. Therefore, there are three and two standard gates in Figures 7.2a and 7.2b, respectively. In order to implement various sizes of logic circuits, a family of master slices is usually prefabricated.

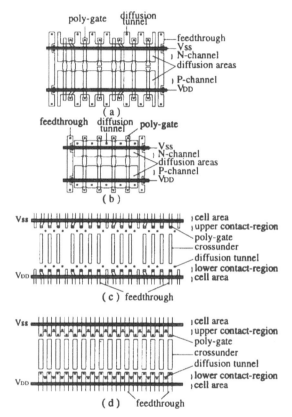

Figure 7.2 Routing channel with fixed width (a) A basic cell consisting of six pairs of transistors (three standard gates); (b) A basic cell consisting of four pairs of transistors (two standard gates); (c) A portion of the channel corresponding to the basic cells of six pairs of transistors; (d) A portion of the channel corresponding to the basic cells of four pairs of transistors.

After the initial processing steps, the basic cells are left unconnected. Like in the standard cell mode, logic function cells are described in the cell library. To implement a logic function, one or several basic cells are specified by *placement* so that a function cell (or macro cell) may be formed. These cells are connected by assigning the appropriate wire patterns from the technology library, and then form the desired circuit type. Figure 7.3 gives examples of implementing logic functions.

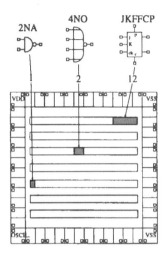

Figure 7.3 Examples of implementing logic functions

Between the two adjacent basic cell blocks is the routing channel with the fixed width. The routing channels corresponding to Figures 7.2a and 7.2b are shown in Figures 7.2c and 7.2d, respectively. The I/O pads are placed all around. The supply power pads are generally placed in corners. An example of suppling power with comb-structure metal nets is given in Figure 7.4.

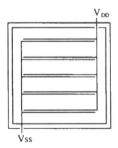

Figure 7.4 An example of suppling power with comb-structure metal nets

One or two metal layers and a fixed polysilicon layer can be used for routing. Because of its low design cost and short design cycle, one-layer metal routing is still in wide use. The channel model with a single-layer metal mask is called the *one-and-half layer channel model*. The "one layer" means the metal layer, and the "half layer" represents polysilicon bars with a fixed length and distance. These polysilicon bars are pre-fabricated on the chip used as the vertical routing resources for transiting the intersections of the nets, simply called *crossunders*. Vias are used to connect crossunders and metal segments. In this chapter, one-and-half layer routing for CMOS gate arrays will be addressed.

Different master slices possess different arrangements of crossunders. In Figure 7.5, an example of the crossunder arrangement and routing result is given [18].

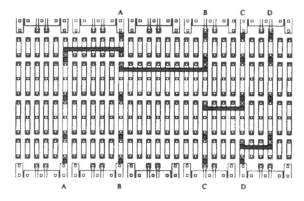

Figure 7.5 An example preplaced four crossunders in a column

There are four crossunders in a column, designated from section 1 to 4. Sections 1 and 4 have two tracks, and sections 2 and 3 have four tracks. So, there are 12 tracks in total. Because of the high resistance of polysilicon, the length of crossunders used in routing should be as short as possible. Although this kind of crossunder construction provides a flexible routing structure, many vias must be used for connections. More vias used will deteriorate the design's performance and reliability. Moreover, the automatic routing for this kind of construction is also very difficult. Therefore, the construction with single crossunder in a column is widely used. Figures 7.2c and 7.2d gives two examples of single crossunder construction. In Figure 7.2c, every three or four crossunders forms a group in which the distance between two adjacent crossunders is two grid units, whereas the distance between the crossunders of adjacent groups is three grid units. In Figure 7.2d, the distance between any two adjacent crossunders is two grid units.

The contacts along each horizontal side of a channel are usually not located at the same row. They change from the poly-gates/feed-throughs to the diffusion tunnels alternatively. For the different technologies, the contacts are arranged differently. For example, in Figure 7.2c, the long contacts and short contacts correspond to the contacts of the diffusion tunnels and poly-gates/feed-throughs, whereas in Figure 7.2d they are arranged oppositely.

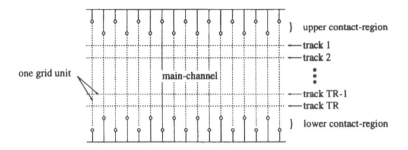

Figure 7.6 Indications for the main-channel and the contact-regions

The region in which the contacts are located is called the *contact-region*. The *main-channel* is composed of the tracks between the top and bottom tracks where the long contacts are located. These tracks in the main-channel are the horizontal routing resources. The number of the tracks is called the *capacity* of the main-channel, denoted by C_a. In general, the larger the size of the gate array chip, the more the number of tracks in the main-channel will be. Figure 7.6 demonstrates the partition of the main-channel and the contact-regions.

7.2 REVIEW OF ROUTING

The routing process is traditionally divided two steps: global routing and channel routing (or detailed routing). However, there are special routing procedures for CMOS gate arrays. Since the contacts along each side of a channel are not located at the same horizontal row, a "contact-region routing" approach is adopted to make full use of all valuable resources. Therefore, the conventional channel routing is divided into two stages: contact-region routing and main-channel routing, and the routing procedure consists of global routing, contact-region routing and main-channel routing.

7.2.1 REVIEW OF GLOBAL ROUTING

Global routing, or loose routing, is the preliminary step of the complete routing process. A global routing approach is essential for a fully automatic design tool to guarantee the 100% routability of all nets. Because, for CMOS gate arrays, the channel capacity is predetermined by the size of the master slice, the main goal of 100% routability can be achieved only if the number of subnets assigned to the channel during global routing does not exceed its given capacity. In addition to routability, the aim of global routing also includes, for example, minimizing the wire length for a minimal chip area or a minimal circuit delay.

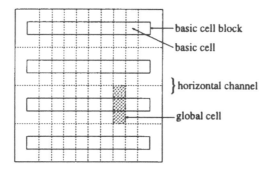

Figure 7.7 Global routing environment

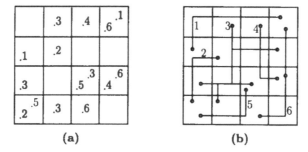

Figure 7.8 An example of global routing

In global routing, the whole gate array chip is usually divided into global cells, as shown in Figure 7.7. The advantage of using global cells is the reduction of problem complexity. The interconnections and exact pin locations within each global cell are ignored. For each global cell, an estimate is made of the number of wires that are allowed to pass through each of its four boundaries, which is called the boundary capacity. An example with 4 × 4 global cells is shown in Figure 7.8a, in which the capacity of each boundary is equal to 1.

The pins in each global cell, designated by the number of nets, are also shown. Pins specified by the same number represent the terminals of the same net and are to be connected. The global routing problem is to determine the path for each net to go through the global cells without violating the capacity requirement at each boundary. A solution to this example is given in Figure 7.8b.

Over the past two decades, many research achievements have been made on global routing of gate arrays. Various techniques have been proposed, such as hierarchical routing, rerouting, simulated annealing, and constructive routing. In [9], a detailed review of the traditional global routing approaches was given. In the following, several recent global routing algorithms for CMOS gate arrays will be introduced.

R. Nair [12] uses the iterative *maze-running* technique to rip up and reroute every net. The advantage of rerouting every net can divert the nets passing through noncongested areas to pass through even less congested areas so as to make rooms for nets in adjacent congested areas. The main innovations of this algorithm are the incorporation of costs on boundaries of cells in addition to the cells themselves, and the inclusion of via costs.

K. Winter and A. Mlynski [18] utilize the hierarchical routing technique to achieve quasi-parallel embedding of all nets. The main procedure consists of the six steps: (1) net modeling for all nets independently; (2) density estimation and routability test; (3) net ordering; (4) hierarchical net partitioning; (5) net decomposition into single channel subnets; (6) feed-through assignment.

C. Chiang et al. [5] provide a two-step greedy approach, which first orders all nets in terms of net length and then assigns the nets one by one on the basis of a Steiner tree with maximum-weight edge minimized. In the first step, a distinct number is assigned to each net. The nets that have the lower order number will be routed first and intuitively will be shorter. The second step is formulated as a graph problem: to obtain a Steiner tree on a weighted graph, whose maximum-weight edge is minimized over all Steiner trees (called Steiner min-max tree). The weight of an edge is a function of current density and capacity and measures *crowdedness* of a border. Each vertex is labeled with *demand* or (potential) *Steiner* depending on whether it is a terminal of the net currently being processed. A Steiner min-max tree dictates a global routing that minimizes traffic in the densest channel.

J. Cong et al. [6] presented a performance-driven global routing algorithm. It constructs a spanning tree with radius $(1 + \varepsilon)R$, where R is the minimum possible tree radius and ε is a non-negative user specified parameter. Since the total wire cost produced by the above strategy can be worse than optimal, another tradeoff method is proposed, which is based on a provably good algorithm that minimizes both maximum delay and total wire length. Given a parameter ε and a set of terminals, the proposed method produces a routing tree with radius at most $(1 + \varepsilon)R$ and with total cost at most $(1+(2/\varepsilon))R$ times the minimum spanning tree cost. In this way, both the total wire length and the maximum delay of the routing are simultaneously bounded by *constant* factors. The whole global routing algorithm works as follows: All net are routed one by one according to their properties. For each net, a bounded-radius minimum spanning tree is constructed. The cost of each edge in the routing graph is a function of wire lengths, channel capacities, and the distribution of current channel densities. After a net is routed, the edge costs in the routing graph are updated. Repeat this process until all nets are routed.

A global routing algorithm based on the analysis of global and local routing resources is proposed by A. Lu et al. [11], which will be discussed in Section 7.3.

7.2.2 REVIEW OF CONTACT-REGION ROUTING

Contact-region routing was first studied by J. Song and Y. Chen. In [14], they proposed a "side-channel routing" and presented an optimal algorithm, whose complexity of computation is $O(n)$ for single contact-region routing.

The contact-region routing is quite different from the usual channel routing, in which only the partial connections of some nets are possible. Therefore, it is hard to evaluate the contact-region routing before the main-channel routing is completed. What is expected here is to provide a good starting point for the main-channel routing based on its requirement.

After the global routing, it is expected that no overflow columns exist in the channel. If not, then it is impossible to achieve 100% routability for this design. On the other hand, even if there are not overflow columns, some nets cannot be routed in the main-channel since routing constraints and constraint cycles may not be solved within the current fixed channel densities. Therefore, one of the goals of the contact-region routing is to reduce the main-channel density, such as Song and Chen's algorithm[14].

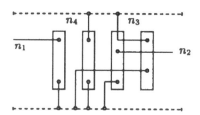

Figure 7.9 An example of using extra horizontal resources for routing

In addition to that, if there are many pins located in one area, the main-channel router may force some nets to utilize extra horizontal resources. This is because some nets in that area can not obtain the vertical resources (crossunders). An example is shown in Figure 7.9, in which the net n_3 uses some extra horizontal resources for connection.

When no extra horizontal resources are available, the dead nets are unavoidable. Therefore, the main-channel router prefers that each net terminal locates at the position where there exists one vertical resource in the channel, and that all pins are evenly placed on the two horizontal boundaries of the main-channel. This can be achieved using the algorithm proposed by A. Lu et al. [11], which reassigns the net terminals in terms of the contact-region resources so that they are evenly located.

7.2.3 REVIEW OF MAIN-CHANNEL ROUTING

Main-channel routing for CMOS gate arrays is similar to the routing in a regular rectangle area. However, it also has some distinctive features. Because the width of the routing channel is fixed, and the fixed-length crossunders used for transit through the intersections of the nets are preplaced, there are severe restrictions in using both horizontal and vertical routing resources in the main-channel routing. The channel routing with 100% routability for this design mode is much more difficult than that for the other modes.

Modern automatic routing methods can be classified as "net-oriented routing" and "area-oriented routing". The net-oriented routers, such as Lee-Moore type "maze running router" and "line search router", work in "a connection-at-a-time" mode whose goal is simply to route the current connection. The maze router and line search router have been put to practical use in gate array's design mode. For example, Hightower's router applies a maze running to gate arrays and can still give a satisfactory routing result. Since the solutions depend on an order of the connections in routing, either an unwise ordering or unforeseen wire patterns could cause routing failure. So the net-oriented routing is not a good method for high-density chips.

The method of *hierarchical routing* was introduced by A. Hashimoto and J. Stevens in 1971 [7]. The hierarchical routing divides whole routing area into several regular channels. The nets are first assigned to every channel using global routing, then the nets are connected in each channel so as to realize the interconnections of nets in whole chip. This is exactly the concept of "area-oriented routing". Many effective approaches of channel routing have been presented, such as "dogleg algorithm", "merging and matching algorithm", "hierarchical routing algorithm" and "greedy router" and so on. These approaches have been successfully applied to standard cell's design mode. Some of them have been applied to gate arrays' mode as well since both have similar constructions. However, the design results are not satisfactory. The complete rates are lower and there are many nets rerouted by user-router interaction. The problem occurs because there are some essential distinctions between them, and that the optimal goals of both modes are totally different.

For standard cell mode, the optimal goals are minimizing the chip's area and minimizing the net's length. Insofar as the present efficient channel routing algorithms are concerned, the number of tracks in a channel is not constrained. When transit is performed through the intersections of the nets, layers can be changed and via holes can be drilled everywhere, if necessary. However, for gate arrays, the optimal goal is the complete rate as high as possible because the chip area has been fixed, the number of horizontal tracks in a channel is constrained, and the positions of vertical routing resources are preplaced. Hence, the aforementioned methods are not suitable for channel routing of the CMOS gate arrays.

For channel routing of the CMOS gate arrays, two new algorithms have been suggested [14, 19]. Both algorithms have proposed some efficient methods for one-and-half layer channel routing in accordance with the CMOS gate arrays' specific characters. In [14], Song and Chen have presented a greedy approach with routability prediction. The approach reasonably uses horizontal routing resources by ordering nets and predicting tracks for nets in a zone. Then, it performs a parallel routing for nets, which adjusts net orders by means of the local "greedy" strategies for the nets to compete for crossunders.

Since the prefabricated vertical routing resources (crossunders) force the intersections at the columns without vertical routing resources to transfer to the nearby columns with crossunders, the intersections among nets will increase the horizontal routing density, and lead to overoccupied horizontal resources (see Figure 7.9). Therefore, the transit through the intersections of the nets in a channel is performed at the expense of increasing horizontal routing density. Furthermore, even if the vertical routing resources are more than the number of intersection columns, an intersection column may only use the adjacent vertical resources for transit through the intersections because the horizontal resources are restricted. The adjacent intersection columns compete directly for the vertical routing resources. The denser the intersection columns, the sharper the competition. Therefore, the number of intersections and the number of intersection columns should be as small as possible, and the intersection columns should be evenly distributed, so that the requirements for the horizontal and vertical resources are as few as possible and the total length of crossunders used in routing is as short as possible. Hence, how to reasonably use the finite horizontal and vertical routing resources becomes the key in raising the completion rate. If the two routing resources could be reasonably assigned before routing, the waste of these limited resources can be reduced, and the blind routing can be avoided, thereby the completion rate can be greatly improved. Based on the above, B. Zhu et al. proposed an algorithm CRABAR [19], **C**hannel **R**outing **A**lgorithm **B**ased on **A**ssigning **R**esources, which has better adaptability and higher completion rate than the algorithms given previously, and has fairly good solved the problem of channel routing for CMOS gate arrays. In Section 7.5.2, this algorithm will be detailedly introduced.

7.3 GLOBAL ROUTING BASED ON THE ANALYSIS OF ROUTING RESOURCES

The basic goal of global routing is to assign nets properly so that all nets can be routed in main-channel routing. To achieve this goal, the global routing must be able to evaluate the routing resources and net densities as exactly as possible. If the global routing model based on global cells is used, the precise locations of pins within the global cell are unknown. This prevents the global routing from evaluating the available routing resources exactly and assigning nets properly.

For the global router discussed below, the global routing model is based on the real master slice. To do so, more information must be taken into account, but it gives a more precise evaluation of routing resources and pin locations.

The global routing algorithm consists of two phases: (1) initial global routing using vertical Steiner tree net; (2) rerouting in terms of the analysis of routing resources.

7.3.1 INITIAL GLOBAL ROUTING

To alleviate the influence of net ordering in the initial global routing, all nets are first assigned using a specific net model. In this way, the terminal locations of a net can easily determine the path of the net. In the gate array master slice, the vertical routing resources for global routing are electrically equivalent pins, feed-throughs, and tracks in left and right vertical routing channels. Because they are usually much less than the horizontal routing resources, the vertical Steiner net model is used in this phase. The main task of the initial global routing is to find a route for each net. The net model used here is called *vertical Steiner tree net*.

Definition 7.1 A vertical Steiner tree net is a subset of Steiner tree nets, the horizontal cut of which is always equal to 1.

An example of a vertical Steiner tree net with 5 terminals is shown in Figure 7.10. Apparently, the vertical Steiner tree net utilizes the minimal vertical routing resources, and thus Proposition 7.1 can be proven.

Proposition 7.1 *The necessary condition that the routing can been successfully completed is that each net can obtain its required vertical routing resources in terms of vertical Steiner tree net.*

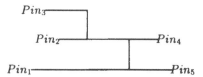

Figure 7.10 A vertical Steiner tree net

Therefore, if initial global routing only makes sure that all nets get their own vertical routing resources and ignores the limitations of horizontal routing resources (tracks available in the channels), the problem of initial global routing can be greatly simplified.

The initial global routing used is very simple but time-efficient, assigning one net at a time using vertical Steiner tree net so that this net takes minimal length. If the initial global routing cannot find vertical routing resources for some nets, another placement strategy has to be used to get a new placement solution. Otherwise, it is impossible to achieve 100% routability.

7.3.2 REROUTING

After initial global routing, all nets have been assigned. However, there may exist some regions in which the horizontal routing resources are less than the number of nets passing through. Therefore, in this phase, a rerouting process is invoked to adjust some nets so that each net can obtain its own horizontal routing resources.

Definition 7.2 A block is defined as a row of logic cells on the chip. *Tgridnu* is defined as the number of horizontal grids in one block. *Chnu* is the number of channels. *Tranu(i)* is defined as the number of tracks in the ith channel.

Definition 7.3 Net Densities:

(1) *Channel density* – Taking a vertical cut at the jth grid column in the channel i, the number of nets crossing the cut (including those nets ending at or starting from the cut line) is defined as the density of column j in channel i, denoted by $CD(i,j)$, $i \in \{1, 2, \ldots, Chnu\}$ and $j \in \{1, 2, \ldots, Tgridnu\}$.

(2) *Vertical density* – The vertical density at the column j is defined as the sum of the densities of each channel at this column, denoted by $VCD(j)$. Thus

$$VCD(j) = \sum_{i=1}^{Chnu} CD(i,j) \tag{7.1}$$

Definition 7.4 Overflow Columns:

(1) *Channel overflow column* – If $CD(i,j) > Tranu(i)$, the jth column in channel i is defined as a channel overflow column. The set of all overflow columns in channel i is denoted by $SOF(i)$. The overflows at the jth column of channel i is represented by $VOF(i,j)$, where

$$VOF(i,j) = CD(i,j) - Tranu(i) \qquad (7.2)$$

(2) *Vertical overflow column* – If $VCD(j) > \sum_{i=1}^{Chnu} Tranu(i)$, the jth column is defined as a vertical overflow column. The set of all vertical overflow columns is denoted by $SVOF$. The overflows at the jth column is represented by $VVOF(j)$. Thus,

$$VVOF(j) = VCD(j) - \sum_{i=1}^{Chnu} Tranu(i) \qquad (7.3)$$

7.3.2.1 REROUTING BASED ON THE ANALYSIS OF GLOBAL ROUTING RESOURCES

In accordance with Definitions 7.2 to 7.4, the Proposition 7.2 can be proven.

Proposition 7.2 *The necessary condition that all nets are able to get their required horizontal routing resources is*

$$VCD(j) \leq \sum_{i=1}^{Chnu} Tranu(i) \qquad (7.4)$$

for $\forall j \in \{1, 2, \ldots, Tgridnu\}$.

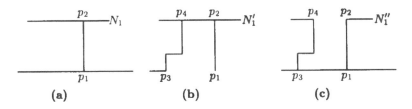

Figure 7.11 Overlap net and its adjustable nets

According to Proposition 7.2, it is essential to eliminate all vertical overflow columns for the purpose of achieving 100% routability. Because all the nets at the moment are the vertical Steiner tree nets, these nets utilize the maximum horizontal routing resources but the minimal vertical routing resources. Therefore it is possible to replace some horizontal net segments using vertical routing resources. An example is shown in Figure 7.11. Figure 7.11a shows a net N_1 before rerouting. If there exist some vertical routing resources (e.g., net segment p_4p_3), N_1 can be equivalently replaced by the net N_1' (as shown in Figure 7.11b) or the net N_1'' (as shown in Figure 7.11c).

Definition 7.5 If a net passes through more than two channels and its horizontal projections of two or more horizontal subnets are overlapped, such a net is defined as the horizontal overlap net. If the overlap net can obtain its required vertical routing resources to replace some horizontal subnets, such a net is defined as an adjustable overlap net. The net shown in Figure 7.11a is an example of a horizontal overlap net.

Therefore, the problem of eliminating the vertical overflow column becomes to find a set of adjustable overlap nets. For an adjustable overlap net, there exist two adjustable subnets (e.g., p_3p_1 in Figure 7.11b and p_4p_2 is Figure 7.11c). The adjustable region D is $D = D_u \cap D_l$, where D_u is the replaceable region in the upper channel of the adjustable overlap net and D_l is the replaceable region in the lower channel. For the example shown in Figure 7.11, $D_u = (p_4, p_2)$, and $D_l = (p_3, p_1)$. Let $S1 = MAX\{CD(i_1, j)|j \in D_u\}$, and $S2 = MAX\{CD(i_2, j)|j \in D_l\}$, where channel i_1 is the upper channel and channel i_2 is the lower channel. In order to determine the subnet to be replaced, the following two rules are used:

(1) If $S1 > S2$ choose D_u, If $S1 < S2$, choose D_l.
(2) If $S1 = S2$ and the length of D_u is larger than the length of D_l, choose D_u, otherwise choose D_l.

To minimize the wire length in the rerouting process, three models are used when forming the adjustable overlap nets as shown in Figure 7.12. Model (a) does not use extra horizontal routing resources, model (b) uses one extra horizontal routing resource, and model (c) uses two extra horizontal resources in both channels. In the cases of model (b) and (c), the precondition of using extra horizontal resources is that it does not cause new vertical overflow columns so as to make the algorithm convergent.

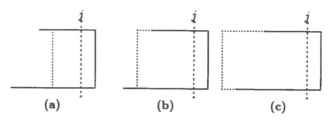

Figure 7.12 Three models of the adjustable overlap nets

Because each adjustable net must use extra vertical routing resources, it is very important to reasonably allocate the limited vertical routing resources to each horizontal overlap net. In order to avoid the net ordering in rerouting process and make the horizontal overlap nets compete the vertical routing resources optimally, several adjustable overlap nets are allowed to use the same vertical routing resource before rerouting. Which one finally uses this resource is determined by finding a minimal cost cover on a dynamic bipartite graph.

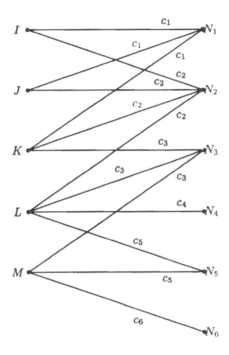

Figure 7.13 Bipartite graph

A bipartite graph is shown in Figure 7.13, on which the left side nodes represent vertical overflow columns and the right side nodes represent the adjustable overlap nets. If, and only if, the adjustable region of a net covers the overflow column, there exists an edge between two corresponding nodes. As shown in Figure 7.13, there are three edges linking nodes $N1$ and I, nodes $N1$ and J, and nodes $N1$ and K, respectively. This means that the adjustable region of $N1$ covers vertical overflow columns I, J, and K. When $N1$ is rerouted, $VVOF(I)$, $VVOF(J)$ and $VVOF(K)$ will all be decreased by 1. The node $N4$ only links to node L, which means that the adjustable region of net $N4$ only covers the overflow column L. Because whether the rerouting process can be successfully completed or not is highly dependent on whether there are enough vertical routing resources, it is intuitively helpful to save vertical routing resources. So, the weight of each edge measures the extra vertical routing resources used by the adjustable net (the cost of the adjustable net), which is equal to the number of blocks between the upper channel and lower channel of the adjustable net being considered. The problem of assigning vertical resources for rerouting becomes that of finding the minimal cost cover of the bipartite graph. The times of covering each left side node are equal to the overflows of the corresponding overflow column. Since the problem of the minimal cost cover on a dynamic bipartite graph is NP-complete, a heuristic algorithm is used here.

Let $ADJNET(j) = \{N_{1j}, N_{2j}, \ldots, N_{mj}\}$ represents the set of adjustable overlap nets whose adjustable regions cover the column j. A set of overflow columns is selected out if they satisfy: (1) column(s) with maximum $VVOF(j)$; (2) column(s) with minimum $|ADJNET(j)|$; (3) column(s) with minimum $(|ADJNET(j)| - VVOF(j))$. Apparently, all these selected overflow columns are more difficult to eliminate than other columns. Therefore, these columns are treated prior to others in the rerouting process so that they have the priority to use vertical routing resources. Thus, the problem of finding minimal cost cover on a dynamic bipartite graph can be approximately translated to a problem of finding a min-cost matching. The set of nodes to be matched was selected above. The min-cost maximum flows described in [9] can be used to find minimal cost matching. By finding a mini-cost matching, a set of adjustable overlap nets can be determined for rerouting. After rerouting, the bipartite graph and the $VVOF(j)$ $(j \in VOF)$ are updated. If for some $j_1 \in VOF$, $VVOF(j_1) \leq 0$, then j_1 can be eliminated from VOF. If for some adjustable nets whose vertical resources have been used up in previous rerouting, these adjustable nets have to be removed from the modified bipartite graph. The above process is repeated until $VOF = \phi$.

If $VOF \neq \phi$ after the whole rerouting process, either another placement strategy or user's interaction has to be invoked to improve the design. If $VOF = \phi$, in order to evenly utilize the routing resources on the whole master slice, an integer parameter $slack$ is introduced for further rerouting some nets based on the global analysis. In this case, if column j satisfies

$$VCD(j) > \sum_{i=1}^{Chnu} Tranu(i) - slack \qquad (7.5)$$

j is also treated as an *overflow column*. The above rerouting process will be used on these columns. Thus, as $slack$ increases, the difference of $\{VCD(j)|j \in \{1, 2, \ldots, Tgridnu\}\}$ will decrease, resulting in an even allocation of nets.

7.3.2.2 REROUTING BASED ON THE ANALYSIS
OF LOCAL ROUTING RESOURCES

After rerouting based on the global analysis, it is assumed that for $\forall j$, $VVOF(j) \leq 0$. However, it does not always mean that $VOF(i,j) \leq 0$, $i \in \{1,2,\ldots,Chnu\}$ and $j \in \{1,2,\ldots,Tgridnu\}$. This implies even if there are no vertical overflow columns, there may still exist some channel overflow columns. Thus, Proposition 7.3 can be proven and it tells how to eliminate channel overflow columns.

Proposition 7.3 *If for $\forall j$, $VVOF(j) \leq 0$; and for a specific channel i, $VOF(i,j) > 0$, there must exist another channel i_1 satisfying*

$$VOF(i_1,j) < 0 \tag{7.6}$$

$i_1 \in \{1,2,\ldots,i-1,i+1,\ldots,Chnu\}$.

Based on Proposition 7.3, it is possible to eliminate the overflow column j in channel i just by assigning nets originally in channel i to channel i_1. Here, the rerouting is processed channel by channel starting from the top and bottom channels simultaneously and ending at the middle channel (it is assumed that the top channel is channel 1 and the bottom channel is channel $Chnu$).

Definition 7.6 (*Upward adjustable nets*) Suppose column j in channel i is a channel overflow column and the net N_1 passes through the column j in channel i. If, in the $(i-1)$th block, there exist two vertical routing resources in both sides of column j and the net N_1 can use these two resources without causing new overflow columns in channel i, N_1 is defined as an upward adjustable net.

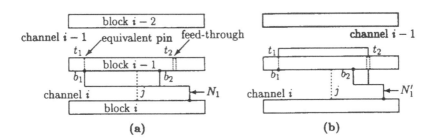

Figure 7.14 Upward adjustable nets

Figure 7.14a shows an example of an upward adjustable net, in which t_1 is an electrically equivalent pin of pin b_1, and t_2 is a feed-through. As shown in Figure 7.14b, if this net is rerouted using t_1 and t_2, N_1 can be equivalently replaced by N_1', resulting in a reduction of $CD(i,j)$.

Definition 7.7 (*Downward adjustable nets*) Suppose column j in channel i is a channel overflow column and the net N_1 passes through column j in channel i. If, in the ith block, there exist two vertical routing resources in both sides of column j and N_1 can use these two resources without causing new overflow columns in channel i, N_1 is referred to as a downward adjustable net.

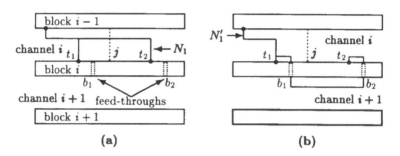

(a)   (b)

Figure 7.15 Downward adjustable net

Figure 7.15a shows an example of a downward adjustable net, in which b_1 and b_2 are two feed-throughs in both sides of column j. If N_1 is rerouted using b_1 and b_2, it can be equivalently replaced by N_1', as shown in Figure 7.15b, resulting in the reduction of overflows in column j. In this example, the prerequisite for using the feed-through b_2 is that $CD(i, j_1) < Tranu(i), j_1 \in (t_2, b_2]$. Otherwise, the rerouted net will cause the new overflow columns in channel i.

According to the above two definitions, both upward and downward adjustable nets related to a channel overflow column can be formed. Since a net can be an upward adjustable net or a downward adjustable net or both, which kind of adjustable nets is used to eliminate the overflow column j is determined based on the following two rules.

Rule 1: If the channel number i satisfies

$$i \le INT[(Chnu + 1)/2] \tag{7.7}$$

upward adjustable nets are considered firstly. At this stage, whether the adjustable net can be rerouted or not is conditional. The prerequisite for rerouting is $CD(i - 1, j_1) < Tranu(i - 1), j_1 \in [t_1, t_2]$. After assigning a net segment in channel i to channel $(i - 1)$, the channel densities in channel $(i - 1)$ and channel i are updated as follows:

$$CD(i - 1, j_1) = CD(i - 1, j_1) + 1, j_1 \in [t_1, t_2] \tag{7.8}$$

$$CD(i, j_1) = CD(i, j_1) - 1, j_1 \in [MAX(t_1, b_1) + 1, MIN(t_2, b_2) - 1] \tag{7.9}$$

After this, another upward adjustable net is tried. Secondly, if there still exist overflow columns in channel i after having tried all upward adjustable nets, downward adjustable nets are used to perform unconditional rerouting. The downward adjustable nets are assigned to the lower channel $(i + 1)$ until all overflow columns in channel i are eliminated or all downward adjustable nets have been rerouted.

Rule 2: If the channel number i satisfies

$$i > INT[(Chnu + 1)/2] \tag{7.10}$$

downward adjustable nets are considered firstly. At this stage, whether the adjustable net can be rerouted or not is conditional. The prerequisite for rerouting is $CD(i+1, j_1) < Tranu(i+1), j_1 \in [b_1, b_2]$. After adjusting a net segment in channel i to channel $(i + 1)$, the channel densities in channel i and channel $(i + 1)$ are updated as follows:

$$CD(i + 1, j_1) = CD(i + 1, j_1) + 1, j_1 \in [b_1, b_2] \tag{7.11}$$

$$CD(i, j_1) = CD(i, j_1) - 1, j_1 \in [MAX(t_1, b_1) + 1, MIN(t_2, b_2) - 1] \tag{7.12}$$

After this, another downward adjustable net is tried. Secondly, if there still exist overflow columns in channel i after having tried all downward adjustable nets, upward adjustable nets are used to perform unconditional rerouting. The upward adjustable nets are assigned to the upper channel $(i - 1)$ until all overflow columns in channel i are eliminated or all upward adjustable nets have been rerouted.

Since there may be more than one upward or downward adjustable nets associated with an overflow column, it is important to select the most suitable ones from the adjustable net set. Like the rerouting process for eliminating vertical overflow columns, a set of adjustable nets are rerouted simultaneously in order to avoid the net ordering. The approach used here is also based on the minimal cost cover of a bipartite graph. The left side nodes of the bipartite graph represent the channel overflow columns, whereas the right side nodes represent the upward adjustable nets (if they are being rerouted upwards) or downward adjustable nets (if they are being rerouted downwards). If, and only if, an adjustable net corresponds to an overflow column, there exists an edge between these two nodes. Because each adjustable net always uses two vertical routing resources and which kind of vertical routing resources is used to form adjustable nets can be controlled by giving the different priorities to different routing resources (in the algorithm, electrically equivalent pins are prior to feed-throughs, and feed-throughs are prior to left side and right side vertical routing channels), the information about the distribution of channel density is extracted and used as the cost of adjustable nets.

Definition 7.8 A column j in channel i is defined as a *crisis column* if $CD(i, j) \geq Tranu(i)$ (for unconditional rerouting) or if $CD(i, j) \geq Tranu(i) - 1$ (for conditional rerouting).

Therefore, the weight of each edge (associated with each adjustable net) on the bipartite graph is equal to the number of crisis columns covered by the net segment assigned to the upper or lower channel.

Example: Figure 7.16 shows an example of selecting the adjustable nets based on the cost of nets. Suppose the conditional upward rerouting is being performed and columns j_1 and j_2 are two overflow columns. In the upper channel (i), there are four crisis columns a, b, c, and d. If N_1 is rerouted using t_1 and t_{11}, it will cover four crisis columns of channel i. Therefore, the cost of N_1 is 4. The cost of N_2 is 1 and the cost of N_3 is also 1. So, N_2 and N_3 will be used for rerouting using t_2, t_{22} and t_3, t_{33} respectively. Actually, if N_1 is rerouted first, N_3 can never be used for rerouting because $CD(i, d) = Tranu(i)$.   ◇

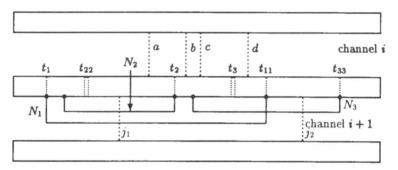

Figure 7.16 Select the adjustable nets based on the net cost

Thus, like the rerouting process for eliminating vertical overflow columns, the problem of determining a set of adjustable nets becomes that of finding a minimal cost cover of the bipartite graph. After the adjustable nets are determined, the rerouting process is performed and the channel densities are updated. Repeat the process until all overflow columns are eliminated or no more adjustable nets are available.

7.4 CONTACT-REGION ROUTING

Generally, there are two tracks in a contact-region, one adjacent to the cell area (or basic cell block), called *near cell track*, and the other adjacent to the main-channel, called *near channel track*. In the following, the attention is paid to routing in the two track contact-region.

In the contact-region routing, the upper contact-region and the lower contact-region are considered as routing areas, and only one metal layer can be used for routing. If no barriers, such as pins or routed nets, exist in a certain interval of a contact-region, some of nets can be partially or completely routed in that interval.

7.4.1 THE APPROACH
AIMED AT MINIMIZING DENSITY

The approach aimed at minimizing density was proposed by Song and Chen [14]. Since the maximum density of nets is a key factor influencing the main-channel routing, the contact-region routing should be able to reduce the density. Thus, one of the goal for the contact-region routing is specified by reducing the density at the columns of high density and filling up the contact-region as much as possible. These requirements will be met by finding a maximum-weight routing.

7.4.1.1 ZONE REPRESENTATION

Let *stripe i* be the area between columns $i - 1$ and i of the channel, and $S(i)$ be the set of nets whose horizontal segments pass through *stripe i*. Since horizontal segments of different nets cannot be overlapped, the horizontal segments of any two nets in $S(i)$ must be routed on the different tracks. However, it is easy to see that only those $S(i)$ that

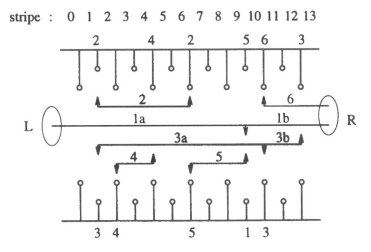

Figure 7.17 The distribution of net's terminals for routing requirement. (Upper-contact-region-routable nets are 1b, 2, 3b, 4, 5, and 6; Lower-contact-region-routable nets are the same.)

are not subsets of another set need to be considered. To achieve this, zones, at which $S(i)$ are maximal, are assigned to the stripes in the sequential number. For the example shown in Figure 7.17, Table 7.1 gives its *stripes*, $S(i)$, and its corresponding zones. Obviously, $S(1)$ and $S(2)$ need not to be considered because the horizontal constraints related to them are included in those of $S(3)$. The zone representation for this example is shown in Figure 7.18.

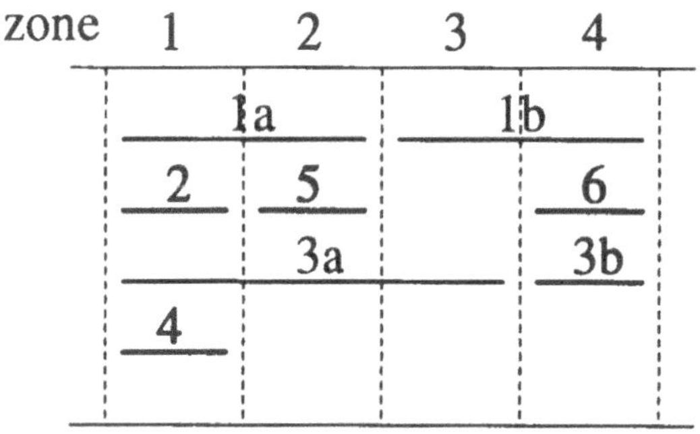

Figure 7.18 Zone representation for the example of Figure 7.17

Table 7.1 *strips*, $S(i)$, and zones

stripe	S(i)				zone
0	1a				
1	1a				
2	1a	2	3a		
3	1a	2	3a	4	1
4	1a	2	3a	4	
5	1a	2	3a		
6	1a	2	3a		
7	1a	3a	5		
8	1a	3a	5		2
9	1a	3a	5		
10	1b	3a			3
11	1b	3b	6		
12	1b	3b	6		4
13	1b	6			

The difference between this zone representation and general zone representation is the use of the term *stripe* instead of *column*, so that there is a chance to route the two nets on the same track of a contact-region in which one starts and the other ends at the same column, e.g., net 2 and net 5 in Figure 7.17 can be routed simultaneously on the same track of upper contact-region as shown in Figure 7.20.

7.4.1.2 CONTACT-REGION-ROUTABLE-NETS AND WEIGHTING STRATEGY

In this section, the basic processing units will be subnets. A *subnet* is a two-terminal net whose ends are determined by two consecutive terminals of the net. The left-terminal column of subnet i is referred to as *starting column* $sc(i)$ of that subnet, and the right-terminal column is referred to as *ending column* $ec(i)$. For the sake of generality, the term *net* will still be used instead of *subnet*. Therefore, the horizontal segment of net i is determined by its left and right terminals.

Definition 7.9 If net i is long enough and no net connects to the near cell track within the interval $(sc(i), ec(i))$, then the horizontal segment of net i can be routed on the contact-region, and net i is called a *contact-region-routable net*.

Thus, there are *upper-contact-region-routable nets*, such as net 2, 5 in Figure 7.17, and *lower-contact-region-routable nets*, such as net 1b, 2 in Figure 7.17.

For the contact-region-routable nets, it does not mean that the horizontal segment of every routable net i can always fill up its span interval $[sc(i), ec(i)]$ on the related contact-region. However, at least a part of the horizontal segment of the contact-region-routable net can be placed on the corresponding contact-region.

Generally, nets with wide spans will have the priority for routing, and so with the large weights. The more the horizontal segment passes through the high density columns, the larger its weight will be. The tie is broken according to the topologies of the nets. Here only the weights of the upper-contact-region-routable nets are discussed.

If every net terminal leading to the left or right boundary of the channel is considered to be connected to the upper contact-region, the topology of a routable net must be the types:

In order to decrease the number of vias and improve the routability of the main-channel routing, the type (a) is given the highest gains in weight, the types (b) and (c) are the second, and the type (d) has no gains or even penalties.

The situation for the lower-contact-region-routable nets is similar, and it is assumed that one contact-region is routed first and then the other.

7.4.1.3 MERGING AND EXTENDING OPERATIONS

Contact-region-routable nets in different zones can be placed on the same track of a contact-region. The procedure linking these nets is called *merging*. In order to optimally determine the merging nets, a merging tree is introduced to record all the possible mergings.

Definition 7.10 A *merging tree* is a directed tree with weight, in which each node represents a net, and a directed edge from node i to node j means that these two nets can be routed on the same track and net i must be placed on the left of net j.

The merging tree can be constructed by the following operations step by step. Let nets i and j be the routable nets on the same contact-region, but not in the same zone. Then, the operation "merging of net i and net j" is to modify the merging tree by drawing a directed edge between net i and net j, from the left net in the zone representation to the right one.

If the operation of merging is limited to the nets between the adjacent zones, the operation *extending of nets* is necessary.

Definition 7.11 If net i is a contact-region-routable net in the zone representation, the operation extending of net to zone k is to update the zone representation by extending the horizontal segment of net i into zone k.

7.4.1.4 ALGORITHM PROCEDURE

After the unroutable nets on a contact-region are specified, all nets in the zone representation are routable nets corresponding to this contact-region. When the weights of the nets are determined, the following algorithm can be used for routing. Since the routing is processed for one contact-region at a time, the algorithm is called a "single-side algorithm". The general flow of the algorithm is as follows:

> begin
>> for $z := 1$ to $zn - 1$ do
>>> $Q := \{$ nets which end at zone z $\}$;
>>> $P := \{$ nets which start at zone $z + 1$ $\}$;
>>> among Q,
>>> find m^* whose weight $(m^*) = \max \{$ weight$(m)|m \in Q$ $\}$;
>>> while $P \neq$ null do
>>>> find n^* among P;
>>>> merge m^* and n^*,
>>>> weight$(n^*) :=$ weight$(m^*)+$ weight(n^*);
>>>> $P := P - \{n^*\}$;
>>> end while;

extend m^* to zone $z + 1$;
end for;
$Q := \{$ nets which end at zone zn $\}$;
among Q,
find m^{**} whose weight $(m^{**}) = \max \{ weight(m)|m \in Q \}$;
end single-side;

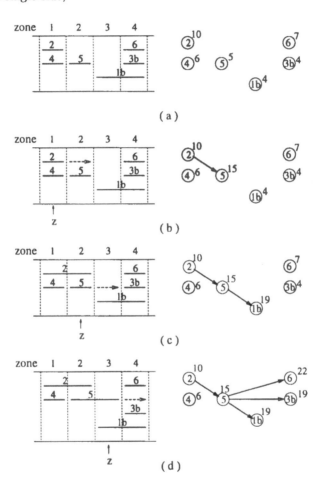

(a)

(b)

(c)

(d)

Figure 7.19 Zone representation and merging tree
(a) Initial status; (b) $z = 1$; (c) $z = 2$; (d) $z = 3$
The set $\{2, 5, 6\}$ is of a maximum weight

The set of the final mode m^{**} is the set of merged nets, and is of the maximum sum of weights. These nets can be placed on the specified track of a contact-region, because no horizontal overlap exists in the zone representation. The algorithm is still a heuristic one since the weighting strategy is locally optimal, although the "single-side algorithm" has been proved to be powerful.

Example: In Figure 7.17, nets 1b, 2, 3b, 4, 5, and 6 are upper-contact-region-routable nets. Assume that the weights of these nets have been calculated to be 4, 10, 4, 6, 5, and 7, respectively. Figure 7.19 illustrates how the set of merged nets with maximum weight on the upper-contact-region is obtained by the algorithm. Having deleted the non-upper-contact-region-routable nets, the zone representation and an initial construction of the merging tree are shown in Figure 7.19a. First, net 2 is merged with net 5 and extended to zone 2, and the weight of net 5 is modified by accumulating that of its direct ancestor. Then, net 5 and net 1b are merged, · · ·. Finally in Figure 7.19d, the set of merged nets {2, 5, 6} with maximum weight is obtained. The result of the assignment of the nets to the upper-contact-region and that of the subsequent lower-contact-region routing are shown in Figure 7.20. Therefore, the problem is simplified by reducing the density from 4 to 2.   ◇

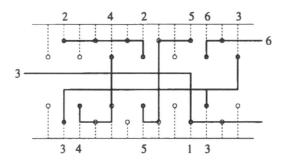

Figure 7.20 A contact-region routing realization for the requirement in Figure 7.17

7.4.2 REASSIGNING PINS EVENLY BY ROUTING IN CONTACT-REGION

The algorithm discussed above emphasizes the reduction of channel densities by routing in the contact-region. Because the area with high channel densities is usually an area with many net terminals located, the contact-region routing resources in that area must be very small, some of which are even not long enough for routing in the contact-region. This means that it is difficult to reduce channel densities for the areas with congested nets. The algorithm discussed below emphasizes the even reallocation of net terminals in the channel so that the areas with congested nets are to be reduced and, in turn, the maximal channel density is decreased.

In one track of the contact-region, a *span* = [*sl, sr*] is defined as a region between two adjacent pins. *LefnetN* is used to express the left-hand side net of a *span*, and *RignetN* is used to express the right-hand side net of a *span*. As shown in Figure 7.21, for *span*$_1$ on the near channel track at the lower boundary, *LefnetN* = n_3 and *RignetN* = n_2. If, in a contact-region, there are two or more pins (or contacts connected with nets) in successive grid columns, these pins are referred to as the continuous pins. If there are n continuous pins in one area, such an area is called the n−continuous pin area. In the lower contact-region of Figure 7.21, the rightmost two terminals of n_3 and n_5 form a 2-continuous pin area.

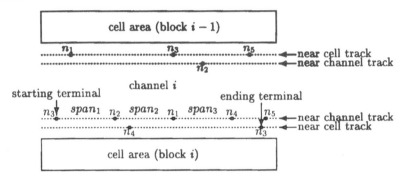

Figure 7.21 Contact-regions environment

From left to right in a channel, if pin j is the first terminal of a net N_1 in the channel, such a terminal is called the starting terminal of N_1; if pin j is the last terminal of a net N_1, such a terminal is called the ending terminal of N_1. Figure 7.21 shows the starting terminal and the ending terminal of net n_3 in channel i. If a net's two adjacent terminals are all located on one block, such a two-terminal subnet is called a *one-block subnet*; if a net's two adjacent terminals are located on two blocks, such a two-terminal subnet is called a *two-block subnet*. In Figure 7.21, the net n_4 is a *one-block net*, while the net n_1 is a *two-block net*.

According to priority, the goals of the algorithm are:

(1) Evenly reallocate net terminals (or reduce the maximal channel density) using the routing resources in the contact-region.

(2) Reassign net terminals to the columns where there exist crossunders in the main-channel.

(3) Reduce the channel densities.

The first goal is to ensure that the nets in the channel are evenly distributed so as to achieve a smaller maximal channel density. To evenly reallocate net terminals, the continuous pin areas should be eliminated. For a n-continuous pin area, if either the leftmost pin of the area is the ending terminal of a net or the rightmost pin of the area is the starting terminal of a net, then routing either the ending terminal net or the starting terminal net in the contact-region results in a $(n-1)$-continuous pin area. if both the leftmost pin of the area is the ending terminal of a net and the rightmost pin of the area is the starting terminal of a net, then routing both nets in the contact-region gives a $(n-2)$-continuous pin area.

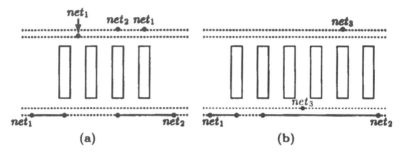

Figure 7.22 Determine the ending columns

The second goal is based on the consideration that each two-block subnet requires at least a crossunder to perform connection. If each net terminal is located at the column where there is a crossunder in the channel, this will provide a good starting point for main-channel routing. The third goal is to make sure that the routing resources in the contact-regions are fully utilized.

The routing process is performed track by track. The near cell tracks are routed first. Then all net terminals in near cell tracks are moved to the same column of the near channel tracks. After this, the near channel tracks are routed. On one track, the routing is performed from left to right.

First, find a $span = [sl, sr]$ on one track with $sr \geq sl + 2$ and determine the $LefnetN$ and $RignetN$ of the $span$. Then determine the net(s) to be routed in this $span$. Let d be the maximal channel density in the $span$ and $CD(i, j)$ be the channel density at column j of channel i. If sl is not an ending terminal of the net $LefnetN$ and one of the following four conditions is satisfied:

(1) sl is located in the continuous pin area.
(2) sl is the starting terminal of $LefnetN$.
(3) $CD(i, sl + 1) = d$, or $CD(i, sl + 2) = d$, or $CD(i, sl + 3) = d$ and there is no crossunder at column $sl + 2$.
(4) there is no crossunder at column sl.

$LefnetN$ can be routed in the $span$. If sr is not a starting terminal of the net $RignetN$ and one of the following four conditions is satisfied:

(1) sr is located in the continuous pin area.

(2) sr is the ending terminal of $RignetN$ in the channel.
(3) $CD(i, sr - 1) = d$, or $CD(i, sr - 2) = d$, or $CD(i, sr - 3) = d$ and there is no crossunder at column $sr - 2$ of channel i.
(4) there is no crossunder at column sr of the channel.

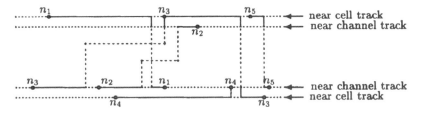

Figure 7.23 Contact-region routing results

$RignetN$ can be routed on the *span*. $LefnetN$ is routed from left to right on the *span* starting at sl and ending at the ending column. $RignetN$ is routed from right to left on the *span* starting at sr and ending at the ending column. The ending column is determined based on the following conditions:

(1) If, in the opposite contact-region of the same channel, there exist the terminals of the same net as the routing net, and suppose these terminals are in the area from sl' to sr' inclusive (if there exists only one terminal of the same net, $sl' = sr'$), the ending column must not larger than the column sl' if $LefnetN$ is being routed or must not less than sr' if $RignetN$ is being routed.

(2) There exists a crossunder at the ending column of the main channel.

(3) There are no net terminals on the adjacent columns of the ending column in the same **contact-region**.

(4) There does not exist a terminal of the different net at the ending column of the opposite contact-region.

(5) Without violating the above conditions, the net is routed which passes through as many maximal density columns as possible.

As shown in Figure 7.22a, the ending columns (the routing lengths) for net_1 and net_2 at the near cell track of the lower contact-region is determined according to conditions (1) and (2). An example of considering conditions (2) to (4), when routing nets net_1 and net_2, is shown in Figure 7.22b.

If neither $LefnetN$ nor $RignetN$ is chosen for routing, select an one-block subnet which passes through the *span* and whose two terminals are located in the same contact-region currently being concerned, and then route this net on the *span*.

For the 5 nets shown in Figure 7.21, the results after routing in the contact-regions are given in Figure 7.23, in which the wires in the contact-region are drawn using solid lines, while the wires in the main channel are drawn using broken lines.

7.5 MAIN-CHANNEL ROUTING

The main-channel routing is a critical step in routing for CMOS gate arrays. It will determine whether the routing is a final success or not. Because there are severe restrictions in using both horizontal and vertical routing resources, it becomes a key how rationally to use both of routing resources. There are two proposed methods, one being a greedy router based on the routability prediction in the practical routing [14], the other being CRABAR algorithm based on assigning the routing resources before the real routing [19].

7.5.1 A GREEDY ROUTER WITH ROUTABILITY PREDICTION

The main phases of the router are:

(1) In a zone, order the nets which pass through or end within the zone. The aim of the ordering is to determine an optimal sequence of the nets in the current zone. An ideal order is used as a guidance in order adjusting.

(2) Predict target tracks for the steady nets and the inserting nets, and the inserting nets are connected to their predicted tracks. A $p-$level prediction is made, which means a trial wiring for the next p zones on the assumption that there are enough crossunders in these zones.

(3) Make a parallel routing for nets from the "pins" at the left side of a zone to those at the other three sides of the zone. The main-channel routing is implemented by routing from left to right and zone by zone.

7.5.1.1 ORDERING FOR NETS

Definition 7.12 A net is called a *rising net* if its right terminal is on the top of the main-channel and within the current zone. A net is called a *falling net* if its right terminal is on the bottom of the main-channel and within the current zone. A net is called a *steady net* if it passes through the zone.

Definition 7.13 A *subnet* is a two-terminal net whose ends are determined by two consecutive terminals of the net. The left-terminal column of subnet i is referred to as *starting column* $sc(i)$ of that subnet, and the right-terminal column is referred to as *ending column* $ec(i)$.

Definition 7.14 Suppose that the main channel is divided into n areas by the columns in a set $S = U_{i \in N} sc(i)$, where N is the set of all nets.

If $zsc(z) \in S, z = 1, 2, \cdots, n$, and $zsc(z_1) < zsc(z_2)$ when $z_1 < z_2$, the area on the interval $[zsc(z), zsc(z+1)]$ is referred to as a *zone* z, whose starting column is $zsc(z)$.

According to the definition, if the channel is scanned from left to right, a new zone starts whenever the left terminal of a new net is encountered. These new nets are called *inserting nets* of the corresponding zone. In the following, the basic processing units will be subnets and zones.

A better order seems to be that, among nets which end within this zone, the earlier the rising net (falling net) ends, the higher track (lower track) the net is placed on, and the steady nets are placed between the rising and falling nets. The order of these steady nets should be consistent with that of the "pins" at the starting column of the next zone whose tracks can be determined by a prediction procedure in a forward-looking approach.

Let L and R represent the sets of the nets to be led to the main-channel from the left and right ends, respectively. If net i and net j form a couple of horizontally related nets which pass through the same column, an *order relation* between them can be specified by the rules in the following sequence:

(1) For the nets which pass through the whole channel from left to right, i.e., net $i \in L$ and $i \in R$, net $j \in L$ and $j \in R, \cdots$, a linear order is given arbitrarily and kept constant in every zone.

(2) When either of the two nets ends within the channel, i.e., net $i \notin R$ or net $j \notin R$, the order between them is determined based on the strategy of avoiding planar intersection ahead.

(3) When both of the nets do not end within the channel, i.e., net $i \in R$ and net $j \in R$, their order is determined to avoid any intersection.

In this way any two horizontally related nets have a unique order which is kept constant in the corresponding zones. The above linear order relation is referred to as an *ideal order* and is represented by vector $I = (i_1, i_2, \cdots, i_s)^T$, where net i_1 should be placed on a higher track than net i_2, and i_2 on a higher track than $i_3, \cdots$, and s is the number of nets to be routed. The order produced by routing at any moment is referred to as a *current order* and is represented by $C = (c_1, c_2, \cdots, c_s)^T$. Net c_k in C is called an *order-consistent net* iff $c_j = i_j$ for which $j = 1, 2, \cdots, k$ or $j = k, k+1, \cdots, s$. Otherwise it is called an *order-conflicting net*. In the succeeding steps, only order-conflicting nets need to be adjusted.

7.5.1.2 PREDICTION APPROACH

Each zone can be considered as a "quasi-switchbox". Pins at the upper and lower sides are given by the main-channel routing requirement;

those at the left side are specified by the routing produced in the left-adjacent zone, and those at the right side can be determined by the prediction approach.

The "quasi-switchbox" routing is carried out from left to right. When routing moves into a new zone, it is always expected that every net should be placed on a suitable track, which enables more nets to be the order-consistent nets and is helpful for the subsequent routing. Therefore, for every steady net and inserting net in the current zone, it is necessary to specify a "target pin" at the starting column of the next zone.

If the nets enter zone k in an ideal order I_{k-1}, together with an inserting net n_t from the top and another inserting net n_b from the bottom, a current order must be $(n_t, I_{k-1}^T, n_b)^T$, which is an imaginary current order. The problem can be described by a matrix M_k, $M_k = (m_{ij})_{\lambda \times w}, \lambda \leq w$, in which

$$m_{ij} = \begin{cases} j & \text{if } i \leq j \leq w - \lambda + i \\ 0 & \text{otherwise} \end{cases} \qquad (7.13)$$

where the columns are arranged according to the sequential number of the tracks, and the rows are arranged according to the imaginary current order. Every row in M_k maps all possible tracks to which a net can be assigned. The net mapped by row i can be assigned to track j if $m_{ij} = j$. In order to simulate a real routing, some operations on M_k are allowed, such as deleting or inserting a row, while the matrix must satisfy the following conditions:

(1) for the leftmost nonzero element in row $i(i \neq 1)$, $m_{ij} = 0$, if $j = 1$ or $m_{i-1,j-1} = 0$;
(2) for the rightmost nonzero element in row $i(i \neq \lambda)$, $m_{ij} = 0$, if $j = w$ or $m_{i+1,j+1} = 0$.

A solution of the problem can be described as follows: *Find a group of nonzero elements* $(m_{1j_1}, m_{2j_2}, \cdots, m_{\lambda j_\lambda})$, *where* $j_1 < j_2 < \cdots < j_\lambda$. The solution will be found by making a multilevel prediction for subsequent zones and removing those bad assignments.

Example: The prediction procedure is illustrated by an example shown in Figure 7.24. Figure 7.24a shows a portion of channel where crossunders are at the odd columns and their length is 5. The purpose of the prediction is to determine the tracks for nets when they pass through the starting column of zone i. So long as the unique solution does not appear in M_i, the following operations will be made:

(1) Determine the initial M_i and prepare for n_3 to intersect (set $m_{11} = 0$, if the uppermost via cannot be over track 2):

$$\mathbf{M}_i : \begin{array}{c} n_3 \\ n_2 \\ n_1 \end{array} \begin{bmatrix} 1 & 2 & 3 & 4 & 5 & 0 & 0 \\ 0 & 2 & 3 & 4 & 5 & 6 & 0 \\ 0 & 0 & 3 & 4 & 5 & 6 & 7 \end{bmatrix} \Rightarrow \begin{array}{c} n_3 \\ n_2 \\ n_1 \end{array} \begin{bmatrix} 0 & 2 & 3 & 4 & 5 & 0 & 0 \\ 0 & 0 & 3 & 4 & 5 & 6 & 0 \\ 0 & 0 & 0 & 4 & 5 & 6 & 7 \end{bmatrix}$$

$$(7.14)$$

(2) Delete n_3 and insert n_3' between n_2 and n_1 and prepare for n_1 to intersect (set $m_{37} = 0$, if no via exists at track 7):

$$\mathbf{M}_i : \begin{array}{c} n_2 \\ n_3' \\ n_1 \end{array} \begin{bmatrix} 0 & 0 & 3 & 4 & 5 & 0 & 0 \\ 0 & 0 & 0 & 4 & 5 & 6 & 0 \\ 0 & 0 & 0 & 0 & 5 & 6 & 7 \end{bmatrix} \Rightarrow \begin{array}{c} n_2 \\ n_3' \\ n_1 \end{array} \begin{bmatrix} 0 & 0 & 3 & 4 & 0 & 0 & 0 \\ 0 & 0 & 0 & 4 & 5 & 0 & 0 \\ 0 & 0 & 0 & 0 & 5 & 6 & 0 \end{bmatrix}$$

$$(7.15)$$

(3) Delete n_1 and insert n_1' over n_2 and n_3':

$$\mathbf{M}_i : \begin{array}{c} n_1' \\ n_2 \\ n_3' \end{array} \begin{bmatrix} 0 & 2 & 3 & 0 & 0 & 0 & 0 \\ 0 & 0 & 3 & 4 & 0 & 0 & 0 \\ 0 & 0 & 0 & 4 & 5 & 0 & 0 \end{bmatrix} \qquad (7.16)$$

Since the solutions on $\mathbf{M}_i$ are still more than one, it is required to make the next level prediction. It is assumed that nets $n_1, n_2,$ and n_3, which have become n_1', n_2 and n_3', get into zone $i + 1$ in an ideal order $(n_1, n_2, n_3)^T$. Therefore

$$\mathbf{M}_{i+1} : \begin{array}{c} n_1' \\ n_2 \\ n_4' \\ n_3' \end{array} \begin{bmatrix} 1 & 2 & 0 & 0 & 0 & 0 & 0 \\ 0 & 2 & 3 & 0 & 0 & 0 & 0 \\ 0 & 0 & 3 & 4 & 0 & 0 & 0 \\ 0 & 0 & 0 & 4 & 5 & 0 & 0 \end{bmatrix} \qquad (7.17)$$

Suppose that $T_k(n_j)$ represents a set of nonzero elements in the row of $\mathbf{M}_k$ which is a mapping of n_j. Then the following "$\circ$" operations are made:

(1) "$\mathbf{M}_{i+1} \leftarrow \mathbf{M}_i \circ \mathbf{M}_{i+1}$"

Since $n_1', n_2,$ and n_3' are common nets both in $\mathbf{M}_i$ and $\mathbf{M}_{i+1}$,

$$T_i(n_1') \cap T_{i+1}(n_1') = \{2,3\} \cap \{1,2\} = \{2\} \qquad (7.18)$$

$$T_i(n_2) \cap T_{i+1}(n_2) = \{3,4\} \cap \{2,3\} = \{3\} \qquad (7.19)$$

$$T_i(n_3') \cap T_{i+1}(n_3') = \{4,5\} \cap \{4,5\} = \{4,5\} \qquad (7.20)$$

Matrix $\mathbf{M}_{i+1}$ is first updated by these results; i.e., the result elements are kept unvaried and the others are set "zero" in the mapped rows. The updated $\mathbf{M}_{i+1}$ is

$$\mathbf{M}_{i+1} : \begin{array}{c} n_1' \\ n_2 \\ n_4' \\ n_3' \end{array} \begin{bmatrix} 0 & 2 & 0 & 0 & 0 & 0 & 0 \\ 0 & 0 & 3 & 0 & 0 & 0 & 0 \\ 0 & 0 & 0 & 4 & 0 & 0 & 0 \\ 0 & 0 & 0 & 0 & 5 & 0 & 0 \end{bmatrix} \qquad (7.21)$$

(2) "$M_i \leftarrow M_i \circ M_{i+1}$"

$$T_i(n_1') \cap T_{i+1}(n_1') = \{2,3\} \cap \{2\} = \{2\} \tag{7.22}$$

$$T_i(n_2) \cap T_{i+1}(n_2) = \{3,4\} \cap \{3\} = \{3\} \tag{7.23}$$

$$T_i(n_3') \cap T_{i+1}(n_3') = \{4,5\} \cap \{5\} = \{5\} \tag{7.24}$$

Similarly, matrix M_i is then updated as

$$
M_i : \begin{matrix} n_1' \\ n_2 \\ n_3' \end{matrix}
\begin{bmatrix}
0 & 2 & 0 & 0 & 0 & 0 & 0 \\
0 & 0 & 3 & 0 & 0 & 0 & 0 \\
0 & 0 & 0 & 0 & 5 & 0 & 0
\end{bmatrix} \tag{7.25}
$$

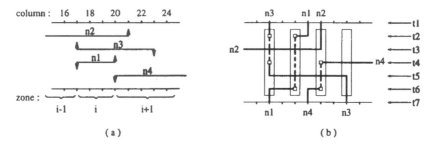

Figure 7.24 Prediction approach.
(a) Routing requirement; (b) Predicted pattern

A unique solution of the example has been obtained after the 2-level prediction. If necessary, further predictions can be made. If $T_i(n_j) \cap T_{i+1}(n_j) =$ null in the above operations, a pair of elements with the least difference in $T_i(n_j)$ and $T_{i+1}(n_j)$ will be chosen. If the unique solution is not found after the specified-level prediction, the tracks to be selected should be as close to the middle of the channel as possible. The routing patterns predicted for the example is shown in Figure 7.24b. ◇

7.5.1.3 GREEDY STRATEGY FOR ZONE ROUTING

The main-channel routing is transformed into the quasi-switchbox routing from left to right and zone by zone. In a zone, nets are routed in parallel fashion from the left side to the three other sides. The key problem is to adjust the order of nets.

Let the ideal order $I = (i_1, \cdots, i_u, \cdots, i_l, \cdots, i_s)^T$ and the current order $C = (c_1, \cdots, c_u, \cdots, c_l, \cdots, c_s)^T$, where the nets corresponding to $(c_u, \cdots, c_l)^T$ are the order-conflicting nets. First, the router will connect the order-consistent nets, corresponding to $(c_1, \cdots, c_{u-1})^T$ and

$(c_{l+1}, \cdots, c_s)^T$ in C, to their pins in a "keep to the edge" way. Then for the order-conflicting nets, the net with the higher greedy priority will be given more opportunities to occupy the current crossunder.

Figure 7.25 illustrates four possible cases in which the order-conflicting nets are adjusted to the order-consistent nets. The sequence "$n_1 - n_2 - n_3 \cdots$" means that net n_1 tries to occupy an encountered crossunder first, and then net n_2 if net n_1 fails, and so on.

Priority sequences:
Case 1 : if $ec(c_u) < ec(c_l)$, the sequence is "$c_u - c_l-$ others". If not, "$c_l - c_u-$ others";
Case 2 : if i_u is rising and $ec(i_u) < ec(c_u)$ and $i_u \neq c_{u+1}$, the sequence is "$i_u - c_u-$ others". If not, "$c_u - i_u-$ others";
Case 3 : symmetric to Case 2;
Case 4 : if i_u is rising and $ec(i_u) < ec(i_l)$, the sequence is "$i_u/c_u - i_l/c_l-$ others", or else if i_l is falling, "$i_l/c_l - i_u/c_u-$ others". If not, "$c_u/c_l - i_u/i_l-$ others".

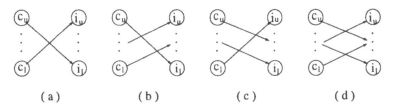

Figure 7.25 Four possible cases for order adjusting.
(a) Case 1; (b) Case 2; (c) Case 3; (d) Case 4

If the order is successfully adjusted, a new current order will be produced. Then the nets are simultaneously moved forward for one step, meanwhile some nets should change their tracks within the limits in order to prepare for the next adjusting.

In the algorithm, (1) if a net has encountered its ending column and is still an order-conflicting one, it can be routed forward since a backward detouring is allowed; (2) if the zone routing is completed but either the order-conflicting nets still exist or some nets do not reach their predicted tracks, this does not mean the routing failure because the difficulties may be solved in the next zone.

7.5.2 ROUTING BASED ON ASSIGNING RESOURCES

The algorithm consists of the following stages: First, the number of intersections among nets and the number of the intersection columns are minimized so that the requirements for the horizontal and vertical routing resources are as few as possible, the total length of crossunders used in routing is as short as possible and the intersection columns are evenly distributed. Then, the horizontal routing resources are assigned to the nets using the active track matrices of the nets. After that, the vertical routing resources are assigned by using the cluster-assignment algorithm of the incidence intersection columns. Finally, the channel routing is implemented by scanning the channel from left to right and column by column. Routing is guided by the assigned routing tracks, and a crossunder will be used only for transit through the intersection column which is assigned to this crossunder column.

7.5.2.1 MINIMIZATION OF REQUIREMENTS FOR ROUTING RESOURCES

The number of intersections among nets and the number of intersection columns will directly affect the requirements for horizontal and vertical routing resources, thereby influencing the completion rate under the condition of limited routing resources. Furthermore, because of the high resistance of polysilicon bars, the length of crossunders used in routing is required to be as short as possible. On the other hand, the densely located intersection columns will cause routing difficulties for some nets because of the local shortages of the vertical resources. Therefore, the number of intersections among the nets and the number of intersection columns ought to be minimized, and the intersection columns should be evenly distributed.

Suppose that there are m nets $n_1, n_2, \cdots, n_m$ and k columns in the channel. If r nets $n_1, n_2, \cdots, n_r$ pass through the same column, an order relation among them in this column can be represented by an order vector $\mathbf{O} = (n_1, n_2, \cdots, n_r)^T$, simply called the order of the nets, where n_i is placed in a higher track than net n_{i+1} for $i = 1, 2, \cdots, r - 1$.

Definition 7.15 If net n has its leftmost terminal in column i, the net n is called the *new net* in column i, and the column i is called the *new net column*.

Definition 7.16 Suppose that $G_c(V, E)$ is a directed graph in which the node v_i is a mapping of the net n_i $(i = 1, 2, \cdots, m)$, where V is the node set and E is the edge set. In an arbitrary new net column j $(j = 1, 2, \cdots, k)$, assume that there are r nets $n_{j1}, n_{j2}, \cdots, n_{jr}$. If there are minimum intersections between nets n_{js} and n_{jt} $(s, t = 1, 2, \cdots, r; s \neq t)$ under the order $\mathbf{O} = (n_{js}, n_{jt})^T$, a directed edge $e = (v_{js}, v_{jt})$ is from node v_{js} to node v_{jt}. This directed graph $G_c(V, E)$ is called the *minimum intersection graph* associated with the channel. If nets n_i and n_j intersect each other, the intersection is denoted by (n_i, n_j).

Definition 7.17 Suppose that $B(V, E)$ is a bipartite graph. The left side nodes of $B(V, E)$ represent column numbers, and these nodes are weighted by the density of this column. The right side nodes of $B(V, E)$ represent the intersections. If there is an intersection (n_s, n_t) in the column j under the condition of minimum intersections, there exists an edge between the left side node j and the right side node (n_s, n_t). The bipartite graph $B(V, E)$ is called the *intersection position graph*.

The approach of minimizing requirements for routing resources consist of three steps:

(1) The initial order relations of nets are given by making the number of intersections among nets the smallest. Meanwhile, the minimum intersection graph $G_c(V, E)$ and the intersection position graph $B(V, E)$ are constructed.

(2) By using $B(V, E)$, the intersections are merged so that a minimum set of intersection columns is obtained. It is the problem of finding a conditional maximum cover on $B(V, E)$, that is, finding a minimum set of left side nodes to cover all the right side nodes with the condition that these right side nodes (or intersections) covered by a same left side node must possess a common net element.

(3) Using $G_c(V, E)$ and $B(V, E)$, the order relations of nets are finally determined in the new net columns. It is the procedure to find a directed path in $G_c(V, E)$ under the following two conditions: (i) The path must cover all the nodes which represent the nets in this column; (ii) If $B(V, E)$ indicates that this column has covered some intersections, the path must possess such a node order that these intersections are engendered in this column.

According to the position and the number of terminals of a new net in the channel, the initial order relation of both nets can be determined in the new net columns, one being the new net, called an *ordering new net*, and the other being the net passing through the current column, called an *ordered net*.

At the new net column i, suppose that net n_A is an ordering new net ending at the column d_A, and nets $n_{B_1}, n_{B_2}, \cdots, n_{B_{PN}}$ are PN ordered nets passing through column i and ending at the columns $d_{B_1}, d_{B_2}, \cdots, d_{B_{PN}}$, respectively. The algorithm of determining the initial order relation of the nets n_A and n_{B_k} can be implemented by the following major steps for $k = 1, 2, \cdots, PN$ (n_A has a upper terminal in column i) or for $k = PN, PN - 1, \cdots, 1$ (n_A has a lower terminal in column i).

Step 1: Determine the closed interval $[i, d]$, where $d = min(d_A, d_{B_k})$.

Step 2: In $[i, d]$, compute the numbers A_t and A_b of the upper and lower terminals of n_A, and the numbers B_t and B_b of the upper and lower terminals of n_B, respectively.

Step 3: There are three order relations:

i. If $A_b + B_t > A_t + B_b$, the order relation is $O = (n_{B_k}, n_A)^T$. When n_A is the new net with lower terminal, the order relation has been determined, and a directed edge is drawn from the node n_{B_k} to the node n_A in $G_c(V, E)$. According to the current order relation, find all columns in which there are the intersections between net n_A and net n_{B_K}. For each intersection, an edge in $B(V, E)$ is linked from the corresponding right side node to the left side node which represents the corresponding column. When n_A is the new net with upper terminal, return to Step 1 to continue determining the order relation between the net n_A and the next ordered net.

ii. If $A_b + B_t < A_t + B_b$, the order relation is $O = (n_A, n_{B_k})^T$. When n_A is the new net with upper terminal, the order relation has been determined, and a directed edge is drawn from the node n_A to the node n_{B_k} in $G_c(V, E)$. According to the current order relation, find all columns in which there are the intersections between net n_A and net n_{B_k}. For each intersection, an edge in $B(V, E)$ is linked from the corresponding right side node to the left side node which represents the corresponding column. When n_A is the new net with lower terminal, return to Step 1 to continue determining the order relation between the net n_A and the next ordered net.

iii. If $A_b + B_t = A_t + B_b$, it is indicated that there are two order relations $O = (n_A, n_{B_k})^T$ and $O = (n_{B_k}, n_A)^T$, which all produce the minimum intersections between both nets. Therefore, two directed edges are drawn in $G_c(V, E)$, one from the node n_A to the node n_{B_k}, the other from the node n_{B_k} to the node n_A, i.e., there is a directed loop between these two nodes. According to each order relation, find all columns in which there are the intersections between both nets. For each intersection, an edge in $B(V, E)$ is linked from the corresponding right side node to the left side node which represents the corresponding column. Then return to Step 1 to continue determining the order relation between the net n_A and the next ordered net.

Example: The terminal distribution of a net set is shown in Figure 7.26a. When column $j = 0$, or left bound, net 5 and net 4 have the order relation $O = (5, 4)^T$, which yields no intersection. In $G_c(V, E)$, a directed edge is drawn from node 5 to node 4. When column $j = 1$, which is a new net column, net 1 and net 3 are two ordering new nets, and net 5 and net 4 are two ordered nets. The order relation between net 1 and net 5 is first determined. Since net 1 is an ordering new net ending at the column 4, and net 5 is an ordered net ending at the column 4 as well, $d = min(4, 4) = 4$. In interval $[1, 4]$, $A_t = 2$, $A_b = 0, B_t = 0, B_b = 1$. Because $A_b + B_t < A_t + B_b$, the order relation $O = (1, 5)^T$, which yields no intersection. In $G_c(V, E)$, a directed edge is drawn from node 1 to node 5. Then, the order relations between net 3 and net 4 and between net 3 and net 5 are determined. Net 3 is an ordering new net ending at the column 3, and net 4 is an ordered net ending at the column 2,

$d = min(3, 2) = 2$. In interval $[1, 2]$, $A_t = 0, A_b = 1, B_t = 0, B_b = 1$. Since $A_b + B_t = A_t + B_b$, there are two order relations $\mathbf{O} = (3, 4)^T$ and $\mathbf{O} = (4, 3)^T$, which all produce one intersection between two nets. In $G_c(V, E)$, two directed edges are drawn, one from node 3 to node 4, the other from node 4 to node 3. Since under the order relation $\mathbf{O} = (3, 4)^T$ there exists an intersection in the column 1, an edge in $B(V, E)$ is linked from right side node $(3, 4)$ to left side node 1. Under the order relation $\mathbf{O} = (4, 3)^T$, there is an intersection in the column 2. Therefore, an edge in $B(V, E)$ is linked from right side node $(3, 4)$ to left side node 2. Similarly, the order relation between net 3 and net 5 can be determined.

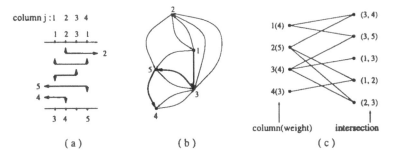

Figure 7.26 An example for intersection minimization
(a) The distribution of net's terminals
(b) The minimum intersection graph $G_c(V, E)$
(c) The intersection position graph $B(V, E)$

The resulting minimum intersection graph $G_c(V, E)$ and intersection position graph $B(V, E)$ are shown in Figures 7.26b and 7.26c, respectively. In the given $B(V, E)$, a maximum cover can be found so that a minimum set of intersection columns is obtained. Because the intersections will increase the density in the intersection column, the cover should begin from the left side node with a minimum weight. Meanwhile, the left side nodes with greater weights should be used as little as possible. The left side node 4 has the minimum weight 3, and covers the sole intersection $(1, 2)$. The left side nodes 1 and 3 have the same weight, 4. The left side node 1 is chosen to cover the intersections $(3, 4)$ and $(3, 5)$, and the left side node 3 is chosen to cover the intersections $(1, 3)$ and $(2, 3)$, respectively. The left side node 2, which has the maximum weight 5, is retained. Then, the final order relations in every new net column can be obtained by finding a directed path in $G_c(V, E)$. For example, in the new net column 1, there are a net set $\{1, 3, 4, 5\}$, and this column is an intersection column including intersections $(3, 4)$ and $(3, 5)$. The only path is found as shown by thick arrowhead lines in Figure 7.26b, which covers the net set $\{1, 3, 4, 5\}$ and satisfies two intersection relations. ◇

7.5.2.2 ASSIGNMENT OF
HORIZONTAL ROUTING RESOURCES

The purpose of assigning horizontal routing resources is to indicate the optimal tracks of nets so that the horizontal resources are reasonably used and the blind routing is avoided. In the procedure, the following aims are emphasized: (1) decrease the unnecessary routing jogs as greatly as possible so as not to occupy more horizontal routing resources; (2) avoid blocking up the new nets; (3) retain enough free tracks so as to transit the intersections.

Because there are density changes in both new net columns and intersection columns, these columns are considered as the *assigned columns*.

Definition 7.18 Suppose that there are L assigned columns in the channel. The *active track matrix* of the nets at the assigned column k is defined as

$$\mathbf{P}_k = [p_{ij}]_{\alpha \times \beta} \tag{7.26}$$

where $\alpha \leq \beta$, α is the number of nets at this column, β is the number of tracks, $i = 1, 2, \cdots, \alpha; j = 1, 2, \cdots, \beta$, and

$$p_{ij} = \begin{cases} 1 & \text{if } i \leq j \leq \beta - \alpha + i \\ 0 & \text{otherwise} \end{cases} \tag{7.27}$$

If there is an intersection which is caused by the upper contact at the assigned column k, then $p_{ij} = 0$ for $j = i$. If there is an intersection which is caused by the lower contact at the assigned column k, then $p_{ij} = 0$ for $j = \beta - \alpha + i$.

Let $H_k(n_i)$ be the element set of row i in $\mathbf{P}_k$, and let n_i represent a net in column k. The operation $\mathbf{P}_k * \mathbf{P}_{k-1}$ is defined as follows: the ith row element set of the matrix $\mathbf{P}_k * \mathbf{P}_{k+1}$ is

$$H_k(n_i) \cap H_{k+1}(n_i), i = 1, 2, \cdots, \alpha. \tag{7.28}$$

When the optimal tracks of all nets in column k are indicated, the following operations are required:

$$\mathbf{P}_k = \mathbf{P}_k * \mathbf{P}_{k+1}, \mathbf{P}_k = \mathbf{P}_k * \mathbf{P}_{k+2}, \cdots \tag{7.29}$$

until either there is only one nonzero element in every row of $\mathbf{P}_k$, or a net passing through the column k is not included in the successive matrices. In the latter situation, if there are more than one nonzero elements in certain row i of $\mathbf{P}_k$, the following heuristics can be used:

i. Maintain the net in the same track as the former assigned column;

ii. Choose the track that is near the channel side of the next terminal of net n_i;

iii. Choose the track that is near the center of the channel.

Example: The distribution of the terminals of a net set and the order relations of these nets are shown in Figure 7.27. In column 3 (the first assigned column), $\alpha = 4, \beta = 7$, and there are intersections caused by the lower contact, giving

$$
\mathbf{P}_1 \;=\; \begin{array}{c} \\ 1 \\ 3 \\ 5 \\ 4 \end{array}
\begin{array}{c}
\begin{array}{ccccccc} 1 & 2 & 3 & 4 & 5 & 6 & 7 \end{array} \\
\left[\begin{array}{ccccccc}
1 & 1 & 1 & 0 & 0 & 0 & 0 \\
0 & 1 & 1 & 1 & 0 & 0 & 0 \\
0 & 0 & 1 & 1 & 1 & 0 & 0 \\
0 & 0 & 0 & 1 & 1 & 1 & 0
\end{array}\right]
\end{array}
\tag{7.30}
$$

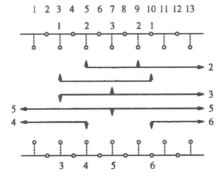

Figure 7.27 The distribution of net's terminals

The second, third, and fourth assigned columns are columns 5, 7, and 10, respectively. When assigning the tracks to the nets in column 3, the operation "$*$" is carried out as follows:

$$
\mathbf{P}_1 = \mathbf{P}_1 * \mathbf{P}_2 =
\begin{array}{c} \\ 1 \\ 3 \\ 5 \\ 4 \end{array}
\begin{array}{c}
\begin{array}{ccccccc} 1 & 2 & 3 & 4 & 5 & 6 & 7 \end{array} \\
\left[\begin{array}{ccccccc}
1 & 1 & 1 & 0 & 0 & 0 & 0 \\
0 & 1 & 1 & 1 & 0 & 0 & 0 \\
0 & 0 & 1 & 1 & 1 & 0 & 0 \\
0 & 0 & 0 & 1 & 1 & 1 & 0
\end{array}\right]
\end{array}
\tag{7.31}
$$

$$
*\;
\begin{array}{c} \\ 2 \\ 1 \\ 3 \\ 5 \\ 4 \end{array}
\begin{array}{c}
\begin{array}{ccccccc} 1 & 2 & 3 & 4 & 5 & 6 & 7 \end{array} \\
\left[\begin{array}{ccccccc}
1 & 1 & 1 & 0 & 0 & 0 & 0 \\
0 & 1 & 1 & 1 & 0 & 0 & 0 \\
0 & 0 & 1 & 1 & 1 & 0 & 0 \\
0 & 0 & 0 & 1 & 1 & 1 & 0 \\
0 & 0 & 0 & 0 & 1 & 1 & 1
\end{array}\right]
\end{array}
$$

where the operator "$*$" gives the ith row element set $H_1(n_i) \cap H_2(n_i)$ for $i = 1, 2, 3, 4$, or

$$
H_1(1) \cap H_2(1) = \{1110000\} \cap \{0111000\} = \{0110000\}
\tag{7.32}
$$

$$H_1(3) \cap H_2(3) = \{0111000\} \cap \{0011100\} = \{0011000\} \qquad (7.33)$$

$$H_1(5) \cap H_2(5) = \{0011100\} \cap \{0001110\} = \{0001100\} \qquad (7.34)$$

$$H_1(4) \cap H_2(4) = \{0001110\} \cap \{0000111\} = \{0000110\} \qquad (7.35)$$

yielding the matrix:

$$
\mathbf{P}_1 \;=\; \begin{array}{c}
 \\ 1 \\ 3 \\ 5 \\ 4
\end{array}
\begin{array}{c}
1\;\;2\;\;3\;\;4\;\;5\;\;6\;\;7 \\
\left[\begin{array}{ccccccc}
0 & 1 & 1 & 0 & 0 & 0 & 0 \\
0 & 0 & 1 & 1 & 0 & 0 & 0 \\
0 & 0 & 0 & 1 & 1 & 0 & 0 \\
0 & 0 & 0 & 0 & 1 & 1 & 0
\end{array}\right]
\end{array}
\qquad (7.36)
$$

Because there are more than one nonzero elements in every row of $\mathbf{P}_1$, the operation "$*$" will be continued.

$$\mathbf{P}_1 = \mathbf{P}_1 * \mathbf{P}_3 =$$

$$
\begin{array}{c}
1 \\ 3 \\ 5 \\ 4
\end{array}
\begin{array}{c}
1\;\;2\;\;3\;\;4\;\;5\;\;6\;\;7 \\
\left[\begin{array}{ccccccc}
1 & 1 & 1 & 0 & 0 & 0 & 0 \\
0 & 1 & 1 & 1 & 0 & 0 & 0 \\
0 & 0 & 1 & 1 & 1 & 0 & 0 \\
0 & 0 & 0 & 1 & 1 & 1 & 0
\end{array}\right]
\end{array}
*
\begin{array}{c}
2 \\ 1 \\ 3 \\ 5
\end{array}
\begin{array}{c}
1\;\;2\;\;3\;\;4\;\;5\;\;6\;\;7 \\
\left[\begin{array}{ccccccc}
0 & 1 & 1 & 1 & 0 & 0 & 0 \\
0 & 0 & 1 & 1 & 1 & 0 & 0 \\
0 & 0 & 0 & 1 & 1 & 1 & 0 \\
0 & 0 & 0 & 0 & 1 & 1 & 1
\end{array}\right]
\end{array}
$$

$$
=\;
\begin{array}{c}
1 \\ 3 \\ 5 \\ 4
\end{array}
\begin{array}{c}
1\;\;2\;\;3\;\;4\;\;5\;\;6\;\;7 \\
\left[\begin{array}{ccccccc}
0 & 0 & 1 & 0 & 0 & 0 & 0 \\
0 & 0 & 0 & 1 & 0 & 0 & 0 \\
0 & 0 & 0 & 0 & 1 & 0 & 0 \\
0 & 0 & 0 & 0 & 1 & 1 & 0
\end{array}\right]
\end{array}
\qquad (7.37)
$$

From this active track matrix $\mathbf{P}_1$, nets 1, 3, 5, and 4 are optimally assigned to the tracks 3, 4, 5, and 6 in column 3, respectively.   ◦

7.5.2.3 ASSIGNMENT OF VERTICAL ROUTING RESOURCES

Assigning the vertical routing resources is to optimally select a crossunder for an intersection column so as to transit the intersections in this column.

Although the algorithm of assigning vertical resources is suitable for any arrangement of crossunders, for convenience, assume that the distances between any two adjacent crossunders are all two grid units.

Definition 7.19 Suppose that b adjacent intersection columns correspond to b nodes $v_1, v_2, \cdots, v_b$ with their corresponding edge weights $x_{12}, x_{23}, \cdots, x_{(b-1)b}$ in the graph $G_L(V, E)$, where the edge weight x_{ij} represents the number of the crossunder columns between two adjacent intersection columns i and j. If an extra node $v_{(b+1)}$ corresponding to a next adjacent intersection column is added, there exist

$$\sum_{\substack{i=1,b \\ j=2,b+1}} x_{ij} \geq (b+1)+2 \qquad (7.38)$$

The graph $G_L(V, E)$ is called the *minimum chain graph* of the incidence intersection columns.

Definition 7.20 A set of all the intersection columns associated with the nodes in the minimum chain graph $G_L(V, E)$ of the incidence intersection columns is called *a cluster of incidence intersection columns*, and is denoted by

$$F_i = \{f_j | j = 1, 2, \cdots, b\}, i = 1, 2, \cdots, k \qquad (7.39)$$

where f_j represents the intersection column, b is the number of intersection columns and k is the number of the clusters of incidence intersection columns in the main channel.

Intuitively, this definition states that all of the intersection columns in a cluster of incidence intersection columns will complete the crossunders in the interval determined by these intersection columns.

Example: In Figure 7.27, there is only one cluster of the incidence intersection columns, and $F_1 = \{3, 7, 10\}$. Its graph $G_L(V, E)$ is shown in Figure 7.28. From this graph, the available number of crossunders in the interval *[3, 7]* is 2, and that in the interval *[7, 10]* is 2 as well. In general, the number of the clusters of the incidence intersection columns is more than one. ◇

Figure 7.28 The minimum chain graph $G_L(V, E)$
of the incidence intersection columns

Definition 7.21 The assignment interval Y_i of the crossunders corresponding to F_i is defined by the two adjacent clusters of the incidence intersection columns F_{i-1} and F_{i+1}. Let left bound column of Y_i be LB_i. Then

$$LB_i = max(SUPR + 1, INFL) \qquad (7.40)$$

where $SUPR$ is the right upper bound column of the crossunder columns assigned in Y_{i-1}, and $INFL$ is the left lower bound column of the crossunder columns located farther from the right bound column of F_{i-1}.

The right bound column RB_i of Y_i is the right upper bound of the crossunder columns, being less than the first element of F_{i+1}.

In order to minimize the influence on the columns in which the crossunders have not yet been assigned, it is desirable to keep routing density sparse in the main-channel when a crossunder is assigned to an intersection column.

Suppose that in order to use the crossunder at column j, the intersection net n at column i ($i \le j$) is required to go through m columns at the locations $i + 1, i + 2, \cdots, i + m$. The original densities at the corresponding columns are $D_i, D_{i+1}, \cdots, D_{i+m}, D_j$, and the increments of the routing density at the corresponding columns are $d_i, d_{i+1}, \cdots, d_{i+m}, d_j$, respectively. Then an assignment matrix $\mathbf{W}$ can be constructed:

$$\mathbf{W} = [w_{ji}]_{s \times t} \qquad (7.41)$$

where s is the number of crossunders in corresponding assignment interval, t is the number of incidence intersection columns ($t \le s$), and the subscripts i and j represent the locations of the columns i and j, respectively. The element w_{ji} can be computed as follows:

$$w_{ji} = max(D_i + d_i, D_{i+1} + d_{i+1}, \cdots, D_{i+m} + d_{i+m}, D_j + d_j) \quad (7.42)$$

If $w_{ji} > C_a$, let $w_{ji} = \infty$, where C_a is the capacity of the main-channel. Hence, w_{ji} describes the maximum routing density from column i to column j by assigning a crossunder at the column j to the intersection net at the column i.

When $i > j$, the matrix W can be constructed in a similar way as follows:

$$w_{ji} = max(D_j + d_j, D_{i-m} + d_{i-m}, \cdots, D_{i-1} + d_{i-1}, D_i + d_i) \quad (7.43)$$

If $w_{ji} > C_a$, let $w_{ji} = \infty$.

To obtain an optimal assignment of the vertical routing resources, the minimum matching in the assignment matrix W can be found according to the following priorities:

(1) Choose the nearest crossunder.

(2) Choose the crossunder whose cost w_{ji} is minimum.

(3) Choose the crossunder whose increment of density is minimum.

(4) Choose a crossunder that does not increase the routing complexity, such as not increasing the jogs of the nets, the new intersection, the length of crossunders used in routing, and so on.

Under the condition of nonincreasing new intersections, the cost w_{ji} of the intersection column i is monotonous and nondecreasing for $j < i$ and $j > i$.

Example: In Figure 7.27, nets 1 and 2 intersect at column 10. If net 1 uses the crossunder at column 11, then $D_{10} = 5$, $D_{11} = 4$, $d_{10} = 1$, $d_{11} = 2$, and $w_{11,10} = max(5 + 1, 4 + 2) = 6$. If net 1 uses the crossunder at column 9, then $D_9 = 4$, $D_{10} = 5$, $d_9 = 2$, $d_{10} = 1$ and $w_{9,10} = max(4 + 2, 5 + 1) = 6$. The assignment matrix is given by

$$
\mathbf{W} = \begin{array}{c} \\ 1 \\ 3 \\ 5 \\ 7 \\ 9 \\ 11 \\ 13 \end{array} \begin{array}{ccc} 3 & 7 & 10 \\ \left[\begin{array}{ccc} 5 & 999 & 999 \\ 5 & 999 & 999 \\ 7 & 7 & 999 \\ 999 & 5 & 7 \\ 999 & 5 & 6 \\ 999 & 7 & 6 \\ 999 & 7 & 6 \end{array} \right] \end{array} \tag{7.44}
$$

where 999 represents infinity. Because both distances from column 10 to columns 9 and 11 are equal, and because the costs and increments of density in columns 9 and 11 are the same, the crossunder at column 11 is chosen in order not to increase the jogs and the new intersections. The resulting assignment is that intersection columns 3, 7, and 10 should use the crossunders at columns 3, 7, and 11, respectively. The routing result is shown in Figure 7.29. ◊

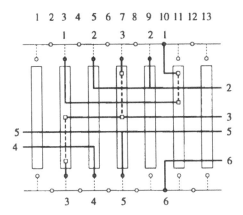

Figure 7.29 A realization for the one-and-half layer channel routing

REFERENCES

1. P. Ammon, *Gate Arrays*, Hüthig, Heidelberg, 1985.

2. M. Burstein and R. Pelavin, "Hierarchical Wire Routing," *IEEE Trans. on Computer Aided Design*, vol. CAD-2, 1983, pp. 215-222.

3. W. K. Chen, *Theory of Nets: Flows in Networks*, John Wiley, New York, 1990.

4. K. X. Cheng, B. Zhu, A. Lu, et al., "LSIS-II : A Flexible and High Density Layout System for Gate Array," *Proc. 2nd Int. Conf. Solid State and Integrated Circuit Tech.*, Nov. 1989, pp. 655-657.

5. C. Chiang, B. Preas, and R. Pelavin, "Global Routing Based on Steiner min-max Trees," *IEEE Trans. on Computer Aided Design*, vol. CAD-9, 1990, pp. 1318-1325.

6. J. Cong, A. Kahng, G. Robins, M. Sarrafzadeh, and C. K. Wong, "Provably Good Performance-Driven Global Routing," *IEEE Trans. on Computer Aided Design*, vol. CAD-11, 1992, pp. 739-752.

7. A. Hashimoto and J. Stevens, "Wiring and Routing by Optimizing Channel Assignment within Large Apertures," *Proc. 8th Des. Auto. Work.*, 1971, pp. 155-169.

8. D. W. Hightower, "A Router for Single Layer CMOS Gate Arrays," *Proc. IEEE Int. Symp. Circuits and Systems*, 1982, pp.1005-1008.

9. E. S. Kuh and M. Marek-Sadowska, "Global Routing," *Layout Design and Verification*, edited by T. Ohtsuki, North-Holland, Amsterdam, 1985.

10. J. Li and M. Marek-Sadowska, "Global Routing for Gate Array," *IEEE Trans. on Computer Aided Design*, vol. CAD-3, 1984, pp. 298-307.

11. A. Lu, X. Wu, W. Zhuang, and W. K. Chen, "New Global Routing Subsystem for CMOS Gate Arrays," *IEE Proceedings on Circuits Devices Systems*, vol. 141, October 1994, pp. 421-426.

12. R. Nair, "A Simple Yet Effective Technique for Global Wiring," *IEEE Trans. on Computer Aided Design*, vol. CAD-6, 1987, pp. 165-172.

13. S. Sastry and A. C. Parker, "Stochastic Models for Wireability Analysis of Gate Arrays," *IEEE Trans. Computer Aided Design*, vol.CAD-5, 1986, pp. 52-65.

14. J. N. Song and Y. K. Chen, "Two-Stage Channel Routing for CMOS Gate Arrays," *IEEE Trans. Computer Aided Design*, vol. CAD-7, 1988, pp. 439-450.

15. M. Terai, "A Method of Improving The Terminal Assignment in The Channel Routing for Gate Arrays," *IEEE Trans. Computer Aided Design*, vol.CAD-4, 1985, pp. 329-336.

16. B. S. Ting and B. N. Tien, "Routing Technique for Gate Array," *IEEE Trans. Computer Aided Design*, vol. CAD-2, 1983, pp. 301-312.

17. M. Vecchi and S. Kirkpatrick, "Global Wiring by Simulated Annealing," *IEEE Trans. on Computer Aided Design*, vol. CAD-2, 1983, pp. 215-222.

18. K. Winter and A. Mlynski, "Hierarchical Loose Routing for Gate Arrays," *IEEE Trans. on Computer Aided Design*, vol. CAD-6, 1987, pp. 810-819.

19. B. Zhu, X. Wu, W. Zhuang, and W. K. Chen, "A New One-and-Half Layer Channel Routing Algorithm Based on Assigning Resources for CMOS Gate Array," *IEEE Trans. Computer Aided Design*, vol.CAD-12, 1993, pp. 250-264.

Chapter 8

RECENT DEVELOPMENTS
IN WIRING AND VIA MINIMIZATION

Paul Molitor

A main feature of VLSI design systems is the placement and routing aspect. A routing problem is given by a routing region, a set of
multiterminal nets (the demands) and the number of available layers.
In this chapter the routing region will always be a planar graph, called
grid. Cross-overs, junctions, knock-knees, bends, and vias may only be
placed on vertices of the routing region. Routing itself typically consists
of two steps. The first step determines the placement of the routing
segments, which is the wire layout. In the second step which is called
wiring or layer assignment, each wire segment of the wire layout has to
be assigned to one of the k available layers so that the segments are
electrically connected in the right way.

In order to state the problem more precisely, we need some definitions
(cf. [9]). The routing region R is given by a connected planar graph (the
grid) which is embedded in the Euclidean plane. The set of multiterminal nets is given by a set $N = \{N_1, \ldots, N_p\}$, where each N_i is a subset of
the grid points, such that $N_i \cap N_j = \emptyset$ for all $i \neq j$. Each N_i is called a
net. The elements of N_i are called its terminals. We assume that there
are k conducting layers $L_1, \ldots, L_k$, each is a copy of the grid. L_{i+1} is
considered to be laid upon L_i, $1 \leq i \leq k-1$. A wire layout of a routing
problem is a mapping that associates each net N_i to a subgraph W_i of
the grid R connecting all terminals of N_i. Such a connected subgraph (or
a part of it) is called wire or wire segment. If W_i does not share an edge
with W_j for all $i \neq j$, the wire layout is a generalized knock-knee mode
wire layout. In literature, knock-knee mode wire layouts are only defined
for square grids, where a square grid is a subgraph of the integer grid
which has vertices $x \in \mathbf{N}_0 \times \mathbf{N}_0$ and edges $\{x, y\}$, where $x = (x_1, x_2)$,
$y = (y_1, y_2)$ and $\mid x_1 - y_1 \mid + \mid x_2 - y_2 \mid = 1$, i.e., which models the
popular constraint that wires can only run horizontally and vertically in
a natural way. If the wire layout is not in generalized knock-knee mode,
it is said to be an overlap mode wire layout. A k-layer wiring of a wire
layout $W = \{W_1, \ldots, W_p\}$ is a mapping that, for each $W_i \in W$, associates each edge in W_i to a layer in $\{L_1, \ldots, L_k\}$. This is done in such
a way that for any $i \neq j$, if there are edges (v_1, v_2) and (v_2, v_3) in W_i
which are assigned to L_s and L_t, respectively, and an edge (v_2, v_4) in W_j
which is assigned to L_u, then $u > \max\{s, t\}$ or $u < \min\{s, t\}$. It follows
that in a wiring all terminals from the same net are made electrically

common – if edges (v_1, v_2) and (v_2, v_3) in W_i are assigned to L_s and L_t, the segments can be connected through a via – and no two distinct nets are electrically connected. A wire layout is called k-layer wirable if there is a k-layer wiring of this wire layout.

In this chapter we first give a historical review on k-layer wiring and its coherent problems (Section 8.1). The remainder sections focus on 2-layer wiring which is discussed in detail. Two-layer wiring is important because the current technologies do wire the signal nets with 2 layers, using either two metal layers or one polysilicium and one metal layer. Section 8.2 discusses the decision problem as to whether a generalized knock-knee mode or overlap mode wire layout is 2-layer wirable, the problem of minimally stretching a wire layout to ensure 2-layer wirability, and the constrained via minimization problem which is the problem of computing a 2-layer wiring with a minimal number of vias. Sections 8.3 and 8.4 discuss new formulations of the wiring problem for VLSI circuits in the case of two layers available. The first one has recently been given by Ciesielski [5] and Kaufmann, Molitor, and Vogelgesang [15]. The objective of this problem is to minimize the interconnect delay by taking into account the resistance and capacity of interconnection wires and vias. The second one arises during hierarchical physical synthesis where the following problem has to be addressed. Let A be a circuit composed of macro cells whose input and output pins lie in certain but fixed layers. Assume that the placement and routing phase is already completed. Find a 2-layer wiring of the wire segments of A such that the pins of the macro cells lie in the preassigned layers and the number of vias is minimal on this condition. A solution of this problem induces a hierarchical bottom-up 2-layer wiring algorithm which preserves the original layout hierarchy of the circuit.

8.1 REVIEW ON WIRING

The results on the decision problem as to whether a given wire layout is k-layer wirable were achieved by Brady and Brown [2] ($k \geq 4$), by Lipski [18] ($k = 3$), and Pinter [27] and Molitor [22] ($k = 2$). Brady and Brown showed that any knock-knee mode wire layout in the square grid can be wired using four layers – the problem is still open for the case of overlap mode and generalized knock-knee mode wire layouts. Lipski proved that it is NP-complete to decide whether a (square) grid based wire layout can be wired using three layers. In [22, 27] the decision problem for 2 layers is shown to be solvable in runtime $O(w)$ for the case of generalized knock-knee mode wire layouts, where w is the sum of the wire lengths of the wire layout. This result can be extended to the case of overlap mode wire layouts. An efficient uniform algorithm for k-layer wiring has been presented by Tollis [30]. It wires each wire layout in two layers if it is 2-layer wirable. Wire layouts which are not 2-layer wirable are wired in three layers in most cases. Of course, some wire layouts that actually need only three layers are wired in four layers because of the NP-completeness of the decision problem for $k = 3$. Nevertheless, to the best of our knowledge, Tollis' algorithm is the best wiring algorithm with respect to the number of layers used for the case that there are no further optimization criterion as, e.g., finding the best k-layer wiring with respect to area, interconnect delay, number of vias and so on.

In the context of wirability, the problem arises of how to minimally stretch a wire layout in one dimension in the square grid to get a 2- or 3-layer wirable wire layout. Brady and Sarrafzadeh [3] proved that the problem in three layers is NP-complete for the case of knock-knee mode wire layouts. It is a straightforward conclusion from Lipski's result on the NP-completeness of the decision problem as to whether a wire layout is 3-layer wirable. They also presented an algorithm for the problem restricted to 2 layers available and knock-knee mode wire layouts. It runs in time $O(A)$, where A is the area of the routing region. This result has been improved by Kaufmann and Molitor [14]. They have presented an algorithm with running time $O(w)$ and showed that the problem is NP-hard for overlap mode wire layouts. When stretching in two dimensions is allowed the problem is already NP-hard for knock-knee mode wire layouts.

If a wire layout is k-layer wirable, one looks for a best k-layer wiring. The goal is to optimize the performance of the circuit and possibly minimize its manufacturing costs. In this context, it is important to minimize the number of vias between conductors on different layers since excess vias lead to decreased performance of the electrical circuit, decreased yield of the manufacturing process and increased amount of area required for interconnections. This problem is called CVM (Constrained Via Minimization problem) (CVM_k) and was first formulated by Hashimoto and Stevens [12]. In [22] it has been shown that CVM_k is NP-complete for any $k \geq 3$ for the case of knock-knee mode wire layouts even when the maximum junction degree $\Delta(W)$ is limited to four, where maximum junction degree is defined to be the maximum number q of wire segments which meet at a grid point and which are to be electrically connected. This result has been extended by Choi, Nakajima and Rim [4]. They proved that CVM_2 is NP-complete when $\Delta(W) = 4$. In 1982, Pinter [27] presented an algorithm for CVM_2 for the case that $\Delta(W)$ is limited to three and the wire layout is in knock-knee mode. This algorithm was based on Hadlock's maximum cut algorithm for planar graphs [11]. Unfortunately, in Pinter's approach for wiring, the planar graph on which maximum cut has to be performed has negative and positive weights associated with its edges. However, Hadlock's algorithm only runs for the case of positive weights (cf. [17]). In 1987, Naclerio, Masuda, and Nakajima [26] and Molitor [22] independently presented a polynomial time algorithm for the above case which runs in time $O(w^3)$. This algorithm also works for overlap mode wire layouts. In 1988, Kuo, Chern and Shih [17] presented a new algorithm based on Pinter's approach which has the time complexity of $O(w^{3/2} \log w)$. Thus, the case of CVM_k ($k \geq 3$) and maximum junction degree $\Delta(W) \leq 3$ has remained the only open problem.

8.2 TWO-LAYER WIRABILITY AND COHERENT PROBLEMS

We first review the results on the decision problem as to whether a given wire layout is 2-layer wirable. They reveal some structural properties of 2-layer wirable wire layouts used in the remainder of this chapter. Secondly, we discuss the problem of how to minimally stretch a wire layout in order to make it 2-layer wirable. Then, we present the main results on the constrained via minimization problem.

8.2.1 TWO-LAYER WIRABILITY

There are two different approaches for 2-layer wiring presented in literature. The first one was designed by Pinter [27]. The second one was independently presented by Naclerio, Masuda, and Nakajima [26] and by Molitor [22]. Here, the second approach – we call it odd-even approach – will be presented as it gives nice insights of the problem which are important for further investigations of wiring. The table below summarizes the results of the decision problem. w denotes the sum of the wire lengths of the wire layout (Table 8.1).

Table 8.1 Decision problem

	knock-knee	generalized	overlap mode
Decision Problem	$O(w)$	$O(w)$	$O(w)$

8.2.1.1 ODD-EVEN APPROACH

Let W be any wire layout. For simplicity, assume that it is a generalized knock-knee mode wire layout. We will show how to extend the results to overlap mode wire layouts at the end of this section. Furthermore, we confine ourselves to the case that at most two different wires run over each grid point of the routing region – otherwise, the wire layout is not 2-layer wirable in any case.

An example illustrates the 2-layer wiring problem. Formal definitions of the notions can be found in [16, 22]. Let W be the wire layout shown in Figure 8.1.

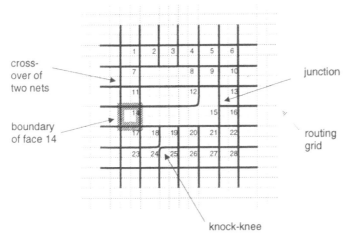

Figure 8.1 Knock-knee mode wire layout W

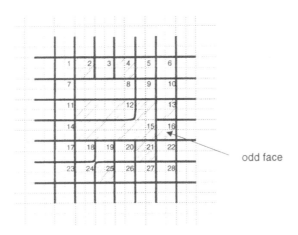

Figure 8.2 Odd faces of wire layout W

W divides the euclidian plane in which the routing region is embedded in elementary faces, 28 inner faces and one outer face. Note that actually the knock-knee nodes are handled as vertices of the graph. The inner faces are enumerated in Figure 8.1. Each face r has a boundary. We call a face r odd if its boundary is 2-layer wirable without inserting a (further) via. Otherwise, the face is called even. In our example face 1 is even because its boundary consists of exactly four cross-overs. At each of these points the layer has to be changed regardless of the wiring. Face 2 is odd because its boundary consists of three cross-overs and one junction. Obviously, a via has to be inserted at one of the five free grid points of the boundary of face 2 in order to wire it. Face 19 is even because there is one cross-over and one knock-knee, where the two adjacent wire segments do not belong to the same net. In Figure 8.2 all the odd faces of W are shaded.

Obviously, it is impossible to find a 2-layer wiring without using a via if there is an odd face. In [22, 26] the converse is shown as well. Hence, there is a 2-layer wiring of W needing no via if and only if each face of W is even. It follows that in order to obtain a 2-layer wiring the odd faces have to be transformed into even ones by inserting vias on their boundaries. Here, we assume that a via is a vertex where the layer has to be changed. In our example, if a via is inserted, e.g., at the free grid point of the vertical wire segment between face 2 and face 3, face 2 which was odd becomes even and face 3 which was even becomes odd. The insertion of a via between face 3 and 4 transforms these two odd faces into even ones. To sum it up, we have transformed the two odd faces 2 and 4 into even ones by joining them by a path of vias. We have married face 2 to face 4.

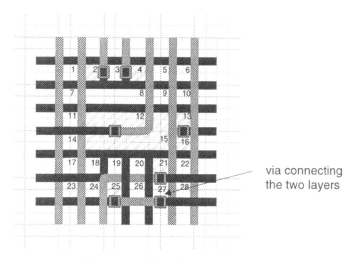

Figure 8.3   Two-layer wiring of W

It is easy to show that there is an even number of odd faces in any generalized knock-knee mode wire layout. Going once around any node of a given wire layout, the layer has to be changed an even number of times regardless of the wiring. Thus, the layer has to be changed an even number of times in the whole wire layout. The even faces contribute an even part to this number. Thus, the odd faces also contribute an even part which is only possible if there is an even number of odd faces in the wire layout.

So, marrying each odd face to exactly one other odd face results in a wire layout where each face is even. Conversely, any 2-layer wiring induces such a marriage of the odd faces. Figure 8.3 shows a 2-layer wiring of W. Here, faces 2 and 4, 13 and 16, 12 and 15, and 21 and 25 are joined. Note that the faces 21 and 25 are joined by a path of vias which runs across the outer face.

Let us now concentrate on the dual graph $G_d = (V_d, E_d)$ of a wire layout W, which is shown for our example wire layout in Figure 8.4. The set V_d of the vertices consists of the faces of W. $\{f, f'\} \in V_d \times V_d$ is an edge of E_d if and only if a via may be located on a grid point adjacent to the faces f and f'. The corresponding wire segments are said to be free. The other ones are called critical wire segments.

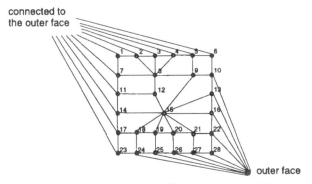

Figure 8.4 Dual graph G_d of wire layout W

The dual graph of the wire layout W of Figure 8.1 is obviously connected (in the graph theoretical sense). Therefore, there are perfect matchings of the odd faces, each defining a 2-layer wiring of W and vice versa. Figure 8.5 shows a matching (given by the bold edges) of the odd faces (marked by black vertices) which corresponds to the 2-layer wiring of Figure 8.3.

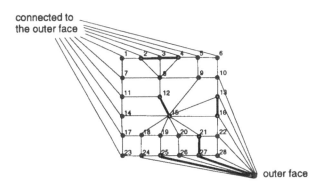

Figure 8.5 Perfect matching of the odd faces of wire layout W

The situation is a little more complex in the case of dual graphs consisting of more than one connected component. However, there is obviously a 2-layer wiring of a wire layout W if and only if each connected component of the corresponding dual graph is even, i.e., if it contains an even number of odd faces. This remark is illustrated by Figure 8.6 where we consider a coarser grid than in the previous figures. The dual graph G'_d of the corresponding wire layout W' consists of a lot of connected components. Two of them are odd so that 2-layer wiring of wire layout W' is not possible.

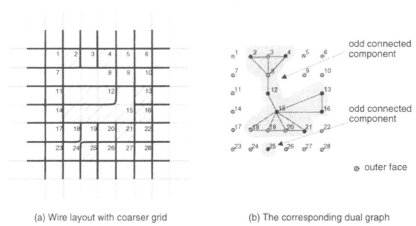

(a) Wire layout with coarser grid   (b) The corresponding dual graph

Figure 8.6   Wire layout W' and its dual graph

Hence, this results in an algorithm to decide whether a generalized knock-knee mode wire layout is 2-layer wirable which runs in time $O(w)$.

8.2.1.2 EXTENSION TO OVERLAP MODE WIRE LAYOUTS

Now, let W be a wire layout in overlap mode. Analogously to the case of generalized knock-knee mode wire layouts, we call a face even if its boundary is 2-layer wirable without using a via. Otherwise it is called odd. By eliminating overlapping wire segments by identifying adjacent grid points p_1, p_2 of the routing region R whenever $\{p_1, p_2\}$ is an edge of the grid over which two wire segments run, it is easy to see that the algorithm proposed above can also be applied in the case of overlap mode wire layouts.

8.2.1.3 EXTENSION TO MACRO CELL DESIGNS

Actually, the designs do not only consist of wire segments but contain basic cells and macro cells. Figure 8.7 shows a design which contains two macro cells A_1 and A_2. Assume that the pins of these macro cells are not preassigned to some layer. (Two-layer wiring with pin preassignments will be discussed in Section 8.4.) Now, 2-layer wiring is performed by deleting the cells first. The remaining design is a wire layout which can be wired as described above: compute the faces which are odd and marry them. The example wire layout of Figure 8.7 contains 5 inner faces and one outer face after deletion of the cells A_1 and A_2. The three inner faces 1, 2, and 4 as well as the outer face are odd. The 2-layer wiring shown marries face 2 to face 4 and face 1 to the outer face.

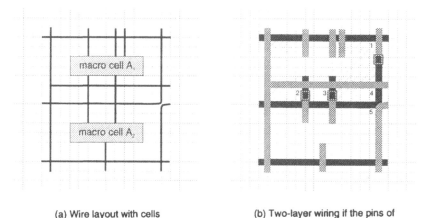

(a) Wire layout with cells

(b) Two-layer wiring if the pins of the cells are not preassigned

Figure 8.7 Two-layer wiring of a macro cell design

8.2.2 STRETCHING TO ENSURE TWO-LAYER WIRABILITY

Assume that a given wire layout in the square grid is not 2-layer wirable. The question which arises is whether it is possible to stretch the wire layout, i.e., to insert new horizontal or vertical grid tracks in order to obtain a 2-layer wirable wire layout. We denote this decision problem by d-stretching$_2$ ($d = 1, 2$), when d-dimensional stretching is allowed. A wire layout is called 2-layer d-stretchable if it can be transformed into a 2-layer wirable wire layout by d-dimensional stretching. Stretching a wire layout increases the area. So, if a wire layout becomes 2-layer wirable by stretching it in appropriate locations, it is desirable to construct a minimum area 2-layer wiring, i.e., to insert the minimum number of new tracks necessary. We call this problem d-minstretching$_2$.

Table 8.2 Decision problem on streching the wire layout

	knock-knee mode	overlap mode
d-**stretching**$_2$	trivial	$O(w)$
1-**minstretching**$_2$	$O(w)$	NP
2-**minstretching**$_2$	NP	NP

Again, Table 8.2 summarizes the results on computational complexities discussed in the following. Here, a decision problem is said to be trivial if either all instances of the problem have as solution the answer 'yes' or they all have the solution 'no.'

Let us concentrate first on 1-stretching$_2$ for the case of knock-knee mode wire layouts, i.e., we allow without loss of generality the insertion of new horizontal tracks in the routing region. The insertion of a new track generates new grid points where vias may be located. Now, call the statements on the structure of 2-layer wirable wire layouts to memory. Obviously, if a new track intersects an odd connected component, this component will be connected in the dual graph to the outer face. If all the odd components are intersected by new tracks, then the new dual graph does not contain an odd connected component any more as there was an even number of odd connected components. Hence, any knock-knee mode wire layout is 2-layer 1-stretchable. So, d-stretching$_2$ is trivial for wire layouts in knock-knee mode. For illustration consider once more the wire layout in Figure 8.6 which is not 2-layer wirable. By inserting two new horizontal tracks 2-layer wirability is ensured. Figure 8.8 shows the new wire layout and its dual graph.

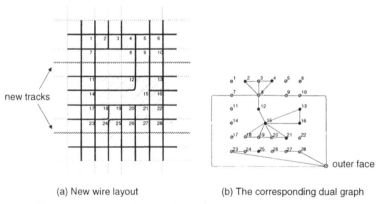

(a) New wire layout   (b) The corresponding dual graph

Figure 8.8 Successful stretching of the wire layout of Figure 8.6

When wire segments are allowed to overlap, an odd connected component intersected by a new track does not have to be connected to the outer face in the new dual graph. Overlapping wire segments prevent the two adjacent faces from being connected in the dual graph even if a new track is inserted at this position. Figure 8.9 illustrates this remark. The overlap mode wire layout shown contains exactly one odd inner face. The only new horizontal track which intersects this face does not connect it to the outer face (which is the other odd face of this wire layout) because of the two overlapping wire segments at the left and the right. Thus, d-stretching$_2$ is not trivial for the case of overlap mode wire layouts. However, it is easy to see that a wire layout is 2-layer d-stretchable if and only if the wire layout which is obtained by maximally stretching it, i.e., by inserting a new track between every two grid tracks of the routing region, is 2-layer wirable. Hence, d-stretching$_2$ is solvable in time $O(w)$.

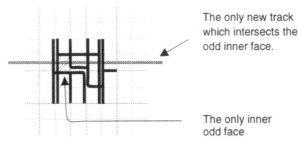

The only new track which intersects the odd inner face.

The only inner odd face

Figure 8.9 Unsuccessful stretching of an overlap mode wire layout

Now, let us discuss the problem of minimally stretching a wire layout to ensure 2-layer wirability. Obviously, d-minstretching$_2$ ($d \in \{1, 2\}$) is equivalent to the problem of finding a minimum number of tracks such that every connected component in the new dual graph is even. Thus, for the case of knock-knee mode wire layouts the problem we have to solve is to find a minimum number of tracks such that each odd connected component is intersected.

Procedure minstretching

---

begin
$\quad M = \{\}$;
$\quad I = I_m$;
$\quad k = \min \{j \mid (i, j) \in I\}$;
$\quad M = M \cup \{k - \frac{1}{2}\}$;
$\quad I = I \setminus \{(i, j) \in I \mid i < k < j\}$;
$\quad$ if $I \neq \{\}$ **then** goto (3);
end

If only vertical stretching is allowed, ignore the horizontal dimension and represent the odd connected components by a set

$$I_m = \{(i, j) \mid 0 \le i < j \le m\} \subset \mathbf{N}_0 \times \mathbf{N}_0$$

of intervals. m denotes the number of horizontal tracks of the routing region. Hence, 1-minstretching$_2$ is equivalent to the problem of finding a minimum set M such that for each $(i, j) \in I_m$ there is an $x \in M$ with $i < x < j$. This clique covering problem on interval graphs is shown to be solvable in runtime $O(|I_m|)$ [14]. An optimal algorithm is given by procedure minstretching. It follows that for the case of knock-knee mode wire layouts 1-minstretching$_2$ is solvable in time $O(w)$.

For the case of overlap mode wire layouts the problem is harder, because an odd connected component intersected by a new track does not have to be connected to the outer face. Recently, Kaufmann and Molitor [14] have shown that for overlap mode wire layouts 1-minstretching$_2$ is NP-hard by reducing 3SAT to it. To each boolean expression E with q variables, they construct a mintrack problem P in overlap mode such that there is a solution for P by inserting q new tracks if and only if E is satisfiable.

When 2-dimensional stretching is allowed, the mintrack problem for knock-knee mode wire layouts is equivalent to a minimum covering of rectangles. It can be formulated as follows: Given a set Q of rectangles in the euclidian plane, find a minimum number of vertical and horizontal lines such that each rectangle of Q intersects at least one of these tracks. Again, 3SAT can be reduced to this covering problem [14]. Hence, 2-minstretching$_2$ is NP-hard even for knock-knee mode wire layouts.

All these results completely close the gap between the polynomially solvable cases and NP-complete cases for the 2-layer stretching problem.

8.2.3 THE CONSTRAINED VIA MINIMIZATION PROBLEM

If a wire layout is 2-layer wirable one looks for a best 2-layer wiring. The goal is to optimize the performance of the circuit and possibly minimize its manufacturing costs. In this context, it is important to minimize the number of vias since excess vias lead to decreased performance of the electrical circuit, decreased yield of the manufacturing process and increased amount of area required for interconnections. This is the most popular version which is called constrained via minimization problem denoted by CVM$_2$ when 2 layers are available.

The table below summarizes the results on CVM$_2$ which close the gap between the polynomially solvable cases and NP-complete cases. The results hold for wire layouts in knock-knee mode and in overlap mode (Table 8.3).

Table 8.3 The results on CVM$_2$

	$\Delta(W) \leq 3$	$\Delta(W) = 4$
CVM$_2$	$O(w^{\frac{3}{2}} \log w)$	**NP**

8.2.3.1 CONSTRAINED VIA MINIMIZATION
IN THE CASE OF 3-WAY JUNCTIONS

As already shown, there is a one-to-one correspondence between 2-layer wirings and marriages of the odd faces of a wire layout W. Hence, CVM_2 for a wire layout W is equivalent to the problem of finding an optimal marriage of the odd faces, i.e., a marriage where a minimal number of grid points are used by the wedding paths.

For the case of $\Delta(W) \leq 3$, i.e., that there is no q-way junction with $q > 3$ in W, it is easy to see that the sum of the lengths of the wedding paths of a marriage of the odd faces equals the number of vias which have to be inserted by the corresponding 2-layer wiring. Therefore, this restricted version of CVM_2 is equivalent to the problem of finding a marriage where the sum of the lengths of the wedding paths is minimal. Formally, let ODD be the set of the odd faces of a wire layout W, the problem is to find a disjoint partition of ODD into pairs $\{f_{1_1}, f_{1_2}\}, \ldots, \{f_{q_1}, f_{q_2}\}$ such that

$$\sum_{i=1}^{q} \text{dist}\,(f_{i_1}, f_{i_2})$$

is minimal. Here, $\text{dist}(g, g')$ is the number of edges of the minimum length path connecting face g and face g' in the dual graph G_d. The best known algorithm for solving this minimum weighted perfect matching problem has worst time complexity $O(w^3)$ [6]. This reduction has been independently presented by Naclerio, Masuda and Nakajima [26] and Molitor [22]. It was the first polynomial time algorithm for CVM_2 for the case $\Delta(W) \leq 3$. At present, the best known algorithm for this restricted version of CVM_2 is that presented by Kuo, Chern and Shih [17] which is based on Pinter's approach for 2-layer wiring [27].

8.2.3.2 CONSTRAINED VIA MINIMIZATION
FOR THE CASE OF 4-WAY JUNCTIONS

Intuitively, when $\Delta(W) = 4$ is allowed, CVM_2 seems to be harder. There are wire layouts such that in every optimal marriage of the odd faces there is a grid point shared by two different wedding paths. An example is given in Figure 8.10. The wire layout shown in Figure 8.10 contains one 4-way junction. Thus, there are four odd faces. Marrying these four faces results in two wedding paths. Thus, the sum of the lengths of these paths is at least 2. Nevertheless, inserting one via at the junction itself converts the odd faces into even ones by marrying face 1 to face 3, and face 2 to face 4. Hence, an algorithm solving the above minimum weighted perfect matching problem does not solve CVM_2 in general.

In fact, Choi, Nakajima and Rim [4] could show that CVM_2 is NP-complete when the maximum junction degree is limited to four. They reduce the vertex-deletion graph bipartization problem (denoted by VDB) to CVM_2. Given a graph $G = (V, E)$ and an integer $q \geq 0$, VDB is the problem of finding a set V' of q or fewer vertices such that the subgraph G' obtained by deleting all vertices of V' (and the adjacent edges) from G is bipartite. The crucial idea behind is based on the fact that in order to obtain a 2-layer wiring, any odd face (a boundary of an odd face is not bipartite) has to be transformed into an even one by inserting a new vertex, namely a via (the boundary is bipartite after this transformation). Note that, in this context, inserting a via on a wire segment is equivalent to deleting the wire segment. In [4], VDB is shown to be NP-complete even when G is planar and the maximum degree is limited to four.

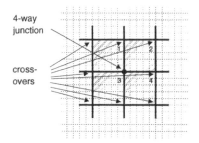
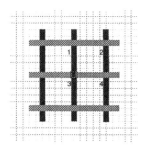

(a) Wire layout which contains a 4-way junction   (b) 2-layer wiring requiring the insertion of one via

Figure 8.10 Optimal 2-layer wiring of a wire layout
containing a 4-way junction

8.3 WIRING FOR INTERCONNECT DELAY MINIMIZATION

Unfortunately, the solutions of the CVM_2 problem are global (cf. [27]), i.e., in the process of minimizing the total number of vias we may burden one net with an excessive number of vias, or embed delay-critical nets in layers with poor conductivity. Since in some VLSI chips the optimization of the electrical performance of interconnections is more important than the minimization of the number of vias, wiring algorithms are needed which minimize signal delays through interconnection lines. This problem – we call it performance driven 2-layer wiring (PDW_2) – was recently formulated by Ciesielski [5]. He investigates the case of two available layers and generalizes Pinter's approach [27] in order to handle the problem. Unfortunately, his reduction results in a maximum cut problem for nonplanar graphs, which is known to be NP-hard. It remained open whether the problem itself is NP-hard.

This section presents some new results on PDW_2. We will discuss two restricted problems of PDW_2. First, we handle the case that all the layers have the same conductivity. Then, we discuss the case of two layers which have different conductivities. We present formal approaches to both problems raising hopes for good heuristics despite their NP-hardness.

8.3.1 TWO LAYERS WITH SAME CONDUCTIVITY

Assume that the two layers have same conductivity. Because a via decreases the performance of the electrical circuit, i.e., increases the corresponding net delay, we have to investigate the following problem in order to solve this restricted version of PDW_2, which we denote by $PDW(=)_2$.

Instance: Let $W = \{W_1, \ldots, W_p\}$ be a 2-layer wirable wire layout and let $v = (v_1, \ldots, v_p) \in \mathbf{N}_0^p$ be a vector consisting of p nonnegative numbers.
Find a 2-layer wiring of W such that for any $i \in \{1, \ldots, p\}$ the number α_i of vias inserted on wire W_i is less than or equal v_i.

From the theoretical point of view, this problem can be formulated as follows using the odd-even approach. We assign a vector $\alpha(e) = (\alpha(e)_1, \ldots, \alpha(e)_p) \in \{0, 1\}^p$ of dimension p to each edge e of the dual graph G_d with $\alpha(e)_j = 1$ if and only if edge e corresponds to a wire segment of W_j. Obviously, one can now assign a vector $\alpha(P) = (\alpha(P)_1, \ldots, \alpha(P)_p)$ to each path $P = (e_1, e_2, \ldots, e_r)$ in the dual graph by adding up the weight vectors of the corresponding edges, i.e.,

$$\alpha(P)_j = \sum_{k=1}^{r} \alpha(e_k)_j$$

Then the problem we have to solve is the following one.

Instance: Let $W = \{W_1, \ldots, W_p\}$ be a wire layout containing $2q$ odd faces and let $v = (v_1, \ldots, v_p) \in \mathbf{N}_0^p$ be a vector consisting of p nonnegative numbers.
Find q simple paths $P_1, \ldots, P_q$ in the dual graph of W marrying the $2q$ odd faces such that for all $j = 1, \ldots, p$ the inequation

$$\sum_{k=1}^{q} \alpha(P_k)_j \leq v_j$$

holds.

As shown in [15], the problem is equivalent to the problem

> **Instance:** Let $G = (V, E)$ be a planar graph and let $\alpha : E \longrightarrow \{0, 1\}^p$ be a function assigning a boolean vector $\alpha(e)$ of dimension p to each edge e. Furthermore, let a, b be two distinguished vertices in V, and let $v = (v_1, \ldots, v_p) \in \mathbf{N}_0^p$ be a vector consisting of p nonnegative numbers.
>
> Find a path P connecting a and b such that $\alpha(P)_j \leq v_j$ holds $\forall\, j = 1, \ldots, p$ for the case $q = 1$, i.e., if W only contains two odd faces.

This problem has been shown to be NP-hard [15]. It remains NP-hard even if the maximum junction degree of the wire layouts is limited to three. Nevertheless, the above formulation of the problem reveals a lot of structural properties of the problem giving hopes for efficient heuristics.

8.3.2 TWO LAYERS WITH DIFFERENT CONDUCTIVITIES

For handling wiring for the case of two layers with different conductivities, we first consider the problem for two layers L_1 and L_2 which have the following two properties. First, the conductivity of L_1 is extremely better than the conductivity of L_2. Secondly, the costs (with respect to signal delays) of a via are much lower than the costs caused by embedding one unit of the corresponding wire in the poor layer L_2. Here, unit length is defined to be the distance between two adjacent grid points, assuming that the routing region is a square grid. Thus, if two different wires W_i and W_j share a common grid point of the routing region, either two units of wire W_i or two units of wire W_j have to be put in the poor layer L_2.

8.3.2.1 THE GENERAL CASE

Let δ be a 2-layer wiring of a wire layout $W = \{W_1, \ldots, W_p\}$. We denote by $u_i(\delta)$ the number of unit lengths of wire W_i which are embedded in the poor layer by δ. Then, the problem – we will denote it by PDW$(\neq)_2$ – we have to handle is equivalent to

> **Instance:** Let $W = \{W_1, \ldots, W_p\}$ be a 2-layer wirable wire layout and let $v = (v_1, \ldots, v_p) \in \mathbf{N}_0^p$ be a vector consisting of p nonnegative numbers.
>
> Find a 2-layer wiring δ of W such that for all $i \in \{1, \ldots, p\}$ the inequation $u_i(\delta) \leq v_i$ holds.

For illustration we need some further definitions (see Figure 8.11). Let a cluster be defined to be a maximal set of mutually crossing and overlapping critical wire segments (cf. [27]). Obviously, in each cluster, once a wire segment is assigned to a certain layer, wiring of the rest of the cluster is forced as no via may be inserted. Thus, there are two possible 2-layer wirings of every cluster of a wire layout, i.e., for any cluster c_q, there are two states which we denote by state 0 and state 1. Each state j of c_q is associated with a weight vector $d_q^{(j)} = (d_{q,1}^{(j)}, \ldots, d_{q,p}^{(j)})$ of dimension p in which the ith component $d_{q,i}^{(j)}$ is the number of units of wire W_i which are embedded in the poor layer L_2 if cluster c_q is in state j.

Actually, this approach can be used to manage the exact relation between the conductivities of the two layers available, too. The definition of the weight vectors for both states of a cluster have to be extended by taking into consideration not only the wire segments assigned to the poor layer L_2 but also those assigned in the preferred layer L_1 and the vias inserted. Free wire segments are considered as trivial clusters.

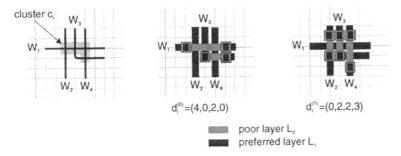

$$d_i^{(0)} = (4,0,2,0) \qquad d_i^{(1)} = (0,2,2,3)$$

▨ poor layer L_2
▦ preferred layer L_1

Figure 8.11 A cluster with its two dual states and weight-vectors

Now, the problem can be stated in the following way.

Instance: Given a wire layout $W = \{W_1, \ldots, W_p\}$ consisting of l clusters $c_1, \ldots, c_l$, and let $v = (v_1, \ldots, v_p) \in \mathbb{N}_0^p$ be a vector consisting of p nonnegative numbers.

Assign a state $j(q)$ to any cluster c_q such that for all $i \in \{1, \ldots, p\}$ the inequation

$$\sum_{q=1}^{l} d_{q,i}^{(j(q))} \leq v_i$$

holds.

As shown in [15], this problem is NP-hard even if the maximum junction degree of the wire layout is limited to three. However, it can be reduced to an integer valued generalized flow problem [7, 13]. The reduction, which is presented in [15], raises hope for good heuristics. First, compute the maximal real-valued flow value. This value can be computed by applying the polynomial time algorithm of Goldberg et al. [7, 8]. Convert this maximal real-valued flow to an integer-valued one by using random-ized rounding. The results of Raghavan [28] who studied this method for maximum multicommodity flow let us hope that we get solutions close to the optimum. For more details, please refer to [15].

8.3.2.2 A POLYNOMIAL TIME ALGORITHM FOR A SPECIAL CASE

Now, let us consider the case that the wire layout is in knock-knee mode and the routing grid is very fine, i.e., that it is possible to place a via between every two adjacent cross-overs and/or knock-knees. Thus, in order to obtain a 2-layer wiring, it is sufficient to construct a bridge for every cross-over and knock-knee by putting two units of one of the two wires in the poor layer L_2; all other wire segments can be embedded in the preferred layer L_1. Figure 8.12 illustrates this remark.

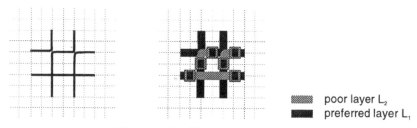

         poor layer L_2
         preferred layer L_1

Figure 8.12 Illustration of bridges

Obviously, this special case of $PDW(\neq)_2$ is equivalent to the problem of not burdening one signal net with an excessive number of bridges. Thus, the problem can be formulated as follows:

Instance: Let $W_1, \ldots, W_p$ be the routed signal nets of a given wire layout W in knock-knee mode, $v = (v_1, \ldots, v_p) \in \mathbf{N}_0^p$ be a vector consisting of p nonnegative integers, and let $c(\{i, j\}) \in \mathbf{N}_0$ be the number of grid vertices commonly shared by W_i and W_j.

Find, for all $i \in \{1, \ldots, p\}$, nonnegative numbers $c^{(i)}(\{i, j\}) \in \mathbf{N}_0$ such that
$$c^{(i)}(\{i, j\}) + c^{(j)}(\{i, j\}) = c(\{i, j\})$$
for all $i, j \in \{1, \ldots, p\}$, and
$$\sum_{j=1}^{p} c^{(i)}(\{i, j\}) \leq v_i$$
for each $i \in \{1, \ldots, p\}$.

$c^{(i)}(\{i,j\})$ $(c^{(j)}(\{i,j\}))$ denotes the number of times wire W_i (W_j) is put in the poor layer L_2 when sharing a common grid point with wire W_j (W_i). Note that $c(\{i,i\})=0$.

This special case of $\text{PDW}(\neq)_2$ can be reduced to a maximum integer network flow problem which can be solved in polynomial running time. We start with some definitions (cf. [20]). A directed network $N=(V,E,cap)$ is given by a directed graph $G=(V,E)$ and a capacity function $cap:E \longrightarrow \mathbf{N}_0$. Let s, $t \in V$ be two designated vertices, the source s and the sink t. A function $f:E \longrightarrow \mathbf{N}_0$ is a legal (s,t)-flow if it satisfies the capacity constraints

$$0 \le f(e) \le cap(e)$$

for all $e \in E$, and the conservation laws

$$\sum_{e \in in(v)} f(e) = \sum_{e \in out(v)} f(e)$$

for all $v \in V \setminus \{s,t\}$ where $in(v)$ $(out(v))$ is the set of edges entering (leaving) v. If f is a legal flow then

$$val(f) = \sum_{e \in out(s)} f(e) - \sum_{e \in in(s)} f(s)$$

is the flow value of f. The maximum network flow problem is to compute a legal flow with maximum flow value. As shown in [8] a maximum flow from s to t can be computed in time $O(ne \log \frac{n^2}{e})$ in any directed network $N=(V,E,cap)$ with $n=|V|$ and $e=|E|$.

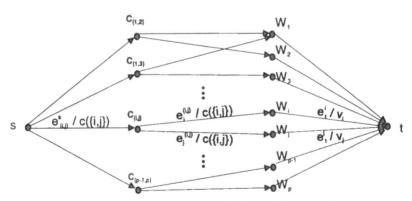

Figure 8.13 Maximum network flow problem to solve

Now, let us reduce our problem to a bipartite maximum network flow problem (V,E,cap). V is partitioned into the set $V_n = \{W_1,\ldots,W_p\}$

representing the signal nets, the set $V_c = \{c_{\{i,j\}} \mid 1 \leq i < j \leq p\}$ and the two vertices $\{s,t\}$, i.e., $V = \{s\} \cup V_c \cup V_n \cup \{t\}$. There is an edge $e^s_{\{i,j\}} \in E$ from s to vertex $c_{\{i,j\}} \in V_c$ with capacity $cap(e^s_{\{i,j\}}) = c(\{i,j\})$. Edges $e_i^{\{i,j\}}$ and $e_j^{\{i,j\}}$ with capacities $c(\{i,j\})$ connect vertex $c_{\{i,j\}}$ to the vertices W_i and W_j, respectively. Furthermore, from every vertex $W_i \in V_n$ there is an edge e_i^t to vertex t with $cap(e_i^t) = v_i$. Figure 8.13 illustrates this construction. The edges are marked by <name>/<capacity>.

It is easy to see that there is a legal flow with flow value

$$\sum_{1 \leq i < j \leq p} c(\{i,j\})$$

if and only if there are nonnegative numbers $c^{(i)}(\{i,j\}) \in \mathbb{N}_0$ such that

$$c^{(i)}(\{i,j\}) + c^{(j)}(\{i,j\}) = c(\{i,j\})$$

and

$$\sum_{j=1}^p c^{(i)}(\{i,j\}) \leq v_i$$

hold for all $i,j \in \{1,\ldots,p\}$ in the given instance of the PDW($\neq$)$_2$ problem. If a maximum flow f with value $\sum_{1 \leq i < j \leq p} c(\{i,j\})$ does exist, then $f(e_i^{\{i,j\}})$ specifies the number of times wire W_i has to be put in the poor layer L_2 when sharing a common grid point with wire W_j.

Actually, if the wires W_i and W_j do not share any common grid point, $c_{\{i,j\}}$ needs not be a vertex of the network as $c(\{i,j\})$ equals 0. Thus, the running time strongly depends on the number q of pairs of signal nets which share a common grid point. By applying the results shown in [8] it is easy to see that the running time of the proposed algorithm is $O(K^2 \log K)$ for $K = \max\{q,p\}$.

8.4 HIERARCHICAL WIRING

Another development in wiring we want to discuss is hierarchical wiring. Today's integrated circuits have up to several hundred thousand transistors. Processing such designs in a naive manner requires very large internal representations with some millions of data so that even linear space (time) optimization algorithms can be too expensive. However, large designs have a regular structure. In such arrangements, there are lots of identical subcircuits so that the designs can be described by small hierarchical representations. During synthesis, it is desirable to handle all instances of a subcircuit identically in order to guarantee identical electrical behavior of all instances, to allow further hierarchical processing and to decrease the running time of the synthesis itself.

Hierarchical physical synthesis is not highly developed yet. Results on hierarchical compaction can be found in [19, 29]. Results concerning the hierarchical physical design of systolic arrays are presented in [21, 23]. Some new results taking advantage of buses during hierarchical wiring appeared in [24].

Here we address the following problem. Let A be a circuit composed of basic cells whose input and output pins lie in a certain but fixed layer, either L_1 or L_2. Assume that the placement and routing phase is already completed. Find a 2-layer wiring of the wire segments of A such that each wire segment is embedded either into L_1 or L_2, the pins of the basic cells lie in the preassigned layers, and the number of vias is minimal on these two conditions. A solution of this problem which we denote by CVMPP (Constrained Via Minimization with Pin Preassignment) induces a hierarchical bottom-up 2-layer wiring algorithm which preserves the original layout hierarchy of the circuit.

We start with some definitions we need in the following. A circuit or a macro cell is said to be of level 1 if it contains no macro cell but only basic cells, and is said to be of level $i + 1$ if it contains at least one macro cell of level i and no macro cell of level greater than i.

Now, assume that A is a circuit of level i composed of basic cells and macro cells $A_1, \ldots, A_q$ which are interconnected by some wire layout. Let δ_{A_i} be a 2-layer wiring of A_i for $i \in \{1, \ldots, q\}$. A wiring δ_A of A is said to be induced by the wirings $\delta_{A_1}, \ldots, \delta_{A_q}$ if δ_A is given by applying δ_X on all occurrences of X ($X \in \{A_1, \ldots, A_q\}$) and then realizing the 2-layer wiring of the interconnecting wire layout. Thus, we have to consider the following problem which is applied as local step by the bottom up wiring algorithm.

> Given 2-layer wirings $\delta_{A_1}, \ldots, \delta_{A_q}$ of the q macro cells $A_1, \ldots, A_q$.
> Find a 2-layer wiring δ_A of circuit A induced by the wirings $\delta_{A_1}, \ldots, \delta_{A_q}$ which has no more vias than any other 2-layer wiring δ'_A of A which is also induced by the wirings $\delta_{A_1}, \ldots, \delta_{A_q}$.

A first suggestion for solving CVMPP in polynomial time has been made by Pinter [27]. His procedure however is only applicable if the preassignments are restricted to exterior pins (located at the outer border of the routing area), a condition normally not fulfilled by hierarchical circuit representations. Grötschel et al. [10] generalized Pinter's procedure to deal with arbitrary pin preassignments. They reduce CVMPP to the maxcut problem for almost planar graphs, i.e., graphs which can be transformed into planar graphs by removal of only one node. As this maxcut problem has been shown to be NP-hard [1], CVMPP is not solved exactly in [10] but attacked by approximation techniques from combinatorial optimization. Grötschel's procedure implied the question as to whether it is adequate to solve CVMPP by heuristics. Or, does there exist a fast exact algorithm? In [25] CVMPP has been proved to be NP-hard, thus shattering hopes for a fast exact solution. The proof is done by showing the opposite direction to Grötschel's reduction, i.e., reducing maxcut for almost planar graphs to CVMPP.

Clearly, the fact that CVMPP is NP-hard in general does not exclude exact polynomial time algorithms for important special cases. Here we will characterize such an important subclass of wiring problems. We restrict ourselves to designs with power supply nets routed in layers L_1 and L_2. The routing of these nets is extremely critical because of voltage drop and current density constraints, and thus, vias on these wire segments are much more expensive than signal wire vias. As a consequence, power and ground wiring is done in advance, without consideration of signal nets. The subsequent signal wiring phase then has to meet the fixed wiring of power supply lines as well as the preassignment of the macro cell pins. This strategy will considerably simplify the solution of CVMPP as shown in [25].

Consider the macro cell design of Figure 8.7. Assuming a bottom-up synthesis process, the pins of the macro cells A_1 and A_2 are preassigned to certain layers. Figure 8.14 shows our example wire layout of Figure 8.7 extended by pin preassignments and wired power supply lines. Note that the pin preassignments of the southern pins of macro cell A_1 induce the wiring of some further wire segments because of the absence of free wire segments, and that the wiring of the power supply nets induces the wiring of each wire segment which shares a common grid point with them.

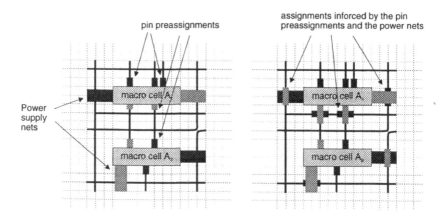

(a) Macro cell design with pin preassignments and wired power supply nets

(b) Inforced wiring of certain wire segments

Figure 8.14 Macro cell design with preassignments

Now, in order to solve the problem, we substitute each macro cell by
two critical wire segments which cross each other. Thus, they have to
be wired in different layers. The first (second) one crosses each of the
signal and power supply pins of the corresponding deleted macro cell
which are preassigned to layer L_1 (L_2); thus, if the wiring meets the
preassignments, this new wire has to be totally wired in layer L_2 (L_1)
as no via is allowed to be inserted. The transformation is illustrated in
Figure 8.15.

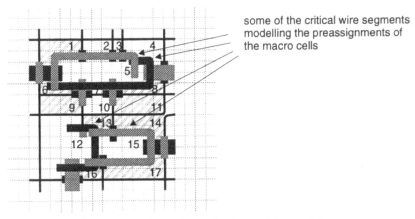

some of the critical wire segments
modelling the preassignments of
the macro cells

Figure 8.15   Transformation to a wiring problem

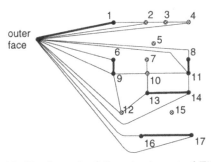

Figure 8.16   Dual graph of the wire layout of Figure 8.15

After this transformation step, the results of Section 8.2 are applied,
i.e., the odd faces, which are marked by shades in Figure 8.15, the corresponding dual graph, and an optimal marriage of the odd faces are
computed. Note once more that the new wire segments modelling the
preassignments are critical, i.e., that no via may be inserted on them.
Figure 8.16 shows the dual graph, the odd faces marked by black vertices, and an optimal marriage represented by bold lines. Obviously,
there exist two different 2-layer assignments to this optimal marriage

which are dual. Both are shown in Figure 8.17. The first one meets all the preassignments whereas the second one does not meet any of the preassignments. It can be proved that this is the general case, i.e., that one of the resulting dual wirings meets all the preassignments, as the preassigned wire segments are connected (in the graph theoretical sense) to each other through critical wire segments in the transformed wire lay‐ out. Those paths connecting wire segments which have to be wired in the same layer (different layers) have even parity (odd parity), i.e., the layers have to be changed an even (odd) number of times on these paths.

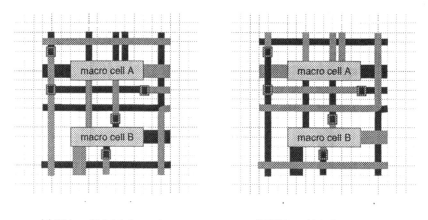

(a) Wiring which totally meets
the preassignments

(b) Wiring which does not meet
any of the preassignments

Figure 8.17 Dual 2-layer wirings corresponding
to the marriage of Figure 8.15.

Because of the discussions of Section 8.2, the number of vias inserted is minimal on the preassignment condition, obviously. The algorithm presented can be applied bottom up to boxes of level 2, 3, 4 and so on. This results in a bottom up hierarchical 2-layer wiring algorithm which locally minimizes the number of vias at each level, and which preserves the hierarchical structure of the specification of the circuit.

Actually, the general problem of hierarchical wiring with respect to via minimization is harder than presented here. It can be formulated in the following manner.

Instance: Given a box A of level i which is composed of macro cells $A_1, \ldots, A_q$.
Find 2-layer wirings $\delta_{A_1}, \ldots, \delta_{A_q}$ of the q macro cells such that there is a 2-layer wiring δ_A of A, which is induced by them, with a minimal number of vias compared to any other 2-layer wiring δ'_A of A which is induced by any other 2-layer wirings $\delta'_{A_1}, \ldots, \delta'_{A_q}$.

This problem is still open.

REFERENCES

1. F. Barahona, "On the Computational Complexity of Ising Spin Class Models," *Journal Phys. A: Math. Gen.*, vol. 15, pp. 3241-3253.

2. M. L. Brady and D. J. Brown, "VLSI Routing: Four Layers Suffice," *Advances in Computing Research*, edited by Fr. Preparata, Jai Press, 1984, pp. 245-257.

3. M. L. Brady and S. Sarrafzadeh, "Stretching a Knockknee Layout for Multilayer Wiring," *IEEE Transactions on Computers*, vol. C-39, 1990, pp. 148-152.

4. H. A. Choi, K. Nakajima, and C. S. Rim, "Complexity Results for Vertex Deletion Graph Bipartization and Via Minimization Problems," *Proc. of 25th Allerton Conference on Computing, Communication and Control*, 1987.

5. M. J. Ciesielski, "Layer Assignment for VLSI Interconnect Delay Minimization," *IEEE Transactions on CAD*, vol. CAD-8, no. 6, 1989, pp. 702-707.

6. H. Gabow, "An Efficient Implementation of Edmonds' Algorithm for Maximum Matching on Graphs," *Journal of ACM*, vol. 23, no. 2, 1973, pp. 221-234.

7. A. V. Goldberg, E. Tardos, and R. E. Tarjan, "Network Flow Algorithms," Technical Report, no. 860, School of Operations Research and Industrial Engineering, Cornell University, Ithaca, NY 14853-7501, 1989.

8. A. V. Goldberg, and R. E. Tarjan, "A New Approach to the Maximum Flow Problem," *Proc. of 18th Symposium on Theory of Computing*, 1986, pp. 136-146.

9. T. S. Gonzalez, and S. Q. Zheng, "Stretching and Three-Layer Wiring Planar Layouts," *INTEGRATION, the VLSI journal*, vol. 8, 1989, pp. 111-141.

10. M. Grötschel, M. Jünger, and G. Reinelt, "Via Minimization With Pin Preassignments and Layer Preference," Technical Report, Department of Mathematics, University Augsburg, Germany, 1987.

11. F. O. Hadlock, "Finding a Maximum Cut of a Planar Graph in Polynomial Time," *SIAM Journal of Computing*, vol. 4, no. 3, 1975, pp. 221-225.

12. A. Hashimoto, and J. Stevens, "Wiring and Routing by Optimizing Channel Assignment within Large Apertures," *Proc. of 8th Design Automation Workshop*, 1971, pp. 155-169.

13. W. S. Jewell, "Optimal Flow Through Networks," Technical Report, no. 8, M.I.T., 1958.

14. M. Kaufmann, and P. Molitor, "Minimal Stretching of a Layout to Ensure Two-Layer Wirability," *INTEGRATION, the VLSI journal*, vol. 12, 1991, pp. 339-352.

15. M. Kaufmann, P. Molitor, and W. Vogelgesang, "Performance Driven *k*-Layer Wiring," *Proc. of 9th Annual Symposium on Theoretical Aspects of Computer Sciences*, 1992, pp. 489-500.

16. R. Kolla, P. Molitor, and H. G. Osthof, "Einführung in Den VLSI Entwurf," *Leitfäden und Monographien der Informatik*, B.G. Teubner, 1989.

17. Y. S. Kuo, T. C. Chern, and W. Shih, "Fast Algorithm for Optimal Layer Assignment," *Proc. of 25th Design Automation Conference*, 1988, pp. 554-559.

18. W. Lipski, "On the Structure of Three Layer Wirable Layouts," *Advances in Computing Research*, edited by Fr. Preparata, Jai Press, 1984, pp. 231-243.

19. D. P. Marple, "A Hierarchy Preserving Hierarchical Compactor", *Proc. of 27th Design Automation Conference*, 1990, pp. 375-381.

20. K. Mehlhorn, "Data Structures and Algorithms 2: Graph Algorithms and NP-Completeness," *EATCS Monographs on Theoretical Computer Science*, Springer Verlag, 1984.

21. K. Mehlhorn and W. Rülling, "Compaction on the Torus," *Proc. of 3rd Aegean Workshop on Computing*, 1988, pp. 212-225.

22. P. Molitor, "On the Contact Minimization Problem," *Proc. of 4th Annual Symposium on Theoretical Aspects of Computer Science*, 1987, pp. 420-431.

23. P. Molitor, "Constrained Via Minimization for Systolic Arrays," *IEEE Transactions on CAD*, vol.. CAD-9, no. 5, 1990, pp. 537-542.

24. P. Molitor, "A Hierarchy Preserving Hierarchical Bottom-Up Two-Layer Wiring Algorithm with Respect to Via Minimization," *INTEGRATION, the VLSI journal*, vol. 15, 1993, pp. 73-95.

25. P. Molitor, U. Sparmann, and D. Wagner, "Two Layer Wiring With Pin Preassignments Is Easier if the Power Supply Nets Are Already Generated," *Proc. of 7th International Conference on VLSI Design*, 1994, pp. 149-154.

26. N. J. Naclerio, S. Masuda, and K. Nakajima, "Via Minimization for Gridless Layouts," *Proc. of 24th Design Automation Conference*, 1987, pp. 159-165.

27. R. Pinter, "Optimal Layer Assignment for Interconnect," *Proc. of International Conference on Circuits and Computers*, 1982, pp. 398-401.

28. P. Raghavan, "Probabilistic Construction of Deterministic Algorithms," *Proc. of 27th Annual Symposium on Foundations of Computer Science*, 1986, pp. 10–18.

29. W. Rülling and T. Schilz, "A New Method for Hierarchical Compaction," *IEEE Transactions on CAD*, vol. CAD-12, no. 2, 1993, pp. 353-360.

30. I. G. Tollis, "A New Algorithm for Wiring Layouts," *Proc. of 3rd Aegean Workshop on Computing*, 1988, pp. 257-267.

Chapter 9

LAYOUT COMPACTION FOR
HIGH-PERFORMANCE/LARGE-SCALE CIRCUITS

Hyunchul Shin

Layout compaction is the process of converting a symbolic layout or a sketch of a topological layout to a design-rule-correct mask layout with minimal area. It can be used in module generators as well as in post-processors for placement and routing systems. Since manual compaction is tedious and error-prone, automatic layout compaction can significantly improve layout productivity and allow designers to explore more design alternatives in symbolic or functional level.

We first define the layout compaction problem and discuss several typical compaction techniques including one- and two-dimensional approaches, in Section 9.1 and Section 9.2. Automatic jog-insertion and "dirty rules" such as conditional design rules which significantly increase the compaction complexity are also discussed in this part.

In Section 9.3, compaction for hierarchical design of large-scale circuits such as datapaths, regular structures, embedded macrocells is reviewed. To construct a compact functional module, one should optimize an assembly of cells rather than each cell. When the layout hierarchy is maintained during compaction, the compacted layout has the same efficient representation as the input. Recently, pitchmatching techniques for connection by abutment were reported.

In Section 9.4, advances in performance-oriented compactors are described, in which stray resistance and capacitance on timing critical paths are minimized. Local transformation of layouts can also be used for performance optimization.

9.1 INTRODUCTION

As the complexity of integrated circuits increases and the first-time-correct design is required, layout compactors become essential for IC design, especially for ASIC design. In addition to minimizing the chip area at the final stage of physical design, a compactor can be used to generate mask layouts for several different technologies (design rules) from a single symbolic cell/module in a library. It can also remove "minor" design rule violations. Several leafcell compactors have been developed and some of them show excellent results as surveyed in [4,8,47].

Layout compaction is the process of conversion from an existing symbolic layout, either manually designed [46] or automatically generated

[22], to a design-rule-correct layout, in which some geometrical aspect such as area is optimized. Figure 9.1 shows symbolic and mask layouts for one of the MCNC compaction benchmark cells, afa [4].

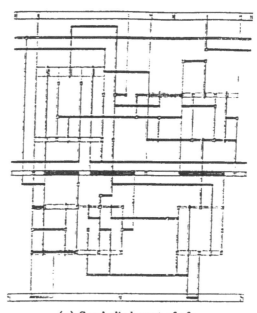

(a) Symbolic layout of afa

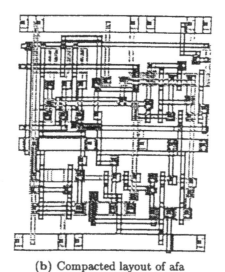

(b) Compacted layout of afa

Figure 9.1 Symbolic layout and compaction

As the process technology develops, compaction can also be used to migrate a layout from one set of design rules to another. This is possible because the input for compaction does not have to be design rule correct. Compaction provides a correct-by-construction process, which enhances designers' productivity. Constant design rule scaling, in which all the transistors and wires are scaled by a constant factor to generate a new layout, can not take full advantage of the latest process technology to reduce the layout area. The design rule migration should take a proven existing layout and transform the design to a new design optimized for a new set of design rules. Efficient and effective design migration allows a designer to reuse his or her proven designs easily and to take full advantages of the latest process technology. Figure 9.2 shows an example of design rule migration for a three-input NAND gate. While wire-width and spacing between wires are scaled by a constant, via-size is not scaled by the same constant.

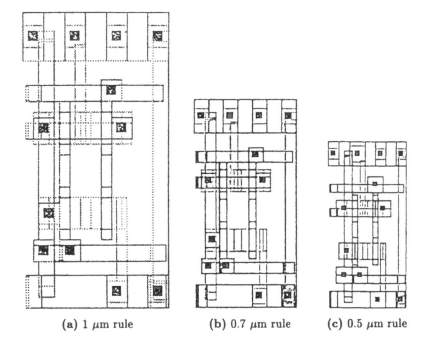

(a) 1 μm rule   (b) 0.7 μm rule   (c) 0.5 μm rule

Figure 9.2 Design rule migration

When compaction is used for library cell generation, much attention should be paid to optimize both of area and performance and to leave feedthrough channels in the layouts of cells and macros [27, 52]. This is because the cells are going to be used many times in various applications. Compaction also can be used for interactive layout optimization [19] or datapath generation [35, 60].

9.1.1 ONE-DIMENSIONAL VERSUS TWO-DIMENSIONAL COMPACTION

The most general form of compaction involves moving the geometries of the design in the x and y coordinates simultaneously. The problem of generating a correct layout of minimum area with this set of moves has been shown to belong to the class of NP-hard problems [51]. For this reason, most of the compactors proposed in the past, e.g., SLIP [13], STICKS [65], CABBAGE [21], SPARCS [5], and MACS [30] decompose this two-dimensional problem into an alternating sequence of independent one-dimensional compaction steps that ignore the information in the direction orthogonal to the axis of compaction. These one-dimensional compaction steps can be solved efficiently with longest path algorithms [21, 29]. (A slightly different approach is virtual grid compaction [8, 18, 40, 59].) Owing to its efficiency, one-dimensional compaction is widely used.

In some cases, however, such algorithms cannot compete with the quality of layouts produced by human designers. Situations where two circuit blocks cannot be shifted past one another because they just catch with their corners cannot be handled by one-dimensional algorithms, whereas a human designer would have no problems to make the necessary adjustments in the other coordinate, as shown in Figure 9.3 [58].

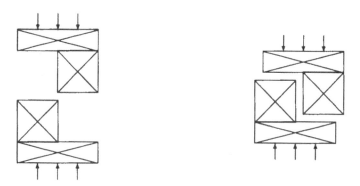

Figure 9.3 Interference that needs to be resolved
with a movement perpendicular to the axis of compaction

9.1.2 COMPACTION OF A HIERARCHICAL LAYOUT

The hierarchical compaction is important in that

- "good" leafcell compactors do not guarantee "good" results for hierarchically described circuits,
- hierarchical compaction can considerably speed up the compaction process, and
- hierarchy-preserving compaction allows efficient representation of the compacted hierarchical layout.

Several hierarchical compaction approaches have been reported. Some methods compact leafcells first and then treat them as rigid objects at the higher level [2, 6]. In these methods, routing is necessary to maintain connections among the rigid subcells. Pitchmatching methods [9, 25] compact leafcells first using a leafcell compactor and then pitchmatch ports of the cells to preserve the connections by abutment. The pitchmatching methods are efficient but they may not generate an identical mask layout for every instance of a cell.

Some other methods partition each cell into clusters of elements and perform hierarchical compaction assuming the clusters as rigid objects [49]. Another method compact the core part of leafcells first [56], and then the wires connected to the ports of the cells are compacted during the higher level compaction.

When the hierarchy is destroyed during compaction, several instances (appearances) of an identical cell before compaction may have different geometric shapes after compaction, which bothers efficient management of the design data. In [33], a powerful hierarchy-preserving hierarchical compactor has been reported. It performs hierarchy-preserving compaction, symmetry-preserving compaction, and wire length minimization using the Simplex method for linear programs. The limitations of this method are in that the Simplex method takes more time and memory than the longest-path method and that automatic wire-jogging and two-dimensional movements of components are not allowed. Note that frequently wire jogging can reduce the layout area significantly, say more than 10%.

When the given layout to be compacted is limited to an array of identical cells, toroidal compaction [15, 36, 38] can be used. With this framework, one-dimensional compaction with and without automatic jog insertion and two-dimensional compaction can be performed.

9.2 COMPACTION TECHNIQUES

In this section, several well-known compaction techniques are described. Then automatic jog insertion techniques are discussed, which are important for area and parasitics minimization. Nontrivial design rules are also discussed in this section.

9.2.1 ONE-DIMENSIONAL COMPACTION

To compact a circuit, layout elements such as wires, contacts, transistors, and subcells are positioned as close as possible without violating process design rules and without affecting electrical connections. One-dimensional compaction in horizontal (vertical) direction requires horizontally positioning the vertical (horizontal) edges of the layout elements and subcells. Most compactors handle this problem by first creating a constraint graph and then by executing longest path algorithms that operate on the constraint graph. Probably the most well-known compaction algorithm for one-dimensional compaction is the one based on longest path computation in a constraint graph [2, 5, 26, 29, 31, 37]. For a cell with n elements, the constraint graph can be constructed in time $O(n^2)$, if one compares every element to every other element. However, it can be more efficiently constructed in time $O(n \log n)$ by using a plane-sweep algorithm for the simplified case where circuit elements are not allowed to swap positions. If, on the other hand, elements of the same net are allowed to swap at most k positions, the run time was shown to be $O(n(k + \log n))$ [12]. Constraint graph generation is the most time-consuming part in graph-based compaction.

In earlier compactors, only lower bound constraints on the position of the geometries were taken into consideration. Several layout problems such as those encountered in the design of analog components require, for electrical reasons, that certain geometries be not separated by more than a prescribed quantity. This requirement amounts to considering upper bound constraints when compacting the layout.

Figure 9.4 shows an example of constraint generation. In the figure, each vertex has a variable weight x_i to represent the horizontal position of its corresponding layout element. Spacing constraints between layout elements are represented by a set of directed edges. Each edge has a fixed weight which represents the minimum constraint distance between two vertex positions. Specifically, consider two edges representing a minimum spacing rule and a maximum spacing rule between two circuit elements i and j:

(i) (minimum spacing rule) $x_j - x_i \geq a$.
This inequality is represented by a directed edge from x_i to x_j with weight a on the directed edge (i, j).

(ii) (maximum spacing rule) $x_j - x_i \leq a$.
This inequality can be rewritten as $x_i - x_j \geq -a$ and is represented by a directed edge from x_j to x_i with weight $-a$ on the directed edge (j, i).

In Figure 9.4, the vertices can be topologically ordered by using the edges. Two possible orders are (x_1, x_2, x_3) and (x_1, x_3, x_2). If we choose the order of (x_1, x_2, x_3), then the first three constraints in Figure 9.4b are called *forward* edges implying minimum spacing rules and the last constraint is called a *backward* edge implying a maximum spacing rule.

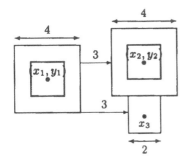

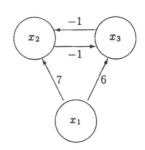

(a) Geometric design rules   **(c)** Constraint graph

$$
\begin{array}{lllll}
x_1 + 2 + 3 & \le & x_2 - 2 & or & x_2 - x_1 & \ge & 7 \\
x_1 + 2 + 3 & \le & x_3 - 1 & or & x_3 - x_1 & \ge & 6 \\
x_2 - 2 & \le & x_3 - 1 & or & x_3 - x_2 & \ge & -1 \\
x_3 + 1 & \le & x_2 + 2 & or & x_2 - x_3 & \ge & -1
\end{array}
$$

(b) Constraint inequalities

Figure 9.4 Constraint generation

Two special vertices X_0 and X_t (Y_0 and Y_t), called the source and the sink, are usually added to represent the left (bottom) side and right (top) side of the layout to be compacted in the horizontal (vertical) compaction. The mathematical formulation of the horizontal compaction problem can be expressed succinctly by the following integer linear program.

Compaction problem :
$$
\begin{array}{rcl}
Minimize & (X_t - X_0) \\
subject\ to & Ax \ge b \\
& x \ge 0
\end{array}
$$

X_0 and X_t are the source and sink positions of the layout and $(X_t - X_0)$ is the width of the layout. The vector x represents positions for all elements and A is the constraint coefficient matrix. Each row A_i of A represents a single constraint. The vector b represents the constraint spacing amounts, where b_i is the value of the constraint A_i. For example, the constraints given in Figure 9.4b can be represented as

$$
\begin{bmatrix} -1 & 1 & 0 \\ -1 & 0 & 1 \\ 0 & -1 & 1 \\ 0 & 1 & -1 \end{bmatrix} \begin{bmatrix} x_1 \\ x_2 \\ x_3 \end{bmatrix} \geq \begin{bmatrix} 7 \\ 6 \\ -1 \\ -1 \end{bmatrix}
$$

In the above leafcell compaction, the constraint coefficient matrix A is totally unimodular (every subdeterminant of A is either +1, -1, or 0) and each constraint A_i has exactly two variables which have unity coefficient. Due to this restricted nature of the constraints, the leafcell compaction problem can be trivially transformed to a simple graph problem, as shown in Figure 9.4c, for which efficient solution method exists.

Now the compaction problem can be solved by using the longest path algorithm. A formal description of Procedure LongestPath is listed in the following. For a given constraint graph G(V, E), let $E_f(E_b)$ be the set of edges representing minimum (maximum) spacing rules. Then E = $E_f U E_b$. Denote the subgraph (V, E_f) of G as G_f. G_f is a single-source, single-sink acyclic graph. Procedure LongestPath takes the graph G_f and the initial values of $V_i(D)$ (or x_i), for $i = 0, 1, \ldots, n$, and calculates each $V_i(D)$ as

$$V_i(D) = \max_j \quad \{initial\ value\ of\ V_j(D) + length\ of\ the$$

$$longest\ path\ of\ G_f\ from\ V_j\ to\ V_i\}$$

where $D \epsilon$ {X, Y} represents direction of one-dimensional compaction and the length of a path is the sum of the edge weights on the path.

> **Procedure** *LongestPath(G_f) [29]*
>   *create a stack;*
>   *push V_0 on stack;*
>   *for i := 1 to n do*
>   *In_Degree(V_i) := the in_degree of V_i in G_f;*
>   *while stack not empty do*
>   *begin*
>   *pop the stack's top node V_t from the stack;*
>   *for each edge (V_t, V_j) in G_f do*
>   *begin*
>   *$V_j(D) = max(V_j(D), V_t(D) + l_{tj})$;*
>   *{l_{tj} is the edge weight}*
>   *In_Degree(V_j) := In_Degree(V_j) - 1;*
>   *if In_Degree(V_j) = 0 then push V_j on stack;*
>   *end;*

> *end;*
> *End(LongestPath)*

After each call to Procedure LongestPath, the result satisfies all the constraints implied by the edges in G_f. Next the compaction algorithm successively accesses each edge (V_i, V_j) in E_b to test if the constraint implied by this edge is also satisfied. If it is not true, then the position of V_j will be increased by the least amount so that the constraint is satisfied. After all the edges in E_b are accessed, the new setup of the node positions may still not satisfy all the constraints. So another pass of reallocation will be executed. This iteration is repeated until all the constraints are satisfied.

Assume that $|E_b| = b$. Then it is proved that the algorithm will yield a minimum solution set and terminate by executing at most $b + 1$ iterations, if the constraints are consistent, i.e., if there exists a solution set [29]. The compaction algorithm can be listed as follows:

> **Procedure** *Compaction(G)*
> > *Count := 0;*
> > $V_0(D) := 0;$
> > *for $i := i$ to n do*
> > $V_i(D) = -\infty;$
> > *Repeat*
> > > *Flag := true;*
> > > *call Procedure LongestPath(G_f);*
> > > *For each edge (V_i, V_j) in E_b do*
> > > > *if $V_j(D) < V_i(D) + l_{ij}$ then*
> > > > > *begin*
> > > > > > $V_j(D) := V_i(D) + l_{ij};$
> > > > > > *Flag := false;*
> > > > > *end;*
> > > *Count := Count + 1;*
> > > *if Count = b + 1 and Flag = false then*
> > > > *begin*
> > > > > *print(inconsistent constraints);*
> > > > > *Return;*
> > > > *end;*
> > *until Flag := true;*
> *End(Compaction)*

9.2.2 TWO-DIMENSIONAL COMPACTION

To obtain competitive layouts from automatic compactors, it is necessary to introduce techniques that have the freedom to move components in both coordinate axes when the compacted circuit is built up. With two-dimensional compaction, the user may have difficulty in predicting the outcome due to the increased degree of freedom during compaction process. However, this cannot be considered a real problem when the physical design procedure becomes completely automatic.

Several approaches have been proposed. Wolf et al. described a method in which cell pitch is minimized by finding the critical path [66, 67], shearing the component pairs on this critical path, and performing one-dimensional compaction in the preferred direction.

True two-dimensional compaction has been attempted with exponential complexity algorithms. One approach [53] is to start with a totally collapsed layout except for topological constraints (base constraints) and then remove the distance violations one by one. For this, a branch and bound search technique is used, keeping only the best layout obtained. Kedem and Watanabe [23] translated the compaction problem into a special form of a mixed-integer programming problem and used a graph-based optimization to solve the resulting problem.

However, formulating a model which is NP-hard does not seem to be very helpful for solving the problem [8]. Weinstein suggested a simple heuristic in selecting between horizontal and vertical constraints; the preference is given to the constraint in the direction where the original distance was larger and this algorithm is strongly dependent on the initial topology of layouts [63]. In [34, 45], automatic overconstraint resolution and jog insertion features are added to two-dimensional layout compaction based on constraint graph solving or based on branch and bound optimization.

Mosteller et al. used a Monte Carlo simulated annealing technique and conjectured that the average complexity of the algorithm is given in polynomial form [39]. However, this algorithm still takes a large amount of CPU time. This approach has been extended in [61] such that wires are treated topologically, i.e., no geometric representation is used for wires during compaction. This is to avoid expensive geometrical manipulations of wires. For simplicity, active devices are described in terms of circular primitives called bubbles in [61].

Another heuristic two-dimensional compaction is proposed based on zone-refining [55]. The zone-refining layout compaction technique bears a strong similarity to a technique used in the purification of crystal ingots. Individual circuit components or small clusters of components are peeled off row by row from the pre-compacted layout, moved across an open zone, and reassembled at the other end of this zone in a denser configuration. In this process, both coordinates of the moved components are altered and jogs are introduced in the connecting wires between them to produce the needed flexibility for placing components into optimal positions. This general approach provides a flexible framework. Without lateral movements of the components, it degenerates to a one-dimensional compactor. At the other extreme, simulated annealing techniques can also be employed within the zone-refining process. This permits tradeoffs of run-time and final layout density.

To adjust aspect ratio (height to width ratio) of a cell, it is desirable to compact the cell to a specified width or height. Figure 9.5 demonstrates that two-dimensional moves in Zorro are effective in width adjustment [58]. Figure 9.5a shows a compacted layout by using a simple one-dimensional compaction, with width×height $= 74 \times 147$. Now the width of the layout is fixed to 85 and greedy compaction by zone-refining is applied to reduce the height. After 10 passes of zone-refining which took 424 CPU seconds on DEC MV8650 running the Ultrix operating system, the result shown in Figure 9.5b is obtained. The final size of the layout is 85×119.

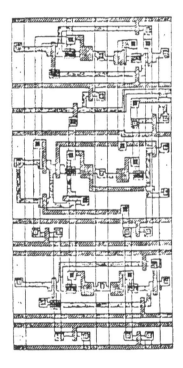
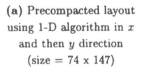
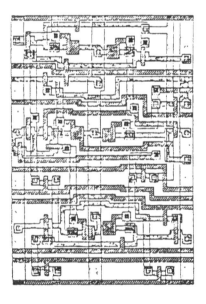

(a) Precompacted layout
using 1-D algorithm in x
and then y direction
(size $= 74 \times 147$)

(b) After 10 passes of
zone-refining with fixed width
(size $= 85 \times 119$)

Figure 9.5 Compaction with a fixed width constraint (From *IEEE Trans. on CAD,*
Vol. 9, No. 2. ©1990 IEEE. With permission.)

9.2.3 JOG INSERTION

Automatic jog creation is an important layout optimization technique used in many compactors and routers to reduce area and to reduce parasitics [28]. Many algorithms such as in [11, 17, 21, 64, 68] focus on how to reduce the cell size and, therefore, jog points are added mostly on the critical paths. However, additional jog points on noncritical paths may reduce parasitic resistance or capacitance.

Force-directed jog insertion was first introduced in [21] and used for many other compactors. Consider the situation where a wire runs through between two objects. If both objects fall on the critical path, then the wire bends or jogs at the midpoint between the objects can reduce the critical path length. When there are multiple wires running between two sets of objects, then appropriate jog points must be determined to produce dense wire bundles as shown in Figure 9.6.

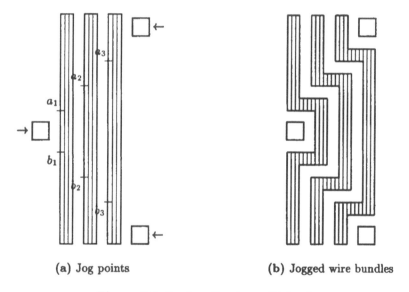

(a) Jog points   (b) Jogged wire bundles

Figure 9.6 Jog insertion for multiple wires

More sophisticated automatic jog-insertion was introduced in [32] and a faster algorithm has been presented in [37]. Maley suggested viewing the wires in a VLSI layout as indicators of the layout topology and to compact the layout maintaining the routability condition [32]. The actual wires are constructed by a homotopic router after compaction. A homotopic router takes a layout sketch consisting of the exact placement of non-wire elements and the topology of the wires as its input and produces a detailed routing of the sketch.

While constraint graphs are usually generated by preprocessing and remain unchanged during compaction, most compaction algorithms with automatic jog-insertion dynamically modifies the constraint graph for jog-insertion. However, a constraint graph which contains constraints for all potential cuts for one-dimension compaction can be generated in time $O(n^2 \log n)$ [16], if necessary.

9.2.4 NONTRIVIAL DESIGN RULES

The design rules for a given fabrication process are constraints placed on the designer by the process capabilities. The rules have a strong impact on the yield and compactness of a chip. The rules are determined such that integrated circuits designed under the rules will have an acceptable yield and compactness. For example, a statistical design rule development method is proposed in [48].

Some of the design rules may not be easily represented by minimum and maximum spacing constraints. For example, MOSIS denser contact rule set 'B' involves considerably more constraints than the simpler rule set 'A'. In addition to this, spacing requirement may depend on the width of the wires involved or on their connectivity. Furthermore, there may be rules that require more information than mask name and net-id. Minimum spacing of an active contact from a poly-Si gate can be less than the spacing from a poly-Si wire [4].

All these rules can be accommodated with extra tests and constraints and the introduction of new pseudo-layers. However many more edges need to be generated between the nodes in the adjacency graph and the constraint generation takes correspondingly longer. While the introduction of these extra rules was somewhat tedious, none of them interfered in any way with constraint graph-based compactors.

Several compactors participated in the ICCD compaction benchmark session used two "dirty" or conditional design rules added to the MOSIS CMOS rules [4]. First, the spacing between a diffusion contact and a gate was made 1 lambda rather than the original 2 lambda. The second "dirty" design rule was a different spacing for a wide metal wire than a thin metal wire. Wires with a width greater than or equal to 9 lambda were spaced 4 lambda apart from other wires, regardless of their width. The normal metal to metal spacing was 3 lambda. Addition of these two conditional design rules roughly doubles the CPU time to run a pass of compaction in Zorro compactor [56].

There also may be slight difference in area of a compacted layout depending on the measure of distance. For example, compaction under Manhattan distance measure usually produces a few percent larger layout than the one produced under Euclidean distance measure. Spacing of two vias under the two different measures is shown in Figure 9.7.

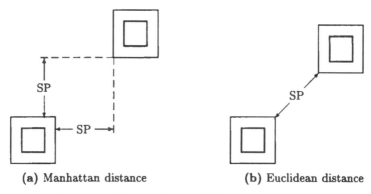

(a) Manhattan distance   (b) Euclidean distance

Figure 9.7 Spacing by two different distance measures

For multiple level metal technology, there may be conditional rules caused by three major factors which are step coverage, photolithographic resolution tolerance, and defect considerations [7]. Figure 9.8 shows a case in which the spacing between two parallel wires of metal 1 has an effect on the step structure. When there is a segment of metal 2 crossing over, the steep steps caused by metal 1 segments create a void space, thus breaking the metal 2 segment as shown in Figure 9.8a. When the spacing of the two metal 1 segments is greater, the void does not exist. When the spacing is much closer, the z-direction topology between the two segments is smooth enough to avoid breaking the metal 2 wire.

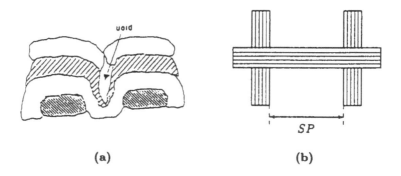

(a)   (b)

Figure 9.8 A segment of metal 2 crosses over two parallel segments of metal 1
(From *IEEE Trans. on CAD.* Vol. 11, No. 4. ©1992 IEEE. With permission.)

Let's assume that an upper level wire crosses over two parallel lower level wires in the orthogonal direction as shown in Figure 9.8. An example of conditional rules is given by

$$SP \geq c_1 \; or \; c_2 \geq SP \geq c_3$$

for the separation width SP between two metal 1 wires, where c_1, c_2, c_3 are constants such that $c_1 > c_2 > c_3$. This condition is called a bridge rule in [7].

Photolithographic resolution is caused by the topological limitation in the planar dimensions (the x and y axes). For example, if two metal 1 wires run in parallel with a length longer than a limit, the etching process tends to leave sleeves that cause shorts between the two wires. Therefore, to control the yield, when the length of the parallel wires is beyond some limit, the spacing needs to be wider.

Figure 9.9 shows an example. Two parallel wires are adjacent for distance L_1. Let the separation width between the two wires be SP. There are three constants c_4 and $c_5 > c_6$ such that

$$SP \geq c_5 \quad if \; L_1 \geq c_4$$
$$SP \geq c_6 \quad otherwise$$

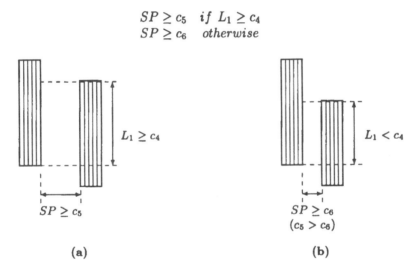

(a)   **(b)**

Figure 9.9 Illustration of the length-dependent rule

To find the optimum solution for the compaction problem with conditional rules, one faces the problem of how to choose the correct edges for each conditional rule so that the constraint graph of the whole layout can generate a legal layout with minimum width or height. This is in fact a very complicated problem. It is proved in [7] that finding the optimum solution for one-dimensional compaction under bridge rules is NP-complete. Since solving an NP-complete problem optimally usually requires either explicit or implicit exhaustive search, one possibility is to find a structure which covers the set of solutions and is small-sized relative to the original problem being considered. Another alternative is, of course, to find and to use a heuristic algorithm.

9.3 HIERARCHICAL COMPACTION

Since most VLSI designs are hierarchical, it is essential that compactors accept hierarchical designs and produce compacted layouts which preserve the input hierarchy. This avoids an explosion in database size, speeds up the compaction process, facilitates characterization, and allows easier modification of the compacted layout.

Some frameworks for hierarchical compaction can be used for several types of layout optimization, such as compaction with jog-insertion, with automatic wire length minimization, and with two-dimensional compaction. However, some others are more restricted and supports only one-dimensional compaction.

Hierarchical compaction works on a hierarchical representation of the input layout. It tries to minimize the area in individual hierarchical block and the interconnections between them. This approach is inherently suboptimal to that of compacting a flat representation of the layout. However, the compaction of a flattened layout takes more CPU time than that of hierarchical one. Therefore, when the CPU time is bounded by a certain value, hierarchical compaction may outperform the flattened compaction in area.

As the complexity of VLSI circuits dramatically increases, the ability to handle large design problems becomes essential. To optimize the overall layout of a chip, full chip compaction is desirable. However, full chip compaction is too expensive in CPU time and memory requirements to perform on a standard workstation.

A true hierarchical compaction can be formulated as an integer linear programming (ILP) problem. However, ILP problems in general belong to the class of NP-hard problems and using brute force ILP techniques to compact even moderate size layouts is prohibitively expensive [43]. To reduce computational complexity, some researchers have used graph-based methods with heuristics to solve the hierarchical compaction problem.

9.3.1 HIERARCHICAL COMPACTION WITH ROUTING

This method divides the hierarchical compaction problem into two parts, leafcell compaction and higher level cell compaction, and then solves the two compaction steps separately. Each leafcell is compacted and represented as an instance in a higher level of hierarchy. The higher level instance can either be abstracted into a polygon for each mask or contain a more detailed description of the leafcell. After all leafcells are compacted and represented as instances, the terminals of instances may not match since each leafcell has been compacted independently. Therefore, a routing phase is necessary to make the required connections between instances. An example is shown in Figure 9.10a. Instead of connection by routing, portions of the compacted cells can be stretched for pitch matching as shown in Figure 9.10b. The stretched regions are shaded in the figure.

The advantage of this approach is that its algorithm is simple since compaction and interconnection are solved as two independent problems. The disadvantages are that too much area may be used for routing and that the regularity of the chip may get lost. For example, when an array of a subcell is compacted, corresponding interconnections between different instances may have different shapes.

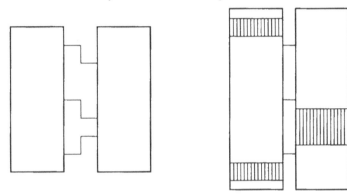

(a) Connection by routing   (b) Connection by stretching

Figure 9.10 Connection in module assembly

9.3.2 LP/ILP-BASED ONE-DIMENSIONAL HIERARCHICAL COMPACTION

True hierarchical compaction is defined as minimizing the area of a hierarchical layout while preserving input hierarchy, design rule correctness, and electrical connectivity between components and subcells. Cell instances that are connected by abutment in the input must remain abutted after hierarchical compaction. To resolve this pitchmatching, the constraints within each of the leafcells must be solved in conjunction with various intercell constraints across the design hierarchy.

Due to the problem complexity, only one-dimensional compaction techniques have been reported for optimal hierarchical compaction [3, 33, 42, 69].

The mathematical formulation of the hierarchical compaction problem can also be expressed by the integer linear program as in the case of leafcell compaction:

$$Minimize \quad (X_t \; - \; X_0)$$
$$subject \; to \quad Ax \; \geq \; b$$
$$x \; \geq \; 0$$

However, the pitchmatching constraints between subcells coupled with the hierarchy preserving constraints result in a general coefficient matrix, i.e., the matrix is no longer unimodular. Hence, the hierarchical compaction problem cannot be transformed into an equivalent problem in a constraint graph. Figure 9.11 shows an example of a hierarchical intercell constraint which cannot be represented by a simple graph. In the figure, vertex position x_2 is with respect to the left edge of subcell A and x_4 is with respect to the left edge of subcell B. Vertex positions x_1 and x_3 are with respect to the left edge of the rootcell.

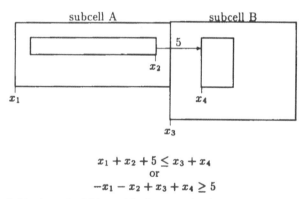

$$x_1 + x_2 + 5 \leq x_3 + x_4$$
$$or$$
$$-x_1 - x_2 + x_3 + x_4 \geq 5$$

Figure 9.11 A typical hierarchical constraint between subcells, A and B

In general hierarchical constraints are linear constraints involving three or more vertices. A further complication with hierarchical constraints is that some of the vertex weights in the constraint may have integer value greater than one, as in the case of mirroring or repeating to form an array. Figure 9.12 shows a spacing constraint between a subcell and the same subcell reflected sideways. In the figure, vertex position x_2 is with respect to the left edge of subcell C. Vertex positions x_1 and x_3 are with respect to the left edge of the rootcell.

Since the hierarchical compaction problem cannot be transformed into an equivalent constraint graph problem, integer linear programming techniques are required to optimally solve the problem. It is easy to see that a hierarchical compaction problem can be written as a linear programming problem, if we assume that a rational solution is acceptable. In [50], it is also shown that any inequality system with rational coefficients can be interpreted as a set of constraints describing an instance of hierarchical compaction. Therefore, the problem of one-dimensional compaction is as hard as linear programming even though a rational solution is allowed.

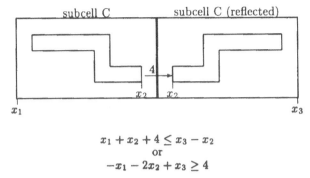

$$x_1 + x_2 + 4 \leq x_3 - x_2$$
$$\text{or}$$
$$-x_1 - 2x_2 + x_3 \geq 4$$

Figure 9.12 A hierarchical constraint between one instance of subcell C and a reflected instance of subcell C

General integer linear programming problems fall into a class of NP-complete problems [43]. Hence, most researchers have used linear programming techniques to solve the problem and perform integer round off if necessary, using heuristics [33]. However, linear programming methods are also expensive in both space and time, even though linear programming problems can be solved in polynomial time in length of the input [44]. Therefore, the size and complexity of design hierarchy that can be handled are severely limited. Because of the large time and space complexity associated with linear programming, the size of the system of linear inequalities must be kept as small as possible. Many researchers tried to solve only a small fraction of the total number of linear inequalities presented by using linear programming. The vast majority of constraints are either factored out or solved by efficient graph algorithms [42, 69].

Regularity in the input is heavily exploited and repetitive instances of a subcell give rise to a single set of equations regardless of the number of instances of the subcell. For example, in Figure 9.12, only a single variable x_2 is used even though there are two instances of the subcell C.

9.3.3 CONSTRAINT GRAPH-BASED APPROACHES

Due to the computational complexity of linear programming techniques, some researchers adopted constraint graph-based approaches with heuristics to handle hierarchical constraints.

9.3.3.1 HIERARCHICAL CONSTRAINTS

During subcell compaction, some objects in a subcell may be moved to positions adjacent to the boundary. Later when the subcells are merged, design rule violation may occur as shown in Figure 9.13.

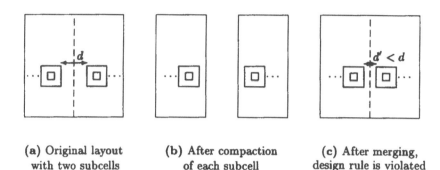

| (a) Original layout with two subcells | (b) After compaction of each subcell | (c) After merging, design rule is violated |

Figure 9.13 Design rule violation
due to independent compaction of subcells

One simple and elegant heuristic approach to resolve the above inter-subcell design rule problem is to enforce an additional design rule, called half design rule, between objects in a subcell and the boundaries of the subcell [69], as shown in Figure 9.14.

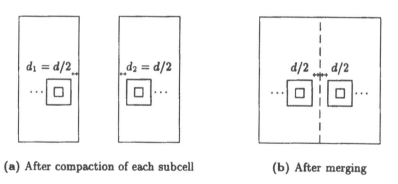

| (a) After compaction of each subcell | (b) After merging |

Figure 9.14 Half design rule constraint

Applying the half design rule to all objects except terminals and wires connected to terminals, simple abutment of individual subcells after compaction can produce correct layouts. This approach is simple and is widely used. However, there may be wasted area since the half design rule is added without considering the availability of unused space for each subcell. To avoid excessive waste in area, the minimum spacing rule between p-well and n-well, which is the largest spacing rule in current process technology, is not usually enforced as a half design rule but is separately checked [10, 69].

A more sophisticated approach attempts to find an optimal distribution of spacing, i.e., d_1 and d_2 ($d_1 + d_2 \geq d$) in Figure 9.14, by performing several iterative compactions [24].

9.3.3.2 MODULE ASSEMBLY BY PITCHMATCHING

The hierarchical compaction or the module assembly problem was established as an integer linear programming problem. Since existing integer linear programming methods are very intensive in both space and time, the complexity of the layout that can be handled is limited. Heuristic methods are used instead to explore layout hierarchy as a practical solution with high speed, low memory and minor area penalties.

In a predominantly butting environment, cells should be stretched and pitchmatched. One problem during stretching pointed out by [14], was the so-called *xy interlock* problem, which was solved by adding more constraints in the leafcell. In particular, constraints in the y direction are added by assuming the x coordinates of objects are in some ambiguity regions. The ambiguity region concept ensures that the cells are *stretchable* under any module assembly constraints. By integrating the leafcell compaction into the module assembly process, the method below requires *no* additional constraints and can achieve a relatively larger degree of free movement [1, 25, 31]. The module assembly algorithm used in [31] is shown in the following:

> **Algorithm** *Module assembly for direction D, $D \in \{x,y\}$*
> *for each leafcell $C_i \in$ module M*
> *{*
>
>   *create graph G_i^x with spacing, connectivity constraints;*
>   *perform automatic jog insertion;*
>   *derive port abstraction g_i^x from G_i^x;*
>
> *}*
> *create module graph G_M^x by*
> *{*
>
>   *instantiate abstraction g_i^x for each C_i;*
>   *add pitchmatching constraints between butted ports;*
>
> *}*
> *solve the module graph G_M^x;*
> *for each leafcell $C_i \in M$*
> *{*
>
>   *restore G_i^x;*

> *add port position constraints to G_i^z;*
> *solve or recompact G_i^z;*
> *perform wire-length minimization on G_i^z;*
> }

For each leafcell C_i in the module M, the spacing and connectivity constraints are formulated. From this cell graph G_i, the port abstraction graph g_i is derived by finding the longest paths between every ordered pair of ports (see Figure 9.15). The problem is a subset of the "finding longest paths between all pairs of nodes" problem which can be solved in $O(n^{1.5} \log n)$ [31].

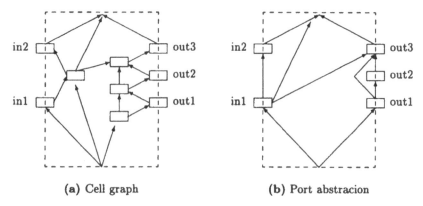

(a) Cell graph   (b) Port abstracion

Figure 9.15 Port abstraction of a cell (From *Proc. of the 27th Design Automation Conf.* ©1990 IEEE. With permission.)

The above pitchmatching technique using port abstraction can be incorporated with two-dimensional optimization based on constraint graphs [57]. This approach can efficiently generate competitive layouts by dynamically resolving constraints in two dimensions, i.e., the size of the layout in one dimension is minimized by moving elements in both (horizontal and vertical) direction. After one-dimensional compaction, all the critical paths which define the height or width of the layout are found and the elements on the paths are rearranged to reduce the size of the layout in the compaction direction. It should be noted that the constraints in the orthogonal direction must be considered during this step. This is to minimize, at least not to increase, the layout size in the orthogonal direction and to enforce pitchmatching constraints to maintain hierarchy during hierarchical compaction.

9.3.3.3 COMBINED HIERARCHICAL COMPACTION

To trade off quality and complexity of hierarchical compaction, other graph-based approaches are suggested, in which terminals are pitchmatched by using partly flexible subcells [49, 56]. differences of the two methods are in that the approach in [56] uses routing, in addition to making the leafcell partly flexible, for pitchmatching and contact/wire sharing, while the one in [49] uses stretching only.

For the special case of an iterative array of identical cells, one may compact the defining cell with suitable cyclic end-around constraints. For example, if a cell is repeated several times in horizontal direction, one can match the terminal positions in vertical direction by adding fixed constraints between terminals of the same height when compacting the leafcell. This will guarantee that the compacted cells can be abutted tightly.

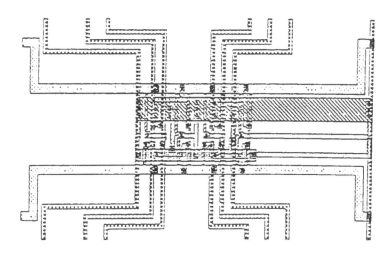

Figure 9.16 A single leafcell with compacted core

The more general case where two cells of different types abut is handled in several steps [56].

(A) First the core of each leafcell is compacted, maintaining the terminal frame in unchanged form as given in the symbolic representation that specifies the topology of the layout. A core is a set of components that are not placed on the boundary of an input layout; a core does not contain any shared terminals. The compacted core remains attached to this shared terminal frame by suitable wires (Figure 9.16).

(B) These core cells with their surrounding frames are then assembled into the desired configuration. A clean-up routine straightens out the wires as much as possible and removes unnecessary wire segments and terminal points (Figure 9.17).

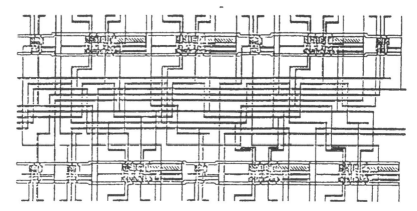

Figure 9.17 Assembly of compacted leafcells

(C) Finally, we apply a second compaction step at the next higher level of the hierarchy, in which the cores of the cells are considered fixed clusters. This compacts the terminal frames, the attached wires, and possible routing channels for all participating cells at this level of the hierarchy (Figure 9.18).

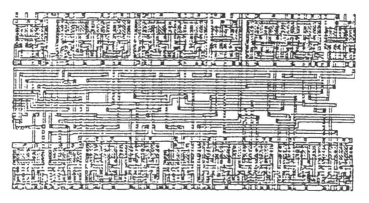

Figure 9.18 Completed hierarchical compaction of cell assembly

This basic strategy, while not as elegant as direct pitchmatching of all terminals between cells during the compaction process, allows us to handle all cases in the benchmarks. The strategic control of the necessary sequence of compaction steps is currently under operator control. For the case of a two-dimensional array of identical cells with some one-dimensional arrays of peripheral cells, one can first compact the cell of the two-dimensional array. Then the peripheral cells are compacted to the respective width or height of the compacted core of this array cell. After execution of the clean-up routine, there is a good probability that the remaining cell cores abut gracefully.

9.3.3.4 PARTITIONING COMPACTION

In [10], simple partitioning compaction is suggested, in which the layout is cut in the vertical (horizontal) direction. All wires running in the horizontal (vertical) direction and cross the cutlines are cut. The cutlines must not intersect any contact or transistor symbols and must not overlap wires which run in the same direction. The individual blocks are then independently compacted only in the horizontal (vertical) direction. This is to maintain the vertical positions of the wires cut the same as before. The compacted blocks are now merged back into the original layout by simply butting them together. The connection by abutment is possible because the vertical (horizontal) positions of the wires cut do not change during compaction. Figure 9.19 shows the procedure of cut, compact, and merge steps. In the figure, three vertical cuts are made to form four blocks. The blocks are independently compacted in the horizontal direction and then merged, resulting reduced horizontal dimension of the layout.

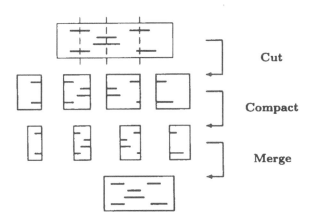

Figure 9.19 The three steps in partitioning compaction

The same sequence of cut, compact, and merge can then be performed in the opposite direction. Although only straight cutlines are shown in Figure 9.19, zig-zag cutlines are allowed for practical applications. Alignment constraints are added to wires which are cut. For those cut wires which run parallel to the compaction direction, only their cut end points are fixed so that their end positions will not be changed after compaction. For cut wires running orthogonal to the compaction direction, the horizontal position as well as the vertical one is fixed with respect to the cell boundary. This allows reconnection of the wires cut by simple abutment when the subcells are merged. But there is penalty in area owing to the alignment constraints. By using this method, a CPU core layout with 28.8K transistors could be compacted on a workstation with 32M bytes of RAM [10].

9.4 PERFORMANCE-DRIVEN COMPACTION

Future needs for layout design automation demand support for high-performance circuits using submicron technology. Designers of high-speed digital circuits or analog circuits must take into account severe constraints to optimize circuit performance as well as the layout area. Most compactors have capability of minimizing the total wire length. Several recent compactors have more sophisticated features for performance optimization.

9.4.1 WIRE LENGTH MINIMIZATION

Long wires use more area and degrade the electrical properties of the circuit. Therefore, a compactor should not only minimize layout area but also minimize the weighted wire length. To reduce the total resistance of electrical nets, the length of wires might be weighted by the resistance per unit length of the layers used. Usually a layout is compacted to minimize area and then the total weighted wire length is heuristically reduced for efficiency. If wire length minimization is not done properly, one may end with poor-quality layouts. Iterative techniques for wire length minimization have been published [28, 54]. Unfortunately, exact wire length minimization may take longer CPU time than that of graph-based compaction. Therefore, many compactors use heuristic approaches.

9.4.2 DELAY MINIMIZATION
ON TIMING CRITICAL PATHS

Minimizing total wire length does not always result in minimal timing delay on critical paths. When delay bounds on critical paths is very important, the delay bounds can be minimized first and then the area is reduced without increasing the delay bounds. Both of these steps are formulated as linear programming problems. Figure 9.20 illustrates the results of three approaches. The critical path is shown by thick lines in the figure [62].

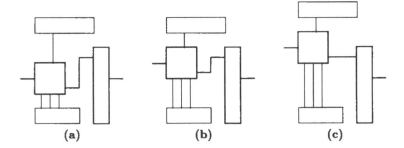

(a)   (b)   (c)

Figure 9.20 Results of three different approaches
(a) The total wire length is minimized
(b) The timing critical path is minimized after area compaction
(c) The path is minimized before area compaction

9.4.3 CROSSTALK MINIMIZATION

The most critical problems other than total wire length are those concerning proximity and separation of the layout elements. It is well known that too much proximate placement of layout elements creates crosstalk, delay, and low yield, and too much separated placement stretches wires and creates nonuniform characteristics of the specific devices [41]. Simple one-dimensional constraint graph compaction pushes all the elements in one direction to place them at their minimum legal locations.

In [41], a one-dimensional compaction algorithm with attractive and repulsive constraints is proposed. The attractive (repulsive) constraint is designed to minimize (maximize) the maximum (minimum) spaces weighted with the given coefficients. This implies that the specified spaces are shrunk (expanded) in proportion to the attractive (repulsive) coefficients as much as possible, as if certain attractive (repulsive) forces with the magnitude of the coefficients were affecting the specified elements. Consequently, certain devices and wires can be placed closely in the attractive case. On the contrary, certain devices and wires can be scattered as much as possible in the repulsive case. This compaction problem can also be formulated as a linear programming problem.

9.4.4 LAYOUT OPTIMIZATION
BY LOCAL TRANSFORMATION

Performance optimization during layout compaction only is rather restricted since the basic topology of a layout cannot be modified by compaction. Therefore, further layout optimization can be achieved by routing and placement transformations. Hojati [20] proposed topology optimization by using routing pattern modification. Each pattern (a small connected piece of routing) is modified as follows. To reduce wire crossings, a pattern is taken and the number of wire crossings is minimized for that pattern. To build better Steiner trees, the pattern is optimized with respect to wire length and the number of corners. To solve the IO terminal alignment problem, routings around IOs are optimized. To perform unconstrained via minimization, the number of wire crossings is reduced first and then the via minimization routine is applied.

9.5 SUMMARY

In this chapter, several typical compaction techniques are discussed. Emphasis is given to hierarchical compaction and performance-oriented compaction techniques. The following capabilities are essential and challenging for state-of-the-art compactors:

- Capability of effectively handling "dirty" design rules including conditional rules.
- Hierarchy-preserving hierarchical compaction capability for general hierarchically designed circuits.
- Two-dimensional compaction or optimization by two-dimensional arrangement of circuit elements.
- Compaction to minimize crosstalks and/or stray parasitic capacitances and resistances.

Many problems pertinent to layout compaction can be formulated by (integer) linear programming problems. Since solution techniques for the (integer) linear programming problems are not efficient enough to deal with the size of current VLSI circuits, various heuristics are used either to reduce the problem size (the number of variables or equations) or to find an approximate solution efficiently.

There are still many problems to be studied for compaction of analog circuits and non-Manhattan style circuits. Analog circuits may yield sophisticated constraints like the constraints for impedance matching or balancing. When diagonal wires (which are not horizontal or vertical wires) or curved wires are used, one-dimensional movements of circuit elements are severely limited since few elements can be moved without affecting the shapes of wires. Therefore, well-known one-dimensional compaction approaches are not effective for non-Manhattan layouts and new efficient and effective techniques should be developed.

REFERENCES

1. B. Ackland and N. Weste, "An Automatic Assembly Tool for Virtual Grid Symbolic Layout," *VLSI '83, Trondheim Norway*, 1983, pp. 457-466.

2. B. W. Bales, "Layout Rule Spacing of Symbolic Integrated Circuit Artwork," *UCB/ERL M82/72*, 1982.

3. S. Bamji and R. Varadarajan, "MSTC: A Method for Identifying Over-constraints during Hierarchical Compaction," *Proc. of the 30th Design Automation Conf.*, 1993, pp. 389-394.

4. DD. Boyer, "Symbolic Layout Compaction Bench-marks," *Proc. IEEE Int. Conf. on Computer Design*, Oct. 1987, pp. 186-191, 209-217.

5. J. Burns and R. Newton, "SPARCS: A New Constraint-Based IC Symbolic Layout Spacer," *Proc. IEEE Custom Integrated Circuits Conference*, 1986.

6. J. Burns and A. Newton, "Efficient Constraint Generation for Hierarchical Compaction," *Proc. IEEE Int. Conf. on Computer Design*, Oct. 1987, pp. 197-200.

7. Cheng, C. K., Deng, X., Liao, Y. Z. and S. Z. Yao, "Symbolic Layout Compaction Under Conditional Design Rules," *IEEE Trans. on CAD*, vol. 11, no. 4, April 1992, pp. 475-485.

8. Y. E. Cho, "A Subjective Review of Compaction," *Proc. of the 22nd Design Automation Conf.*, 1985, pp. 396-404.

9. W. Crocker, R. Varadarajan, and C. Lo, "MACS: A Module Assembly and Compaction System," *Proc. IEEE Int. Conf. on Computer Design*, Oct. 1987, pp. 205-208.

10. J. Dao, N. Matsumoto, T. Hamai, C. Ogawa, and S. Mori, "A Compaction Method for Full Chip VLSI Layouts," *Proc. of the 30th Design Automation Conf.*, 1993, pp. 407-412.

11. D. N. Deutsch, "A Compacted Channel Routing," *Proc. IEEE Int. Conf. on CAD*, 1985, pp. 223-225.

12. J. Doenhardt and T. Lengauer, "Algorithmic Aspects of One-Dimensional Layout Compaction," *IEEE Trans. on Computer Aided Design*, 1987, pp. 863-878.

13. A. E. Dunlop, "SLIP: Symbolic Layout of Integrated Circuits with Compaction," *Computer Aided Design*, Nov. 1978, pp. 387-391.

14. P. A. Eichenberger, "Fast Symbolic Layout Translation for Custom VLSI Integrated Circuits," *Technical Report No. 86-295, Stanford Univ.*, April 1986.

15. P. Eichenberger and M. Horowitz, "Toroidal Compaction of Symbolic Layouts for Regular Structures," *Proc. IEEE Int. Conf. on CAD*, 1987, pp. 142-145.

16. S. Gao, M. Kaufmann, and F. M. Maley, "Advances in Homotopic Layout Compaction," *Proc. Symp. Parallel Algorithms and Architectures*, 1989, pp. 273-282.

17. A. Herrigel, J. Kamm, and W. Fichtner, "Macrocell-Level Compaction with Automatic Jog Introduction," *Proc. IEEE Int. Conf. on Computer Design*, 1989, pp. 536-539.

18. D. D. Hill, Fishburn, J. P. and M. P. Leland, "Effective Use of Virtual Grid Compaction in Macro-Module Generators," *Proc. of the 22nd Design Automation Conf.*, 1985, pp. 777-780.

19. R. Hojati and D. P. Chen, "Transformation-Based Layout Optimization," *Proc. IEEE Custom Integrated Circuits Conference*, 1990, pp. 30.5.1-30.5.3.

20. R. Hojati, "Layout Optimization by Pattern Modification," *Proc. of the 27th Design Automation Conf.*, 1990, pp. 632-637.

21. M. Y. Hsueh, "Symbolic Layout and Compaction," *UCB/ERL M79/80*, 1979.

22. M. Ishikawa, T. Matsuda, T. Yoshimura, and S. Goto, "Compaction-Based Custom LSI Layout Design Method," *IEEE Trans. on CAD*, vol. CAD-6, no. 3, May 1987, pp. 374-382.

23. G. Kedem and H. Watanabe, "Graph-Optimization Techniques for IC Layout and Compaction," *IEEE Trans. on CAD of ICAS*, vol. 3, no. 1, Jan. 1984.

24. W. Kim, J. Lee, and H. Shin, "A New Hierarchical Layout Compactor Using Simplified Graph Models," *Proc. of the 29th Design Automation Conf.*, 1992, pp. 323-326.

25. C. Kingsley, "A Hierarchical Error-Tolerant Compactor," *Proc. of the 21st Design Automation Conf.*, 1984, pp. 126-132.

26. A. A. J. Lange, J. S. J. Lange, and J. F. Vink, "A Hierarchical Constraint Graph Generation and Compaction System for Symbolic Layout," *Proc. IEEE Int. Conf. on Computer Design*, 1989, pp. 532-535.

27. J. F. Lee, "A Layout Compaction Algorithm with Multiple Grid Constraints," *Proc. IEEE Int. Conf. on Computer Design*, 1991, pp. 30-33.

28. J. F. Lee and C. K. Wong, "A Performance-Aimed Cell Compactor with Automatic Jogs," *IEEE Trans. on CAD*, vol. 11, no. 12, Dec. 1992, pp. 1495-1507.

29. Y. Liao and C. K. Wong, "An Algorithm to Compact a VLSI Symbolic Layout with Mixed Constraints," *IEEE Trans. on CAD*, vol. CAD-2, no. 2, April 1983, pp. 62-69.

30. C. Lo, R. Varadarajan, and W. Crocker, "Compaction with Performance Optimization," *Proc. IEEE Int. Symp. Circuit Syst.*, 1987, pp. 514-517.

31. C. Lo and R. Varadarajan, "An $O(n^{1.5} \log n)$ 1-D Compaction Algorithm," *Proc. of the 27th Design Automation Conf.*, 1990, pp 382-387.

32. F. M. Maley, "Compaction with automatic jog-insertion," *Proc. Chapel Hill Conf. on VLSI*, 1985, pp. 261-283.

33. D. Marple, "A Hierarchy Preserving Hierarchical Compactor," *Proc. of the 27th Design Automation Conf.*, 1990, pp. 375-381.

34. J. L. Martineau, G. Bois, and E. Cerny, "Automatic Jog Insertion for 2D Mask Compaction: A Global Optimization Perspective," *Proc. EDAC*, 1992, pp. 508-512.

35. N. Matsumoto, Y. Watanabe, and K. Usami, "Datapath Generator Based on Gate-Level Symbolic Layout," *Proc. of the 27th Design Automation Conf.*, 1990, pp. 388-393.

36. K. Mehlhorn and W. Rülling, "Compaction on the Torus," *Proc. IEEE Int. Conf. on Computer Design*, 1987, pp. 201-204.

37. K. Mehlhorn and S. Näher, "A Faster Compaction Algorithm with Automatic Jog Insertion," *IEEE Trans. on CAD*, vol. 9, no. 2, Feb. 1990, pp. 158-166.

38. D. Mehlhorn and W. Rülling, "Compaction on the Torus," *IEEE Trans. on CAD*, vol. 9, no. 4, April 1990, pp. 389-397.

39. R. C. Mosteller, A. Frey, and R. Suaya, "2-D Compaction - A Monte Carlo Method," *Proc. the 1987 Stanford Conference*, 1987, pp. 173-197.

40. L. S. Nyland, "Improving Virtual-Grid Compaction Through Grouping," *Proc. of the 24th Design Automation Conf.*, 1987, pp. 305-310.

41. A. Onozawa, "Layout Compaction with Attractive and Repulsive Constraints," *Proc. of the 27th Design Automation Conf.*, 1990, pp. 369-374.

42. P. Pan, S. K. Dong, and C. L. Liu, "Optimal Graph Constraint Reduction for Symbolic Layout Compaction," *Proc. of the 30th Design Automation Conf.*, 1993, pp. 401-406.

43. C. Papadimitriou and K. Steiglitz, *Combinatorial Optimization, Prentice Hall*, 1982.

44. R. G. Parker and R. L. Rardin, *Discrete Optimization, Academic Press*, 1988.

45. T. Pérez-segovia and A. F. Joanblanq, "CACTUS: A Symbolic CMOS Two-Dimensional Compactor," *Proc. EDAC*, 1990, pp. 201-205.

46. S. D. Posluszny, "SLS: An Advanced Symbolic Layout System for Bipolar and FET Design," *IEEE Trans. on CAD*, vol. CAD-5, no. 4, Oct. 1986, pp. 450-458.

47. B. Preas and M. Lorenzetti, *Physical Design Automation of VLSI Systems*, The Benjamin/Cummings Pub., 1988, pp. 211-281.

48. R. Razdan and A. J. Strojwas, "A Statistical Design Rule Developer," *IEEE Trans. on CAD*, vol. CAD-5, no. 4, Oct. 1986, pp. 508-520.

49. M. Reichelt and W. Wolf, "An Improved Cell Model for Hierarchical Constraint Graph Compaction," *Proc. IEEE Int. Conf. on CAD*, 1986, pp. 482-485.

50. M. Rülling and T. Schilz, "A New Method for Hierarchical Compaction," *IEEE Trans. on CAD of IC&S*, vol. 12, no. 2, Feb. 1993, pp. 353-360.

51. S. Sastry and A. Parker, "The Complexity of Two-Dimensional Compaction of VLSI Layouts," *Proc. IEEE Int. Conf. on Circuits and Computers*, 1982, pp. 402-406.

52. T. Sasaki, K. Kawakyu, T. Seshita, A. Kameyama, T. Terada, Y. Kitaura, K. Ishida, and N. Uchitomi, "Cell-Shifting Compaction of Building-Cell Methodology for High-Speed GaAs Standard-Cell LSIs," *Proc. IEEE Custom Integrated Circuits Conference*, 1991, pp. 28.3.1-28.3.4.

53. M. Schlag, Y. Z. Liao, and C. K. Wong, "An Algorithm for Optimal Two-Dimensional Compaction of VLSI Layouts," *Integration, VLSI J.*, 1983, pp. 179-209.

54. W. L. Schiele, "Improved Compaction by Minimized Length of Wires," *Proc. of the 20th Design Automation Conf.*, 1983, pp. 121-127.

55. H. Shin, A. L. Sangiovanni-Vincentelli, and C. H. Séquin, "Two-Dimensional Compaction by 'Zone Refining'," *Proc. of the 23rd Design Automation Conf.*, 1986, pp. 115-122.

56. H. Shin, A. Sangiovanni-Vincentelli, and C. H. Sequin, "Two-Dimensional Module Compactor Based on 'Zone-Refining'," *Proc. IEEE Int. Conf. on Computer Design*, 1987, pp. 201-204.

57. H. Shin and C. Lo, "An Efficient Two-Dimensional Layout Compaction Algorithm," *Proc. of the 26th Design Automation Conf.*, 1989, pp. 290-295.

58. H. Shin, A. L. Sangiovanni-Vincentelli, and C. H. Séquin, "Zone-Refining: Techniques for IC Layout Compaction," *IEEE Trans. on CAD*, vol. 9, no. 2, Feb. 1990, pp. 167-179.

59. D. Tan and N. Weste, "Virtual Grid Symbolic Layout 1987," *Proc. IEEE Int. Conf. on Computer Design*, 1987, pp. 192-196.

60. K. Usami, Y. Sugeno, N. Matsumoto, and S. Mori, "Hierarchical Symbolic Design Methodology for Large-Scale Datapaths," *Proc. IEEE Custom Integrated Circuits Conf.*, 1990, pp. 30.3.1-30.3.4.

61. J. Valainis, S. Kaptanoglu, E. Liu, and R. Suaya, "Two-Dimensional IC Layout Compaction Based on Topological Design Rule Checking," *IEEE Trans. on CAD*, vol. 9, no. 3, March 1990, pp. 260-275.

62. L. Y. Wang, Y. T. Lai, B. D. Liu, and T. C. Chang, "Layout Compaction with Minimized Delay Bound on Timing Critical Paths," *Proc. IEEE Int. Symp. Circuit and Syst.*, 1993, pp. 1849-1852.

63. J. Weinstein, "A New Two-Dimensional Compaction Algorithm for Symbolic Layout," *Proc. IEEE Custom Integrated Circuits Conf.*, 1987, pp. 605-609.

64. S. J. Wei, J. Leroy, and R. Crappe, "An Efficient Two-Dimensional Compaction Algorithm for VLSI Symbolic Layout," *Proc. EDAC*, 1990, pp. 196-200.

65. J. D. Williams, "STICKS: Graphics Editor for High-Level LSI Design," *Proc. National Computer Conf.*, 1978, pp. 289-295.

66. W. Wolf, "An Experimental Comparison of 1-D Compaction Algorithms," *Proc. Chapel Hill Conf. on VLSI*, 1985, pp. 165-179.

67. W. H. Wolf, R. G. Mathews, J. A. Newkirk, and R. W. Dutton, "Algorithms for Optimizing, Two-Dimensional Symbolic Layout Compaction," *IEEE Trans. on CAD*, vol. 7, no. 4, April 1988, pp. 451-466.

68. X. M. Xiong and E. S. Kuh, "Nutcracker: An Efficient and Intelligent Channel Spacer," *Proc. of the 24th Design Automation Conf.*, 1987, pp. 298-304.

69. S. Z. Yao, C. K. Cheng, D. Dutt, S. Nahar, and C. Y. Lo, "Cell-Based Hierarchical Pitchmatching Compaction Using Minimal LP," *Proc. of the 30th Design Automation Conf.*, 1993, pp. 395-400.

Chapter 10

RIP-UP AND REROUTE STRATEGIES

Manuela B. Raith

Automated routing of printed circuit boards (PCB) and very large-scale integrated (VLSI) chips is a major task of the physical design process. The goal of routing is to realize all logical signal connections without violation of any rules required by a particular technology. Unfortunately, because of the enormous complexity of the routing task, up to now there exists no fully automatic router capable of meeting all requirements for all routing problems. Given complex circuit demands, the sequential processing of the signal connections frequently leads to blocking situations. Since previously routed connections constitute obstacles for later connections, they prevent the completion of the routing task. As a consequence, we may have a blocking situation and an unconnect is leftover. Thus, a human designer has the tedious task of manually routing leftover unconnects, a process, which requires valuable time and additionally is a source of errors. Automated rip-up and reroute strategies constitute an answer to this problem and will be introduced and discussed in this chapter.

10.1 BASIC CONCEPTS AND STRATEGIES

There exist several different concepts and strategies to solve routing blockades which will be discussed in this chapter. First, in Section 10.1.1, the problem of blocking situations will be defined. In Section 10.1.2 we will discuss current approaches. Section 10.1.3 will point out the limitation of the different strategies and in Section 10.1.4 we will formulate requirements for future approaches.

10.1.1 PROBLEM DESCRIPTION

An example for a typical blocking situation is shown in Figure 10.1a. In this small routing problem two layers are available and between all tracks a spacing of one grid point is required. The task is to complete the connection between s and t. It is easy to see that due to the previously routed connections no valid path can be found. However, there exists a routing solution which is shown in Figure 10.1b.

The relocation of some tracks allows completion without any violation of spacing rules (see Figure 10.1b). In the next section we will see that there exist procedures to achieve such a solution automatically.

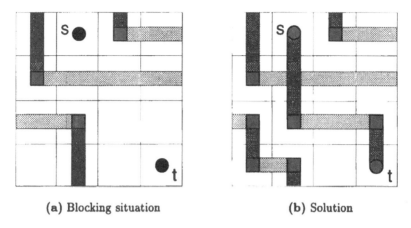

(a) Blocking situation   (b) Solution

Figure 10.1   Incomplete routing and a rerouting solution

10.1.2   CURRENT APPROACHES

Only a few rip-up and reroute approaches have been published in the last years [2 - 13]. They have been developed to automatically reorganize existing routes to generate room for leftover connections. All these methods and strategies essentially work in three steps:

(1) Rip-up of one or more connections completed previously to eliminate a blocking situation.
(2) Routing of the blocked connection.
(3) Rerouting of all connections ripped-up in step (1).

In particular, the first point, the selection of routed connections to be ripped up, must be handled very carefully, since this determines the success of the whole rip-up process. A wrong selection in step (1) leads to new blocking situations in step (3) and an additional rip-up becomes necessary. Because of this iteration, the rip-up procedure may get out of control and an oscillation occurs. In such a case, the number of leftovers can increase instead of being reduced. The known algorithms can be classified into two main categories, according to the strategy they use. We distinguish between the method of "push aside" and the "strong rip-up".

10.1.2.1   PUSH ASIDE

This technique implies a local relocation of the existing wiring, while maintaining all existing connections, in order to create room for the failed connection. Thus, this approach does not cancel previously wired connections before finding an alternative path. The "push aside" technique is a very popular rip-up and reroute strategy and has been discussed

in numerous papers [4, 5, 6, 11]. It locally shifts tracks which block
the routing process into neighbored areas. This strategy is illustrated
in Figure 10.2. In this example only a single layer is available. Three
connections have already been routed successfully but now the way for
the forth connection between s and t is blocked. Temporarily, in Figure
10.2b an additional grid has been introduced on which a path can be
found by using, e.g., a Lee-type router. This process of searching for
a path on a finer grid than the predefined ("real") grid, is known as
"squeeze through technique".

The path on the dummy grid must then be realized on a neighbored
real grid. This can be done by pushing tracks into free areas, shown in
Figure 10.2c. In the last step the previously blocked connection can be
routed.

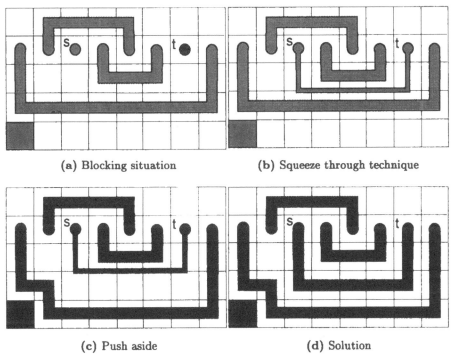

(a) Blocking situation   (b) Squeeze through technique

(c) Push aside   (d) Solution

Figure 10.2 A push aside technique

During this procedure a relocation of existing tracks has been done
before the blocked connection has been completed. This means that
with this method the number of unconnects can not be increased. Very
often, in order to generate new room, it is not enough to push only one
connection. In Figure 10.3a we will assume that the marked point must
be freed. This demand influences 24 other grid points marked by a circle.

At the end of such a propagation it can be recognized if a push aside will be successful. The circles at the border of the propagation pyramid are all located on free grid points, which indicates success. In our example four tracks must be relocated. This method of rip-up and reroute seems to be very elegant. However, during implementation we will encounter many unexpected problems. While for a human being it is very easy to recognize even complicated push aside possibilities, a computer must run through a multitude of different decision processes. Nearly always there exist several possibilities to squeeze through. Remember, in practice you mostly have more than one layer available. Then you need to decide in which direction you want the algorithm to push. The number of possibilities increases with the number of layers and with the number of propagation steps. Very soon the problem gets too complex to calculate all possibilities. Additionally, this method follows very strict regulations because relations like "left from" or "right from" are untouched.

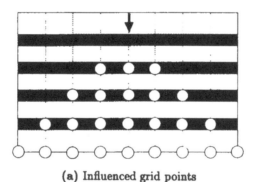

(a) Influenced grid points

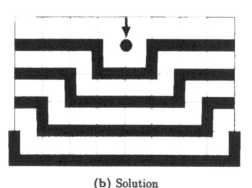

(b) Solution

Figure 10.3 Point propagation for pushing a set of paths

10.1.2.2 REFLECTION

Sometimes "push aside" is combined with a reflection technique [6, 11]. The advantage of this variant is that the topology can be changed. Figure 10.4 shows a blocking situation based on two layers. By reflection of one via the blocking situation can be solved.

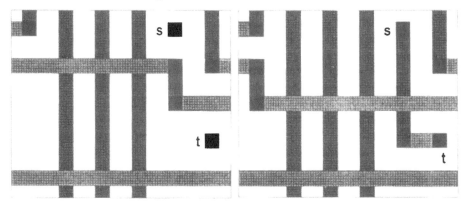

Figure 10.4 A reflection technique

In general, there are two possibilities for the implementation of this strategy. One is to allow the push aside procedure to push track segments over routed nets and to tolerate temporary errors. The second method checks in advance if there is enough capacity to include the reflected pattern of the blocking segments. The advantage of so-called "weak" rip-up techniques like push aside and reflection is that reroutability never becomes critical because it is always checked in advance. The main drawback is its strong local view. Since only local reorganization of the wiring is considered, often no solution is found and this approach can only be used as a first attempt. In case of a failure another strategy must be adopted.

10.1.2.3 STRONG RIP-UP

This technique is based on a (mostly very poor) analysis of the current blocking environment. In every iteration step, it removes a single connection till the failed connection can be routed successfully. To our knowledge, all published strong rip-up algorithms for detailed routing remove laid out connections sequentially and thus, in each step a connection is selected without considering future and past rip-ups. Due to this unstructured progression, these algorithms often delete connections which do not contribute to the overall goal of deleting a "minimal" set of reroutable connections which allow the currently blocked connection to be routed. In general, this results in a removal of a too large set of connections, which have then to be rerouted again, leading possibly to an increased number of final unconnects. This fact may be explained by a simple example: The router has identified a blocking situation. The rip-up and reroute procedure decides to remove connection A, but does not know whether the blocked connection will be routable after this step. It might be necessary to additionally rip-up connections B, C and D to eliminate the blocking situation. Altogether the rip-up procedure has removed four connections. For the same problem, a human being might have recognized connection E as an alternative and probably better solution. As you see, a "wrong" decision in the beginning can endanger the whole rip-up and reroute process. To counteract this, most programs make use of control mechanism to stop the rip-up process when there is an oscillation danger. Two more papers should be mentioned. Marek-Sadowska introduced an interesting non-sequential rip-up strategy based on cross-capacity limitations but applicable for global routing only [7]. A non-sequential router was also presented by Rosenberg [10], however, the rip-up and reroute process is based on a sequential greedy heuristic.

10.1.3 LIMITATIONS

The drawbacks of the current approaches are visible in the example shown in Figure 10.5. For this example the routing task is limited to one layer; the connection between s and t is blocked.

Figure 10.5b shows a result which has been achieved after the rip-up of connections 1 to 4. The previously blocked connection between s and t could be finished successfully. But, for three of the ripped-up connections no valid path can be found any more. Connection 2, 3 and 4 are now leftover. The original problem with one unconnect has been changed to a new problem with three unconnects. Indeed, lots of rip-up strategies would choose this "solution" because they rip-up one connection in every iteration step without taking the resulting situation into consideration. Starting from the source s they find connection C_2 obviously interrupting the routing process. After the removal of C_2 they get stuck in front of C_3 and so on. This sequence of ripping up connections leads to a cancellation of an unnecessarily great number of connections. Since all cancelled connections must be subsequently rewired, the overall result with respect to the completion rate can definitely be deteriorated by this excessive cancelling of previously laid out connections.

That there exists a solution with 100% completion rate is shown in Figure 10.5c. The rip-up of only two connections, C_1 and C_5, allows the routing between s and t and additionally also for the ripped-up connection an alternative path can be found.

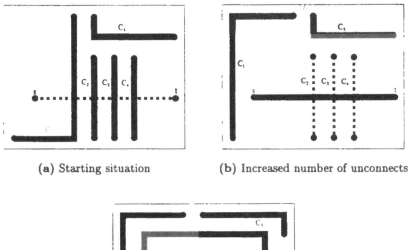

(a) Starting situation   (b) Increased number of unconnects

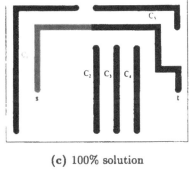

(c) 100% solution

Figure 10.5 Reroute connections crossed by an incomplete connection

10.1.4 REQUIREMENTS FOR FUTURE STRATEGIES

Because of the previously mentioned drawbacks, the aim must be the development of "intelligent" rip-up and reroute strategies on the basis of a specific analysis of the given routing blockade. Very important for the success of any strategy seems to be the following assumption: *An automatic rip-up and reroute procedure must be based on a strategy which provides powerful "look-ahead features".*

The "look-ahead feature" stands for the capability to consider in advance routing problems which may occur in the future. The removal of connections must directly depend on the particular routing problem, and not on the order of rip-ups as it used to be in the above-discussed strong rip-up strategies. It definitely does not make sense to remove, for instance, four connections without even knowing that there would also be a solution by ripping up only one connection. Thus, any rip-up and reroute approach should analyze the current blocking situation very much in detail. The time spent for this analysis can easily be absorbed by avoiding unnecessary rip-up and reroute steps. A prerequisite for every successful rip-up and reroute strategy is the generation of useful information about each specific blocking situation. Only then does there exist a chance of performing the "right" rip-ups in each blocking situation. What is meant by "useful information"? Probably there exist more than one answer and it is worthwhile thinking about it. An example of a "useful information" is the following: Suppose, all different "minimal" combinations of connections whose rip-up will eliminate the blocking situation can be calculated in advance. By "minimal" we mean that for each of those rip-up sets, no proper subset exists which would solve the blocking situation as well. It may be assumed that after knowing all possible minimal rip-up sets, we will have a better base to decide which one will fit best. The next section will show that such an approach exists.

10.2 THE METHOD OF MINIMAL RIP-UP SETS

We now present a brand new rip-up and reroute strategy based on a hypergraph model [9]. The novelty of this approach is that given a blocking situation, a minimal set of connections to be ripped-up is simultaneously calculated and removed. The selection of the best rip-up set is based on its reroutability potential. Results have shown that not only a lower number of rip-ups is produced, but also better results compared to an industrial rip-up and reroute based router are obtained. For more details on routing results see [8].

10.2.1 MINIMIZATION PROBLEM

A blocking situation occurs when at least one connection is leftover. The reason for the failed completion is that previously routed connections do not leave enough space for further routing. Let (s, t) be the failed connection. Then there exists no legal path to lay out a track from s (source) to t (target) without violation of any design rules.

Definition 10.1 A set $RS(s, t) = \{c_1, c_2, ..., c_n\}$ of previously routed connection is called a *rip-up set* of (s, t) if their removal allows to route a valid path between s and t.

To eliminate the blocking situation we need to find a rip-up set as defined before. Note, that the conditions that make up a rip-up set are still very weak, because removing all wired connections will always do the job. Let us think about similar problems. In general, the rip-up and reroute problem is a path-finding problem under specific conditions. Suppose there is a city with some streets which can be passed and others which are blocked by barriers. The rip-up problem then can be formulated as a path-finding problem, where a "cheapest" path between two houses must be found. Differently from a normal path-finding problem, we additionally have to get rid of blockades which keep us from moving through a street. The removal of a blockade increases the total costs and thus a path with lots of blockades is more expensive than one with only a few blockades. This problem could be easily solved using for example the normal Dijkstra algorithm. But there is one additional fact to be considered: There might be several blockades due to the same source. In such a case it would be sufficient to spend some cost units to get rid of one blockade because simultaneously some other barriers would disappear as well. Remember, there might be one and the same connection which blocks a net in several different areas. Different from a normal path-finding problem, the algorithm needs to consider that there exist several possibilities to reach the same point. Suppose that half of the way between s and t has been realized and you now have to decide whether you want to remove connection C_1 or C_2. This decision can be done only if you know all possible paths leading to this specific point. There might be one path where C_1 already has been deleted in an earlier stage. Then no additional costs would be necessary. As you see, it is not enough to consider only the lowest costs with which you can reach a specific point. A successive cost-based path-finding method would lead to the fatal situation that the cheapest possibility might not be considered because in the beginning phase it seemed to be too expensive. In most cases there exists more than one rip-up set whose removal eliminates the blocking situation. The problem is to find a "good" solution, so that not only the blocked connection can be routed, but also the ripped-up connections. It seems to be very natural to ask for a rip-up set containing as few connections as possible. The removal of too many connections can always end in an oscillation which leads to the possibility that after the rip-up and reroute procedure more connection are leftover than before (see Figure 10.5b). In order to give every connection different cost units we introduce a function F.

$$F : C \rightarrow \{1, 2, 3, ...\}$$

F indicates the supposed difficulty for finding an alternative pass, in case this connection will be ripped up. Then we have to find the rip-up set $RS_{min}(s, t)$ with the constraint condition:

$$\sum_{c_i \in RS(s,t)} F(c_i)\xi(c_i) \rightarrow min \qquad (10.1)$$

where $\xi(c_i) = 1$ if c_i has been selected for a rip-up, else 0.

This seems easy, but the difficulty is hidden in the demand of checking whether a set is a rip-up set or not. In a real routing example there very often exist thousands of connections and thus it is impossible to

calculate all sums and to check if it is a rip-up set. A detailed analysis of
the blockade problem and the development of alternative solutions are
necessary.

10.2.2 THE NEW STRATEGY

To develop an alternative strategy we use the following characteristic
of the function F: Since $F(C_i)$ can never be a negative number (see
definition), we can find the minimum sum by calculating all minimal
rip-up sets as defined in Definition 10.2:

Definition 10.2 A rip-up set $RS(s,t) = \{c_1, c_2, ..., c_n\}$ is called a minimal rip-up set $(MRS(s,t))$ if no proper subset exists which eliminate
the blocking situation as well.

If we know all minimal rip-up sets to a given blocking situation then
we can easily calculate the sums as claimed in Equation 10.1 to eventually
find the minimum. We mentioned earlier that strong rip-up approaches
usually find a rip-up set by sequentially deleting connections; thus, they
may generate many superfluous rip-ups. Our goal will be to find a rip-
up set (1) simultaneously and (2) of minimal cardinality (see Definition
10.2). In order to be able to calculate all minimal rip-up sets (MRS)
responsible for a specific blocking situation, we need to build a structured
global view of its environment. Therefore, we look for the smallest region,
called critical area (see Section 10.2.2.1), where all useful rip-ups have to
be located and we then analyze this critical area (see Section 10.2.2.2).
In Section 10.2.2.3 we introduce our hypergraph model for finding all
MRSs. Finally, in Section 10.2.2.4 we describe how to select the best
MRS with respect to reroutability.

10.2.2.1 CRITICAL AREA LOCATION

Before we can proceed, we need two more definitions:

Definition 10.3 An island $I(v_i)$ is a set of free grid points which can
be reached by a path-finding algorithm starting from v_i. By reachable
we mean that there is a path between any grid point of the island and
the vertex v_i. $I(v_i)$ is maximal so that there exists no vertex which can
be added.

In particular, we will very often need the islands I_s and I_t. The sets
of points on the borders are called I_s blocking points $(I_s BP)$ and I_t
blocking points $(I_t BP)$, respectively (see Figure 10.6).

Definition 10.4 The critical area, (s,t), is the region between the two
islands $I(s)$ and $I(t)$.

Figure 10.6 explains the idea of I_s, I_t and the critical area. Again, I_s is the area reachable from s (respectively, I_t). The critical area lies between the two areas I_s and I_t and is shaded.

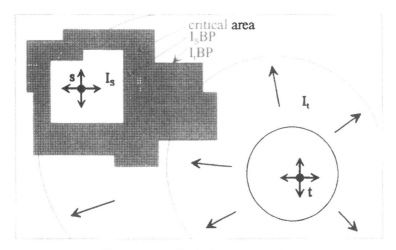

Figure 10.6 Critical area generation

To each boundary of an island corresponds a set of connections such that the removal of anyone of them will reduce the critical area. We call the set of connections related to the source/target islands respectively I_s blocking connections (I_sBC) and I_t blocking connections (I_tBC). To route a connection from s to t we cannot avoid crossing the critical area and its boundaries as you can easily see in Figure 10.6. This leads to the following:

Theorem 10.1 *All connections belonging to any MRS(s,t) use at least one vertex of the critical area.* (The Proof is given in [9].)

This theorem is fundamental for our algorithm. It states that searching for rip-up candidates only makes sense inside the critical area.

10.2.2.2 CRITICAL AREA ANALYSIS

In order to find all MRSs which are candidates for the simultaneous ripping process, the critical area will be constructed further into the islands $I_1, ..., I_n$. Similarly to I_s and I_t, the set of points belonging to the border of I_i is called the I_i blocking points (I_iBP), and the set of connections whose removal expands the island I_i is called the I_i blocking connections (I_iBC), $(i = 1, ..., n)$.

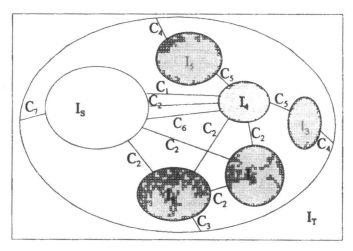

Figure 10.7 Islands of the critical area

As a result of the described assignment, we have detected all useful connections being candidates for a rip-up, since

(1)   a connection contained totally in one island is not useful for a rip-up because it would not allow us to escape into another area;

(2)   connections between the islands permit us to walk from one island to another and are therefore ideal candidates to be ripped-up.

The situation is visualized in Figure 10.7. Each useful connection C_k is represented by an edge between I_i and I_j if it is an element of $I_i BC \cap I_j BC$ $(i, j = s, t, 1, ..., n)$.

10.2.2.3 HYPERGRAPH MODEL

Since several connections may appear between two islands and a connection can run multiple times between different islands, we solve the problem of finding all MRSs on a hypergraph model. First we give some definitions from graph theory which we need in order to formulate our hypergraph model [3].

Definition 10.5 Let $X = \{x_1, x_2, ..., x_n\}$ be a finite set. A hypergraph on X is a family $H = (E_1, E_2, ..., E_m)$ of subsets of X such that

(1) $E_i \neq \emptyset$ $(i = 1, ..., m)$, and

(2) $\sum_{i=1}^{m} E_i = X$

where the elements $x1, x2, ..., x_n$ of X are called *vertices* and the sets $E_1, E_2, ..., E_m$ are called *edges of the hypergraph*.

Note, we do not consider isolated points to be vertices (see condition (2)). A hypergraph H may be drawn as a set of points representing the vertices. The edges are represented by a continuous curve joining the two elements if $|E_i| = 2$, by a loop if $|E_i| = 1$ and by a simple closed curve enclosing the elements if $|E_i| \geq 3$, respectively. Two vertices are considered to be *adjacent* if they are joined by an edge. Two edges are considered to be *incident* if they have the same vertex. A *path* of length n from s to t is a sequence of distinct vertices $v_0, ..., v_n$ whose initial vertex v_0 is s, whose final vertex v_n is t and where the vertices v_i and v_{i+1} are adjacent ($i = 0, ..., n-1$). A hypergraph H is connected if there exists a path in H between any pair of vertices. In our problem each of the islands $I_s, I_t, I_1, ..., I_n$ is represented by a vertex of the hypergraph. An edge $C_k = \{I_1, ..., I_m\}$ is created if there exists a connection C_k with $C_k \in I_iBC$ ($i = 1, ..., m$). In Figure 10.8 the hypergraph $H = (C_1, ..., C_7)$ represents the routing problem of Figure 10.7.

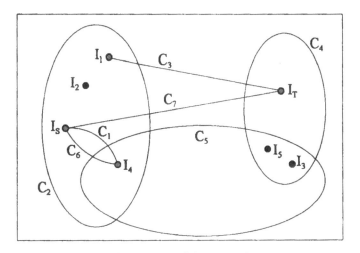

Figure 10.8 A hypergraph

Our goal is to find all MRSs in the hypergraph. First, we discuss the simplest case of finding all MRSs of cardinality one. Then we will show the general case of calculating all MRSs of any cardinality.

10.2.2.3.1 MRS OF CARDINALITY ONE

All sets of MRS of cardinality one are represented by the edges on the subhypergraph $H_1 = \{C_i | I_s, I_t \in C_i; i = 1, ..., n\}$. In the following we will sketch a justification for this statement. An edge of H_1 exists only if $I_sBC \cap I_tBP \neq \emptyset$, or in other words if there exists a common connection between I_sBC and I_tBC. If we remove such a connection we achieve a path from I_sBP to I_tBP crossing the critical area. A path from s to $I_s BP$ and from t to $I_t BP$ exists by definition. Thus we create room for accomplishing the failed connection from s to t. Connections represented by edges which do not include both I_s and I_t never can cross the critical area and therefore they cannot be a MRS of cardinality one. A similar approach has been presented by Ohtsuki [12]. However, his separator approach presumes that there exists a connection reachable from source and target, and thus just a special case of our concept of a critical area is considered.

10.2.2.3.2 MRS OF ANY CARDINALITY

To find all MRSs of any cardinality we first compute all alternating sequences of vertices and edges $S = I_0, C_1, I_1, C_2, I_2, C_3, ..., I_n$, fulfilling the following conditions:

(1) the sequence starts with I_s and ends with I_t, i.e. $I_0 = I_s$ and $I_n = I_t$;

(2) C_1 is incident to I_{i-1} and I_i $(i = 1, ..., n)$;

(3) $C_1, ..., C_n$ and $I_0, ..., I_n$ are pairwise unequal;

(4) the sequence is minimal relating to the set of C_i (there exists no sequence having the same C_i a subset).

Condition (1) ensures creating a path crossing the critical area. In (2) we require that there exists a single connection to be ripped-up, which allows going from I_{i-1} to I_i. Finally, (3) and (4) guarantee the minimality because we can visit the same island only once and no proper subset of the MRS is sufficient to eliminate the blocking situation. In our example the alternating sequences are

$S_1 = \langle I_s, C_7, I_t \rangle$
$S_2 = \langle I_s, C_2, I_1, C_3, I_t \rangle$
$S_3 = \langle I_s, C_1, I_4, C_5, I_3, C_4, I_t \rangle$ or $\langle I_s, C_1, I_4, C_5, I_5, C_4, I_t \rangle$
$S_4 = \langle I_s, C_2, I_4, C_5, I_3, C_4, I_t \rangle$ or $I_s, C_2, I_4, C_5, I_5, C_4, I_t \rangle$
$S_5 = \langle I_s, C_6, I_4, C_5, I_3, C_4, I_t \rangle$ or $\langle I_s, C_6, I_4, C_5, I_5, C_4, I_t \rangle$

As MRSs we get five sets $\{C_7\}$, $\{C_2, C_3\}$, $\{C_2, C_5, C_4\}$, $\{C_1, C_5, C_4\}$ and $\{C_6, C_5, C_4\}$. Each set is contained in one of the sequences.

10.2.2.4 SELECTION OF THE BEST MRS

Up to now we have not considered the reroutability factor. This aspect is taken into account during the selection of the best MRS. Based on heuristics we will estimate for each MRS its reroutability potential and the MRS with the best potential will be ripped-up. Note that the calculation of all MRSs is exact and only the selection of the best MRS is based on a heuristic. Therefore, we will be able to guarantee that after removal of the selected MRS the failed connection will be routable and that no subset of the selected MRS would suffice.

10.2.3 THE ALGORITHM

We will now discuss the algorithm in more detail. The pseudocode in Figure 10.8 sketches our approach.

PROCEDURE *fIND_BEST_MRS;*
begin
    if no_oscillation then begin   { *see* **10.2.3.5**}
    {*avoidance of endless oscillation*}
        best_mrs := ∅;
        critical_area_analysis;   { *see* **10.2.3.1/2**}
        while (not complete_hypergraph) and (best_mrs = ∅) do begin
        {*stop if hypergraph modeling is finished or*
        all MRSs of lowest cardinality are found}
            create_edges_in_hypergraph;   { *see* **10.2.3.3**}
            {*create or update the hypergraph*}
            estimate_reroute_potential;   { *see* **10.2.3.4**}
            update_vertex_info;   { *see* **10.2.3.3**}
            find_best_mrs;   { *see* **10.2.3.3**}
            {*find the best MRS of lowest cardinality*}
        end;
    end;
    if best_mrs ≠ ∅ then rip_up (best_mrs)
    else create_final_unconnect;
end;

The procedure is called whenever the basic router gets stuck in a blocking situation. It
(1) calculates all MRSs of lowest cardinality,
(2) selects an MRS with the best reroutability potential and
(3) performs the rip-up process.

Points (1) and (2) stress the reroutability of ripped-up connections. In addition, (1) leads to a speed-up in creating the hypergraph. After each rip-up process, the basic router reroutes the previously blocked connection and tries to reroute the MRS just removed.

10.2.3.1 CRITICAL AREA GENERATION

We will determine the boundary of the critical area, $I_s BP$ and $I_t BP$ by executing a bidirectional wave expansion starting from source and

target. All reachable grid points are marked. The elements of I_sBP
(I_tBP) must fulfill the conditions of:
(a) being not reachable on either of the layers and
(b) having at least one reachable grid point and at least one point of
     the critical area as neighbor.

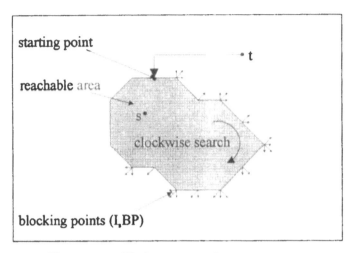

Figure 10.9 Blocking points of source expansion

For instance, in case of detecting I_sBP, we find the boundary incrementally beginning from a starting point (to be defined later) on the
boundary and by looking for neighbors satisfying the boundary conditions mentioned above. All possible directions which have to be checked
in order to find a neighbor, fulfilling conditions (a) and (b), are shown in
Figure 10.9. As starting point we will select the last non-reachable grid
point detected by a rectilinear beam which is sent from t to s. Note, that
a beam starting in s and ending in the first point which is not marked
as reachable would not be sufficient since there may be some unreach-
able islands which have nothing to do with our critical area boundary.
Finally, we calculate all connections I_sBC corresponding to elements of
I_sBP which cause the generation of these blocking points. We will use
them later in the hypergraph model.

10.2.3.2 STRUCTURING THE CRITICAL AREA

A structuring of the critical area as introduced in Section 10.2.2.1 is
necessary for calculating all MRSs of cardinality higher than one. The
goal is to assign grid points reachable from each other to the same island.
To achieve this, we scan the critical area and for each non-assigned free
grid point a wave expansion is started toward either s or t. All grid

points reachable from this seed point form a new island. Additionally, we consider the space between two neighbored connections as an island, if there exists no grid point between them.

10.2.3.3 HYPERGRAPH OPERATIONS

In this section we discuss how to (1) create and update the hypergraph and (2) select the best MRS. We create the hypergraph bidirectionally from s and t till we find all MRSs of the lowest possible cardinality. By using a MRS of the lowest cardinality we avoid creating the complete hypergraph.

10.2.3.3.1 HYPERGRAPH CREATION

Starting in s and t we create two trees, *Tree(s)* and *Tree(t)*, having s and t as their roots and being a subgraph of the hypergraph. In each step, alternately one more level of one of the trees is produced.

Figure 10.10 gives an example: two levels of each of the two trees have been created. Up to now the graph is not connected and thus there exists no MRS of cardinality ≤ 4. Let us denote the vertices connected in the last step that belong to *Tree(t)* with I_i^t (I_j^s, respectively). Vertices not yet connected are called I_k^{unc}.

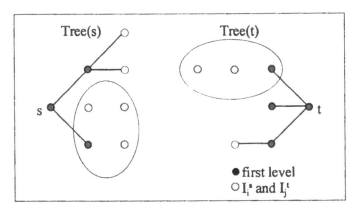

Figure 10.10 Hypergraph creation

Suppose it is the turn of *Tree(t)* to expand. For each connection C^* being an element of
(1) $I_i^t BC \cap I_j^s BC$ or
(2) $I_i^t BC \cap I_k^{unc} BC$.
we could create a new edge C^* (see Section 10.2.2.3 for edge conditions) connecting at least one more vertex I^* with *Tree(t)* in order to com-

plete the hypergraph. To simplify the hypergraph by avoiding creating unnecessary edges, we

(1) estimate the reroutability potential (see Section 10.2.2.4) for connection C^*;
(2) calculate the complete reroutability potential (see Section 10.2.3) of all connections being part of the path starting in t and ending in the vertex I^*;
(3) check if the vertex I^* is already connected; if so, we keep only the best edge (with respect to the reroute potential calculated in (2));
(4) feed the vertex I^* with the potential rerouting data for the whole path calculated in (2).

We stop the process if (1) the level of the current tree is finished and (2) the two trees are merged to one connected graph.

10.2.3.3.2 SELECTION OF THE BEST MRS

Using the information calculated during the creation of the hypergraph, we easily find the best MRS. For each vertex I' being an element of both I_i^t and I_j^s we estimate the complete reroutability potential for the whole path from s to t crossing the island I'. Since we know the potential for the two subpaths starting in s and t and ending in I', the complete potential is calculated in one step. The vertex I' with the best reroutability potential relating to the whole path from s to t is selected. All edges visited by walking from I' to s and t are the elements of the best MRS.

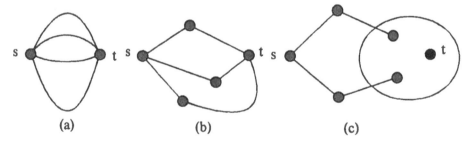

(a)   (b)   (c)

Figure 10.11 Hypergraphs having MRS of different cardinalities

Figure 10.11 shows an example of a hypergraph having an MRS of cardinality one (a), of cardinality two (b), and of cardinality three (c).

10.2.3.4 HEURISTIC FOR
THE REROUTABILITY POTENTIAL

The reroutability heuristic is mainly based on the density of the routing environment. A density percentage for each single connection of the

MRS is determined by scanning its corresponding surrounding rectangle. The environment around the source and target component is weighted higher because reaching these areas is crucial, while other dense areas can be avoided by a detour. Furthermore, the history of a connection already ripped-up several times will get a lower probability to be ripped-up. For estimating the complete reroutability potential of a rip-up set of a cardinality higher than one, we use a quadratic formula to achieve a higher penalty for the connections having a poorer value.

10.2.3.5 OSCILLATION AVOIDANCE

If the rerouting process fails all the time, we run the risk of getting stuck into an endless iteration. To avoid this we use a control procedure which enables us to terminate the process and to create a final unconnect. The procedure keeps count of the current rip-up and reroute iteration. If the number of removed and routed connections is balanced or negative (more ripped-up than routed connections), we are still iterating the same problem. Otherwise the algorithm will have proceeded to a new blocking situation which is not a consequence of the former iteration steps. We restrict the maximum iteration number between 2 and 10. In each new blocking situation this value is recalculated depending on the potential of successfully solving the current blocking situation. This value is again estimated by a density analysis.

10.2.4 EXAMPLE

We will explain the new rip-up and reroute strategy based on a small example. It is the example which we know already from Figure 10.3. Instead of graphs and vertices we will now talk of routing areas and grid points. First we start with a bidirectional search starting from s and from t. All free grid points we can visit belong to I_s (I_t, respectively). The only connection on the border of I_s is connection C_1 and the border of I_t is built out of C_5 and C_1. In addition we will color the region between I_s and I_t, calling them I_1, I_2, I_3 and I_4. After doing the coloring it is time to build up the hypergraph. Each island becomes a vertex in the hypergraph and we include an edge between two or more islands if the connection touches the border of the island. For instance, C_3 lies on the border of I_2 and I_3 and thus a connection is generated in the hypergraph. Connection C_5 is very special. It touches the borders of I_1, I_2, I_3, I_4 and I_t. As you remember, an edge of a hypergraph joining six vertices is represented by a loop. Let us now consider all MRSs of this small example. In the hypergraph we can find two paths to get from s to t. One is $\{C_1, C_2, C_3, C_4\}$. This is the "solution" which produced an oscillation as shown in Figure 10.5b. The other solution identified very quickly with help of the hypergraph is $\{C_1, C_5\}$ which probably produces lower costs than the other one. Removing this MRS will lead to the 100% solution seen in Figure 10.5c.

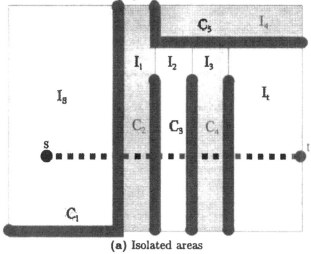

(a) Isolated areas

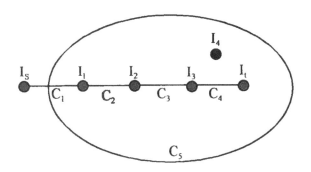

(b) Hypergraph

Figure 10.12 Rip-up rerouting with hypergraph

10.3 CONCLUSIONS AND OPEN PROBLEMS

We proposed a strong rip-up based strategy which tries to avoid the drawbacks of previous approaches and which can be used whenever "weak" rip-up strategies, like push aside, fail. The new aspect is a simultaneous selection of a "minimal" set of connections to be ripped-up based on a global look-ahead analysis of the situation. Our strategy is based on a hypergraph model and consists of calculating all possible "minimal" sets of rip-up combinations, also called minimal rip-up set (MRS), each of which enables us to route the failed connection. By "minimal" we mean that no proper subset of a rip-up set exists that is capable of fulfilling the above-mentioned condition. The novel feature of this approach is that, due to the simultaneous removal of a rip-up set, the success does not depend on the order of the connection removals and only a necessary and sufficient number of rip-ups is performed. An open problem remains the selection of the best MRS, taking into account the reroutability. Up to now we used a heuristic approach, but there might be a better solution.

REFERENCES

1. M. Bartholomeus, W. Weisenseel, "A Routing Concept for Large Sea-of-Gates Designs," *Proc. Euro ASIC*, 1990, pp. 225-229.

2. M. Bartholomeus, M. Raith, "A New Graph Theoretical Approach for the Selection of Rip-Up's," *Euro ASIC*, 1991.

3. C. Berge, "Hypergraphs," *Combinatorics of Finite Sets*, North-Holland Mathematical Library, 1989, pp. 45.

4. M. Bollinger, "A Mature DA System for PC Layout," *Proc. Int. Printed Circuit Conference*, 1979, pp. 85-99.

5. W. A. Dees, Jr., P. G. Karger, "Automated Rip-Up and Reroute Techniques," *Proc. 23rd ACM/IEEE Design Automation Conference*, 1986, pp. 32-439.

6. K. Kawamura, M. Umeda, H. Shiraishi, "Hierarchical Dynamic Router," *Proc. 23rd ACM/IEEE Design Automation Conference*, 1986, pp. 803-809.

7. M. Marek-Sadowska, "Global Router for Gate Array," *Proc. of the IEEE Int. Conf. on Computer Design*, VLSI in Computers, 1984, pp. 332-337.

8. M. Raith, M. Bartholomeus, "A New Hypergraph Based Rip-Up and Reroute Strategy," *28th ACM/IEEE Design Automation Conf.*, 1991, pp. 54-59.

9. M. Raith, "Analyse von Routing-Blockaden mit Hilfe von Hypergraphen (German)," thesis, Ludwig-Maximilians-University Munich, 1994.

10. E. Rosenberg, "A New Iterative Supply/Demand Router with Rip-Up Capability for Printed Circuits Boards," *Proc. 24th ACM/IEEE Design Automation Conference*, 1987, pp. 721-726.

11. H. Shin, A. Sangiovanni-Vincentelli, "A Detailed Router Based on Incremental Routing Modifications: Mighty," *IEEE Trans. on Computer-Aided-Design*, Vol. CAD-6, No. 6, 1987, pp. 942-945.

12. K. Suzuki, Y. Matsunaga, M. Tachibana, T. Ohtsuki, "A Hardware Maze Router with Application to Interactive Rio-Up and Reroute," *IEEE Trans. on Computer-Aided-Design*, Vol. CAD-5, No. 4, 1986, pp. 466-476.

13. T. Watanabe, Y. Sugiyama, "A New Routing Algorithm and its Hardware Implementation", *23rd ACM/IEEE Design Automation Conference*, 1986, pp. 575-580.

Chapter 11

PARALLEL ROUTING

Tae Won Cho

Routing on the parallel machines can be applied either on the special purpose hardwared parallel machine or on the general purpose multiprocessor computer. Architecture of hardwared parallel machine designed specially for the maze routing would use a meshlike grid plane with many thousands of processors in a two-dimensional array. Host computer controls the processors in the system by an array, dealing with instruction and data. Application of hardwared machines are in general confined to the maze routing problems, which is inherently suitable to map on an array. Thus, hardwared machines are not flexible to the variety of problems such as channel and switchbox routing problems. Channel and switchbox routing problems are mostly implemented on the general purpose multiprocessor system such as shared memory [16], pipelined [20] and neural network systems [26].

Parallel routing concepts were found in special cellular arrays by J. Soukup [71] as early as 1978. He has introduced some very original heuristics for parallel routing in software implementation. They are based on "divide and conquer" efforts to improve performance of hardware and software and on the observation that parallel algorithms will run better on parallel hardware. Although he suggested the parallel algorithm concepts only by software, his idea can easily be implemented on the hardware machine. The quest for improved performance has been focused on avoiding the limitations and bottlenecks characteristic of sequential machines, such as contention for system resources and degraded performance due to parallel algorithms imposed on a sequential environment. A strong catalyst for development of parallel machines is the fact that a large subset of computational problems, including many routing problems in design automation, are inherently parallel in nature. Interest in parallel processing has also been heightened, in recent years, by the availability of low-cost, high-performance microprocessors which, in turn, have served as excellent vehicles for experimentation with parallel machine configurations.

In the following section, design guides for parallel routing will be discussed in greater detail. In Section 11.2, maze routing algorithms on the hardwared machines will be discussed. Parallel routing on the multiprocessor environment is described in Section 11.3 and finally future development issues of parallel processing will be described.

369

11.1 DESIGN GUIDES FOR PARALLEL PROCESSING

Early experiments in parallel routing were attempts to do more work within each cycle of a meshlike array processors designed specially [37, 41, 43, 53, 68, 83, 86, 87]. Hardware implementations separated the functions of instruction fetch and instruction execution into many subparts and assigned these tasks to separate processor elements. In this manner, processing of several parts of different instructions in a sequence could be overlapped or "pipelined" to achieve improved performance for vector data, parallel paths through computational logic, and special hardware for array processing [82]. Early motivation for this kind of improved performance seems to have limitations in the complex routing problems such as channel and switchbox problems. In contrast, performance gains in the commercial data processing arena have been achieved through the use of complexes of interconnected computer systems referred to as multiprocessor systems or attached-processor systems. Microprocessor development, in recent years, has led to a variety of sophisticated machine architectures and building blocks for larger systems [48, 58, 73].

Today's microprocessors offer a rich set of functions at a low enough price so that it is now cost-effective to quickly and cheaply build parallel processing systems for experimentation on architectures, control programs, and algorithms. Architectural options range from two or more processors sharing a common bus, to networks of processors and memory. Control programs to support this environment range from unique, special purpose, highly optimized control programs to modification of such widespread operating systems as UNIX. Easy availability of parallel hardware has also triggered an increased level of research into parallel routing [11, 16, 26, 63, 64, 89]. In recent years, parallel neural algorithms [26] and genetic approaches in a general purpose parallel machine in VLSI circuits [14, 15, 17, 50] have been developed.

11.1.1 HARDWARE OPTIONS FOR PARALLEL ROUTING

Developers of applications for parallel routing must choose between implementing their ideas in special hardware or general purpose hardware [28]. The primary trade-off is performance gain versus flexibility in implementing applications. Special hardware usually yields higher performance for specific applications while general purpose hardware can be programmed for a variety of applications. Both approaches have their adherents in the parallel routing community [7, 28, 54, 84, 85]. Before discussing specific implementations, it will be useful to introduce the architectural components of a parallel processor. They are

- Processors. The implementation may be a processor capable of executing only a few specific, application-related instructions. Alternatively, the processors may be as complex as complete general purpose computing systems. A survey of available parallel machines shows that their processors fill the spectrum between these two extremes [3].

- Memories. These are the data storage facilities in the system, and within the context of this tutorial, it is assumed that they are made of random access memories (RAM) only.
- Interconnection Networks. These provide the means for connecting processors to memories of other processors. A variety of interconnection networks have been implemented in hardware [45, 48]. Several examples will be discussed.

In succeeding sections, examples of hardware embodying many of the above constructs will be discussed.

11.1.2 INTERCONNECTION SCHEMES

As noted above, there are a variety of ways in which processors and memories can be interconnected to build practical parallel machines. Figure 11.1 shows two aspects of single bus architectures. Figure 11.1a shows separate processors and memory attached to the bus; this is termed a shared memory system since all memory attached to the bus is accessible to each processor. A distributed memory system occurs when the memory resources are local to the processors (Figure 11.1b).

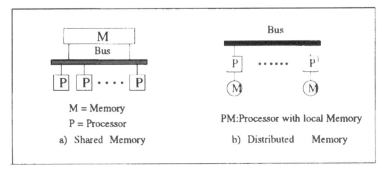

Figure 11.1 Single bus architectures (From *23rd Des. Auto. Conf.* ©1986 IEEE. With permission.)

If the ends of the common bus are joined, the interconnection network becomes a ring. The common bus can become a critical resource if there is a requirement for frequent communication between processors or between processors and memories. Degradation in performance due to contention on the bus can be minimized by providing each processor with a large local memory and by limiting the number of processors and memories that can be connected to the common bus. Bus operation may be synchronous or asynchronous. While synchronous busses may operate at relatively high speeds, data transfer and bus access can be quite complex. Data transfer often requires fixed length data packets with routing information included. Processors must monitor bus activity to receive data packets addressed to them. They must also monitor bus activity in order to determine when to transmit data packets to other processors. These activities constitute an "overhead" associated with data transfer and communication. Asynchronous busses, by contrast, are less complex in operation but may operate somewhat more slowly. An arbitration unit assigns bus usage to a particular processor.

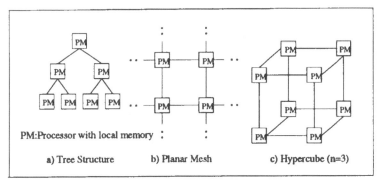

Figure 11.2 Multiport architectures (From *23rd Des. Auto. Conf.* ©1986 IEEE. With permission.)

A finite number of "hand-shaking" signal lines permit a processor to gain control of the bus. A typical scenario is to request use of the bus, wait for permission, use the bus for data transfer, and release the bus. Because of the asynchronous nature of operation, a processor may transfer unlimited amounts of data when it gains control of the bus. Comparison of bus architectures, then, shows that synchronous busses have high data transfer rates and complex transmission and reception of data while asynchronous busses have slower data transfer rates but more flexibility in terms of bus access and the amount of data that can be sent at any given time. A variety of interesting, and more complex, interconnection topologies may be implemented when individual processing and memory elements have multiple external connection ports. Hierarchical, multiple bus structures can be built when two ports are available. Planar mesh structures can be implemented when processing and memory elements each have at least three ports. If four or more ports are available, structures with more than two dimensions can be created. Examples, including a variety of hierarchical topologies, such as, trees, pyramids, and hypercubes are shown in Figure 11.2. Hypercubes are particularly interesting because, in these architectures, the processors are mapped to the nodes of an n-dimensional geometric figure called a hypercube [53]. As Figure 11.2b shows, each processor is linked to its nearest neighbors. As the dimension of the hypercube grows, there is an increase both in the number of processors in the system and the communication links from each processor to its nearest neighbors as in Figure 11.2c. One attribute of this architecture is its compactness; contrary to other network topologies which exhibit longer path lengths, the longest path across a hypercube is "n" links. In other words, communication path length increases linearly with increase in the dimension of the hypercube. Hypercubes with up to 1024 processors (i.e., $n = 10$) are currently being marketed. Providing a control method to pre-define usable interconnection links allows a variety of ring tree on pyramid structures to be mapped within a hypercube.

11.1.3 SPECIAL PURPOSE HARDWARE

Special hardware for design automation has appeared from a number of researchers in recent years. A majority of these machines are designed as accelerators. The control processor maintains and distributes global clock data and provides special services needed to simulate memory array. During the course of routing, signals that are generated in one processor and used in another are transmitted on the common bus. Some of the machines use a cross-point switch for communication between processors. Processor memory is used for storage of data and signal values that are needed only within a processor. Switch memory, by contrast, is used when routing data must be transmitted from one processor to another. A multistage network called a router cell is used to support communication between processors. Hardware engines have been designed, built and demonstrated for maze routing [34, 37, 38, 71].

11.1.4 GENERAL PURPOSE HARDWARE

General purpose computing systems that exhibit varying degrees of parallelism have been available for many years [25, 55]. The earliest systems were characterized by two or more high-speed processors sharing access to a common memory. Even with the performance offered by high-speed processors, there were problems that could not be solved due to limitations in memory capacity and in communication bandwidth between processors or between processors and memory. High-performance microprocessors now make possible the construction of highly parallel machines to relieve these limitations. Microprocessor-based parallel machines built according to several of the architectural schemes shown above are available in the marketplace. These machines are different from special purpose parallel machines because they can be programmed and because they incorporate control programs to provide a service interface between application programs and the hardware. Creation of efficient application programs for parallel machines requires high-level language compilers that incorporate constructs in support of parallel environments. The functions provided by control programs in a parallel machine composed of general purpose processors are essentially independent of the machine architecture. It is important though, that there be a high degree of synergism between the control program and its underlying hardware in order to achieve effective parallel performance. As an interface to the hardware, important control program tasks include scheduling work among the processors [47], monitoring system performance levels [52], and supporting communication between processors [24, 46].

Task management (e.g., scheduling the work among processors) includes such functions as task identification and retrieval, task distribution among processor, and task shifting between processors. These are basic functions in terms of getting useful work done on any machine and also provide the information necessary for effective performance monitoring.

In addition to task management and performance monitoring, control programs often include storage management functions. These functions are necessary in order to prevent fragmentation and to provide sequential access to shared memory areas.

Efficient communication between processors and with the "outside world" is required when tasks running on separate processors must share application data or control information. Message passing and forwarding between processors also requires an efficient communications system. When hardware problems are detected, control program services must be available to sense the problem, isolate the failing components and allow the balance of the machine to continue operating in a degraded manner. To realize the control functions outlined above, an existing uniprocessor control program can be modified. Otherwise, a complete, new, specially tailored control program can be designed and implemented. Modifying an existing program allows many existing functions to be used, but the result may not achieve maximum performance due to limitations in the original uniprocessor design. A completely new control program can avoid these limitation and can be implemented to take advantage of any unique characteristics of the parallel machine for which it is written.

There is a high probability that a special control program will be smaller and provide better performance than an existing program that has been modified. The resources required to develop a special control program, however, may be greater. To enable a developer to design and implement effective application programs for parallel environments, high-level languages that have parallel constructs are needed [22, 29, 33]. The language syntax, for instance, must have functions to support message sending and receiving, task or process creation, and task or process deletion. A storage management interface to the control program is also necessary. This means that it must be possible to obtain and release blocks of storage, and to differentiate between blocks of storage that are private and those that may be shared.

11.2 HARDWARED ROUTING FOR LEE'S ALGORITHM

After J. Soukup [71] introduced the fast maze router with cellular array, many attempts to achieve parallel router emerged. Iosupovici [37, 39] described a modular design with a VLSI chip of $n \times n$ array in O(L) time where L is a shortest path length. He used N pins in an array of $n \times n$, with $N = 4k + 14$ and $k = \log(n - 1)$ to reach the pin limitation problem. The VLSI chips can be expanded to form a larger array with a control unit.

Breuer and Shamsa's L-machine [5] consists of a control unit that communicates with a host computer and sequences the operations of an array of simple processing elements, called L-cells. M. Tachibana, T. Ohtsuki, et al. [41] presented a twisted torus architecture which requires N^2 processors to implement the Lee algorithm on an $n \times n$ grid plane. A control unit and each processing element has its own local memory. He introduced a unique processors-to-cells mapping method. T. Watanabe et al. [82] introduced a maze router on an SIMD cellular array processor showing 230 times faster than a software router on a uniprocessor for 256×256 grid.

Won and Sahni [86] investigated the implementing of Lee's maze routing algorithm on an MIMD hypercube multiprocessor computer. A similar experiment was performed by Olukotun and Mudge [53] in 1987. The parallel routers on the hypercube are classified in the special purpose hardware systems because of its inherent nature of easy mapping characteristics on the array, though it can be dealt with on the general purpose multiprocessor system.

11.2.1 FAST MAZE ROUTER

This method [71] may be the first attempt to use cellular array for maze routing problem. The method combines two techniques: A line search is first directed toward the target. When the line search hits an obstacle, an expansion technique typical of the Lee-type algorithms is then used to "bubble" around the obstacle. The line search can then continue toward the target again.

For the sake of the simplicity, the algorithm is described for an orthogonal one-layer array. For each cell, two values are maintained:

S   **signal definition and a trace back pointer,**
    3 bits required.
C   **reach flag, 2 bits required.**

Two working arrays store coordinates of the cells on the old and new wavefront:

RO is the old reach set (2 coordinates for each entry);
RN is the new reach set (2 coordinates for each entry).

The following codes are used for variables S and C:

```
S = 0 other layer (not used here)
S = 1 left
S = 2 right
S = 3 down
S = 4 up
S = 5 starting point (or subnet)
S = 6 target point (or subnet)
S = 7 other signals or obstructions
C = 0 not reached yet
C = 1 reached through Lee expansion only
C = 2 reached through line search
```

Before the algorithm starts, C is set to 0 for all cells. The algorithm for the fast maze router can be expressed as follows:

(1) Enter the start cell as the first entry into stack RN. Set C = 2 for the starting point.

(2) Move RN into RO; empty RN; reverse order of RO.

(3) One by one, proceeding from the end of the array, take the entries of RO. For each entry, check values of S and C for its neighbors:

```
if C = 2 or S = 7, then do nothing.
if S = 6, then go to (8).
if C < 1 and the neighbor is in the direction
   toward the target, then go to (5).
if C = 0, then do:
   - add the neighbor to the end of RN.
   - set C = 1 for the neighbor.
   - if S < 4, then set S to the traceback code.
```

(4) If RN is empty, then EXIT - there is no connection, otherwise go to (2).

(5) Move RN into RO after the last unused entry.

(6) Add the neighbor to the end of RO. Set C = 2 for the neighbor. If $S \leq 4$, then set S to the traceback code.

(7) Find the next neighbor continuing in the same direction. Check its values of S and C:

```
if C = 2 or S = 7, then go to (3).
if S = 6, then go to (8).
if not closer to the target, then go to (3).
Go to (6).
```

(8) A connection has been found, traceback to the starting point, using S.

11.2.2 ITERATIVE ARRAY MAZE ROUTER

A. Iosupovici and A. Vachidsafa [39] presented a number of new procedures for the simultaneous routing of k nets, which are well traced to parallel processing. A special purpose array processor the PRIAP, has been developed for effective implementation of these routing procedures. Iosupovici [37] has designed a special-purpose two-dimensional array processor router (based on Lee's shortest-path algorithm [44]) dedicated to finding the shortest path on a grid in time of $O(L)$ (i.e., processing time is a linear function of the length L of the shortest path). The major advantage of that router, in comparison with other proposed hardware routers, is its modularity.

General routing techniques involve the sequential processing of nets. If routing completion fails then usually some of the nets are ripped-up and then rerouted [23]. The rip-up and reroute approach may, however, fail. Parallel routing is the problem of simultaneous routing of k disjoint pairs of grid points. The problem is NP-complete for single layer as well as for two-layer routing. W. Dees et al. developed a number of new heuristics for parallel routing [23], which have been tailored for straightforward implementation on a hardware array router called the PRIAP. The first heuristic is based on the detection of bottleneck cells. The definition of a bottleneck cell is as follows.

Definition: A bottleneck cell for a net is a cell whose blockage prevents the routing of that net. Any route for a net must pass through all its bottleneck cells. Furthermore, if the same cell is a bottleneck for nets N_i and N_j, then routing completion is impossible. The following is an effective procedure for listing the bottleneck cells of net N_i:

(1) Find a shortest path for N_i.

(2) Test each cell on the shortest path as a candidate bottleneck cell by marking it as a blocking cell and attempting to route the net. The cell is a bottleneck cell if and only if this attempted routing is unsuccessful.

The time complexity of this procedure is of $O(L^2)$ on the PRIAP. The following bottleneck (BN) iterative procedure, finds the bottleneck cells of nets $N_i, \cdots, N_k$. Its time complexity is of $O(L^4)$.

(1) Derive the cell maps of nets $N_i, \cdots, N_k$. In the cell maps, if a cell is used in net Ni it is marked as a blocking cell for all nets $N_j, j \neq i$.

(2) Repeat for all N_i, $i = 1, k$: If N_i has a new blocking cell, derive its list of bottleneck cells. If N_i and N_j have a common bottleneck cell, go to step 4.

(3) For each cell map which has new bottleneck cells, mark those cells as blocking cells in all other nets and go to step 2. If no cell map has a new bottleneck, go to step 5.

(4) Stop. The problem is not routable.

(5) Stop. Find a shortest path for each of $N_i, i = 1, k$. If the k shortest paths are pairwise disjoint then a solution has been found. Otherwise a solution has not been found as yet.

The bottleneck detection procedure could have one of three outcomes: solution found, solution does not exist, or solution not found though one could exist. For the last outcome two additional procedures are employed, the blocked-gulf detection procedure and the cell assignment procedure. Figure 11.3 shows an example for two nets where rip-up and reroute does not lead to a solution whereas the BN procedure does. For convenience, the two nets are shown on separate cell maps.

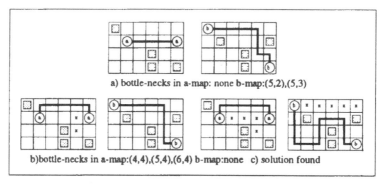

a) bottle-necks in a-map: none b-map:(5,2),(5,3)

b)bottle-necks in a-map:(4,4),(5,4),(6,4) b-map:none   c) solution found

Figure 11.3 Bottleneck detection procedure (From *Proc. IEEE Int. Conf. on CAD-85.* ©1985 IEEE. With permission.)

The purpose of the blocked-gulf detection procedure (BG) is the reduction of the size of the area under consideration for routing by eliminating areas into which routes should not enter. A blockage for given pair of pin cells is defined as a collection of cells whose blockage has no effect on the ability to connect the two [39]. The cell assignment (CA) technique is employed whenever the bottleneck detection procedure is ineffective and following the application of the blocked-gulf detection procedure. It consists of assigning an unused cell to one of the nets according to a number of criteria which tend to improve the probability of a correct assignment while reducing the time complexity of the procedure. Note that only one of the unused cells adjacent to a pin cell belongs to the same net as the pin cell. Therefore, by assigning randomly to a net one of the unused cells adjacent to one of its pins, the probability of the assignment being correct is between 25 to 100 percent, depending on the number of unused cells adjacent to that pin cell. That is a much better probability than random assignment of a cell which is not adjacent to

a pin cell. Furthermore, all the other unused cells adjacent to that pin cells can be turned into blocking cells since they will not be used for that net. The following criteria are applied in the given order in selecting a pin cell whose neighbor is to be assigned to a net:
(1) Pin cells with the lowest number of unused cells adjacent to them.
(2) Pin cells with the lowest total number of unused cells adjacent to their pin cells.
(3) Pin cells which have two unused cells in the X or Y direction but not both.

Rules 1 and 2 select pin cells which gave a higher probability of correct assignment. Rule 3 selects pin cells, which increase the probability of the cell assignment by creating new blocked-gulfs and does not lead to the appearance of bottleneck cells. On the other hand, the addition of bottleneck cells could lead to new blocked-gulfs. The time complexity in selecting a cell for assignment is of $O(L^2)$. Thus, the three procedures can be combined into a procedure dictated by the result of marking blocked-gulfs. The following is the parallel routing algorithm with L as the iteration level:
(1) Initialize L to 0.
(2) Apply the BN procedure. There are three possible outcomes:

```
A. Solution found.    Stop.
B. Solution is not possible.    Go to step 4.
C. Solution has not been found as yet.
   Continue with step 2.
```

(3) Apply the BG procedure.
(4) Let $L = L + 1$. Apply the CA procedure. Go to step 1.
(5) If $L = 0$, The problem is not routable; Stop. If $L < 0$, let $L = L+1$; cancel the last cell assignment. Is there another choice for a cell assignment?

```
A. Yes.   Go to step 3.
B. No.    Go to step 4.
```

The time complexity of each iteration is dominated by the blocked-gulf procedure, and thus is of polynomial. But the number of iterations is not polynomial since the problem is NP-complete. The architecture of the PRIAP is similar to that of the array router described in [37].

11.2.3 MESH CONNECTED MULTIPROCESSORS

The Iterative Array Maze Router described in the previous section could be classified as mesh connected processors to apply Lee's algorithm. S. Hong and R. Nair [34] had examined three special purpose machines with a rectangular array of mesh connected processors: highly special L-machine [5], slightly more general SAM [4] machine, and the WRM [35] designed with fully general microprocessors in the array.

The L-machine consists of a control unit that communicates with a host computer and sequence the operations of an array of simple processing elements, called L-cells. The Synchronous Active Memory (SAM) machine proposed by T. Blank, M. Stefik, and W. van Cleemput [4] is also aimed at a compact design suitable for subarray packing in a VLSI ship. One main difference between this machine and the L-machine is that this was designed with a somewhat broader range of applications in mind.

The Wire Routing Machine (WRM) described by Hong, Nair, and Shapiro [35] is also a mesh-connected multiprocessor complex, as the two machines described above. The major difference in the design between this and others lies in the power and scope of the node processing element. The WRM system consists of three major parts: the processing array, the control processor, and the array control unit that interfaces the array and the control processor. The control processor connects to the host computer, disk units, console, printer, and the array. This section only describes the L-machine that transfers the Lee-Moore Algorithm into a parallel Lee-Moore Algorithm.

An important and often employed simplification is to find a shortest path on the rectilinear grid between two given points. The length of a shortest path is often longer than the rectilinear distance between two points because of blockages. The basic technique proposed by E. Moore and C. Lee [44], commonly referred to as the Lee-Moore (LM) algorithm, can be briefly stated as follows. For simplicity of arguments we take a single wiring plane. The technique can be adapted easily to situations where multiple planes are involved. Let one of the points be called the source and the other the sink. We make use of two lists of nodes called OLD and NEW and two status markers per node for recording whether the node has been visited, and if so from which direction.

Consider a mesh connected complex of simple processing elements, one per track intersection of the chip being wired. Suppose each processing element can perform the following basic operations (among some other equally simple operations) in parallel.

Lee-Moore Algorithm (LM):
(1) Initialization: Mark source node as visited OLD ← source node.
(2) Propagation: Starting with empty NEW list for each node in the OLD:
     new ← the neighbor nodes of the current node
     that are not visited and not blocked.
     Mark *new* nodes as visited and the direction visited from.
     If sink ∈ NEW, go to LM 3, else append *new* to NEW.
     Let OLD ← NEW (if NEW is empty, the path does not exist, i.e.,

an overflow)

Go to LM 2.

(3) Backtrace: Starting from the sink node, follow the directions noted on the nodes to the source node.

Parallel Lee-Moore Algorithm (PLM — Kernel of Propagation):

(1) Receive wavefront token from neighbors, if any.

(2) Ignore the token if blocked or already visited.

(3) Mark it visited and mark the direction from where the token was received.

(4) If sink node, signal halt, else send wavefront token to all four neighbors simultaneously.

M. Breuer and K. Shamsa's L-machine [5] is the first published design of this nature. The L-machine consists of a control unit that communicates with a host computer and sequences the operations of an array of simple processing elements, called L-cells. This is expressly designed to implement the parallel Lee-Moore (PLM) algorithm. The L-cells are simple (about 75 gates) and, hence, many of them can be laid out on a VLSI chip to form a subarray. The machine is capable of performing the following tasks:

(1) Initialization: This involves the loading of source, sink, and blockages information into the L-cells.

(2) Parallel Propagation: This essentially implements the PLM described earlier. In addition, a BUSY-status signal is raised by those L-cells which are active during a step of processing the wavefront. The controller receives the wired-OR of BUSY long as some cell is active. If the controller sees the BUSY signal go down before the sink is reached, an overflow is indicated.

(3) Backtrace: This process determines the wire path by following the stored direction flags from an activated sink back to the source. The X, Y coordinates of each node on the path of the wire are output by the machine.

(4) Clear: Cells along the backtraced wire path are marked as blockages for subsequent wire routing. The internal status of all other L-cells is cleared to an idle state.

Each L-cell communicates with its four neighbors through bidirectional lines, one per neighbor. These lines are used during the propagation and the backtrace. In addition, there are 7 more signal I/O's and a clock input line per cell:

• Global Broadcast Bus (3 bits of control signals) from the unit in wired-OR lines.

• Global Status Bus (2 bits) from all cells to the control unit in wired-OR lines.

• X and Y (1 bit each) Select/Response wired-OR bidirectional lines along each column and each row of L-cells.

The control signal lines are used to sequence the loading of blockage, source and sink status (in conjugation with X, Y selection), the forward propagation, the backtrace, and clearing operations. The two status bits informs the control unit of the busy status of the wavefront, whether the sink is reached during the propagation phase, and whether the source is reached during the backtrace phase. The X and Y select/response lines are used for communicating the cell addresses between the control unit and the array. The control unit selects a specific L-cell, through the X and Y addresses of the cell. Theses addresses are decoded by X and Y decoders, the outputs of which activate the appropriate column and row signal lines. When an L-cell raises its X and Y lines, column and row encoders translate these signal to X and Y addresses for the control unit or for other I/O lines of the system. The machine is capable of processing an entire wavefront of propagation in one clock cycle. Contrast this with tens of instructions necessary to process just one node of the wavefront for the Lee-Moore algorithm in a conventional serial computer. For two wiring planes, the number of L-cells double and each L-cell has five neighbor connection lines. To accommodate the two wiring planes in three dimensions, the authors propose to interweave rows of L-cells of top and bottom layers in a plane. While the arrangement is perfectly capable of finding a shortest path, the current design can neither accommodate preferred wire directions in the wiring planes nor treat via as anything other than one unit of length, should such technology constraints exist.

For an $n \times n$ track wiring surface, the wiring array contains n^2 L-cells, each of which contains about 75 gates. The total number of pins of the array including two power pins and the row and column decoder and encoder is $4log_2n + 8$ (this can be reduced to $2log_2n + 8$ if the row and column select/response address signals share one bidirectional bus).

The corner and side chips realize portions of encoder and decoder structure besides the subarray. The mid-array chip now needs $4m$ neighbor lines, $4m$ X, Y lines plus six global wires.

To reduce the pin counts, Iosupovicz [37] suggests a scheme involving serial transfer of information between subarray boundaries. That is, all neighbor communication between a subarray boundary are serially transmitted by encoding the active cell positions, and then passed on to the next chip which, in turn, serially decodes the correct neighbor cell position on the boundary. A substantial reduction in the chip pin count, as well as a complete modularity of subarray chips is accomplished at the expense of lengthening the operation of the machine. The processing time is still more or less proportional to the length of the path. The array size of the L-machine must be as large as the wiring surface (twice as large for two-layer wiring). If a large enough machine is built, perhaps all practical problems can be handled. But no problem larger than the physical size if the machine can be processed. The L-cell processor is compact and very fast. However, it is inflexible and limited to a very narrow field of applications; namely, finding a shortest path between two points.

11.2.4 THE WAVEFRONT MACHINE

In the previous section, mesh connected machines require N^2 processors to implement the Lee algorithm on an $N \times N$ grid plane, which is not practical for the large N. Thus, M. Tachibana et al. [41] proposed an architecture that requires only $O(N)$ processors to find a path in $O(N)$ time for the $N \times N$ grid plane. A key consideration for realizing an economic hardware Lee router is that the grid-cells (processors) need to be activated only when they are on the current wavefront. As the average length of wavefronts for an $N \times N$ grid plane is $O(N)$, a machine with $O(N)$ processors should be enough to find a path in $O(N)$ time provided that they are pertinently mapped into the grid-cells. Such a hardware Lee router is henceforth called a *wavefront machine* [41].

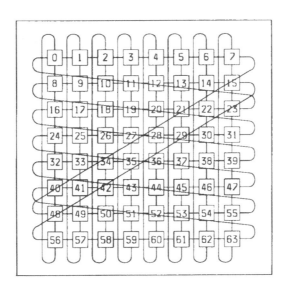

Figure 11.4 Interconnection of processor elements (twisted torus) (From *Int. Conf. CAD-85.* ©1985 IEEE. With permission.)

The actual number of processor must be determined by the trade-off between running speed and resource utilization economy. There are two ways of assigning processor elements to grid-cells; "dynamic" and "static" assignment. A dynamic assignment can provide an optimal processor-to-cell mapping by means of on-line calculation for each particular instance of wave propagation. However, it tends to make hardware configuration complicated, and the optimal assignment algorithm itself. Therefore, a static assignment is employed.

The method of processor-to-cells mapping exploits that the wavefronts consist of diagonal line segments, i.e., the mapping of processors to any diagonal lines on the grid plane is optimized. Then the same processors appears at every $N/2$ processors along any diagonal lines on the $N \times N$ grid plane. Note that it is optimal because, when a processor is used to send a labeling message, another processor must be simultaneously used to receive the labeling message.

For example, if the machine has 64 processors, the maximum degree of parallelism is 32. In order to achieve an optimal assignment with respect to diagonal lines, the processor elements are interconnected in a "twisted torus" configuration [1] as in Figure 11.4.

11.2.5 MAZE ROUTING ON A HYPERCUBE COMPUTER

Due to the two-dimensional nature of the maze routing problem, the most common processor configuration proposed for parallel maze routing hardware has been that of the two-dimensional mesh of processors as discussed in the previous sections. However, other architectures, such as raster pipeline subarrays have also been used [67]. Most approaches to the parallelization of maze routing aim to reduce the worst case time complexity of the time-consuming wavefront expansion step from $O(L2)$ to $O(L)$.

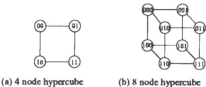

(a) 4 node hypercube   (b) 8 node hypercube

(c) 16 node hypercube

Figure 11.5 Hypercube topology (From *Int. Conf. Paral. Proces.* ©1987 IEEE. With permission.)

In 1987, Olukotun et al. [53] and Y. Won et al. [87] independently developed a maze routing on a hypercube computer with 64 nodes. They partitioned the $N \times N$ routing grid into k parts and assigned one partition to each of the k node processors. Then, each processor utilizes a maze routing algorithm, such as a wavefront algorithm, to route within grids of a part, while the results of routing from each processor will be shared by the adjacent processors to make a routing problem solvable.

The NCUBE is an example of a general purpose parallel computer with a hypercube processor interconnection topology which can accommodate up to 1024 nodes in a 10-dimensional hypercube. Hypercube topologies are shown in Figure 11.5. It is well known that every 2^d node hypercube has embedded in it a $2^{(2/2)} \times 2^{(2/d)}$ two-dimensional mesh. Figure 11.6 shows such an embedding for 4, 8 and 16 node hypercubes.

11.2.5.1 GRID PARTITIONING AND MAPPING

As remarked in the previous section, we utilize only an embedded mesh of the hypercube. Further, we assume that the number of grid cells, n^2, is significantly greater than the number of node processors k. Since the node processors from a hypercube, k is a power of 2, i.e., $k = 2^d$. d is called the *dimension* of the hypercube.

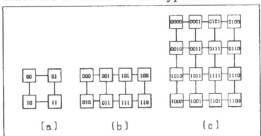

Figure 11.6 Meshes embedded in a hypercube (From *Int. Conf. Paral. Proces.* ©1987 IEEE. With permission.)

In discussing the grid partitioning and mapping strategy, we make the simplifying assumption that n is a power of 2, $n = 2^p$, and $p \geq d/2$. Suppose that $n = 8$ and $k = 4$. The 8×8 grid may be partitioned and mapped onto the 4 processors as shown in Figure 11.7a. Figure 11.7b shows a possible partitioning and mapping when $k = n = 8$. Each partitioning is labeled with the processor to which it is assigned. As can be seen, the neighbor partitions of any partition are assigned to node processors that are adjacent in the hypercube connection. This partitioning strategy may formally be defined as below. *Partitioning Strategy 1:* Cover the $n \times n = 2^p \times 2^p$ with rectangles of size $2^{(d/2)} \times 2^{(d/2)}$. Each rectangle in the cover defines a partition of the grid. The partitions are mapped to processors in such a way that partitions that are adjacent in the grid are mapped to processors that are adjacent in the hypercube. A grid cell is said to be on a partition boundary if at least one of its neighbor cells (i.e., north, south, east, west) is in a different partition. It is easy to see that only boundary cells have a potential to cause interprocessor communication during frontwave expansion. In an effort to reduce interprocessor communication one may attempt to reduce B, the number of boundary cells. Each node is a powerful custom 32-bit microprocessor with 128 K-bytes of memory and the ability to perform floating point arithmetic [30]. The connection between the nodes is by dedicated point-to-point bit-serial DMA channels. The hypercube is managed by a host processor (an Intel 80286). The experiments were performed using a 64 node version of the NCUBE, the NCUBE/six. A simple three step maze routing algorithm has been implemented on the NCUBE/six with significant speed-up over the same algorithm running on a large mainframe. Furthermore, due to the power and general purpose nature of the node processors, it is possible to implement a routing algorithm for the NCUBE with a considerable degree of sophistication in order to improve routing quality or to adapt the algorithm to a specific technology. The speed and flexibility make the NCUBE hypercube both a practical and a cost-effective solution to the maze routing problem.

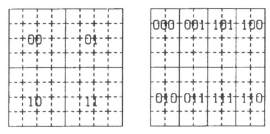

Figure 11.7 Grid partitionings (From *Int. Conf. Paral. Proces.* ©1987 IEEE. With permission.)

11.2.5.2 MAZE ROUTING ALGORITHM

One of the most important aspects in parallel algorithm design is the decomposition of the problem so that it maps efficiently onto the parallel architecture. Typically, this requires that the mapping does not result in the undue loading of a subset of the processing nodes, as this imbalance would degrade the overall performance of the algorithm. The decomposition should also minimize the ratio of interprocessor communication time to the computation time within the processors, as this will result in high efficiency for the overall computation. Unfortunately, there is no optimum solution to the decomposition of the maze routing problem. To decompose the maze routing problem we have chosen the most obvious approach, which is to divide the grid into as many square regions as there are node processors, N. If the grid is A square units in area, and there are $N = 2d$ nodes, where d is the dimension of the hypercube used, then each node will be responsible for a partition or cell of the grid containing $A/2^d$ square units. At the edges of each cell there are crossings which connect the cell to its adjacent cells. Each edge has a fixed capacity of crossings in which nets can be routed. A node keeps the number of crossings available at the eastern and southern boundaries of the cell assigned to it. Given this decomposition of routing problem, the parallel routing algorithm consists of the following steps:

(1) Global routing.
(2) Boundary crossing placement.
(3) Detailed maze routing.

Each of these steps will be described in detail in the sections that follow. The speed-up of the algorithm over a conventional maze routing algorithm rests on the fact that the above steps can all be performed in parallel. As noted, a 2-dimensional mesh may be mapped onto a hypercube connected parallel computer such as the NCUBE, and the parallel routing algorithm described above implemented. The algorithm does not make any use of the hypercube interconnection topology beyond that of a 2-dimensional mesh; however, the preliminary work described here is intended to set the stage for further experimentation to develop algorithms that utilize the full power of the hypercube interconnection scheme.

11.2.5.3 GLOBAL ROUTING

The first phase of the routing algorithm is a goal routing to assign nets to routing regions (cells) rather than detailed grid points. A global routing phase should increase the routing quality by globally optimizing net placement [66]. Global routing also serves as the first step in the decomposition of the routing problem into N independent routing problems. This phase requires communication among adjacent processors to perform the routing and to update the values of the edge capacities. The global routing phase is performed under the following assumptions:

• all nets consist of two points (source-target points),
• all nets can be wired within their minimum global bounding rectangle.

The first assumption is a simplification of the general multipoint net routing problem, and will be relaxed in later work. The second assumption relies on the fact that in practice most nets are routable near their minimal lengths, and do not detour much beyond the minimum bounding rectangle. In all cases the area enclosed by the global minimum bounding rectangle is always greater than or equal to that of the actual minimum bounding rectangle.

If we wish to connect grid points A and B, that reside in a global rectangle of n units in the x-direction and m units in the y-direction, there are G possible minimum global length routes from the cell containing point B, where G is given by

$$G = \binom{n+m-2}{n-1} \tag{11.1}$$

To explore these routes and find the one with the least boundary crossing cost, a global expansion phase is performed, using global routing messages, that is analogous to the wavefront expansion step of maze routing. A global routing message consists of seven fields, these are

- *The net number* identifies which net a global route message belongs to.
- *The source processor* indicates where the message originated from.
- The *destination* or target processor indicates where an expansion message should terminate.
- The *direction* gives the direction of the destination processor in relation to source processor. The direction has a field to specify the x-direction, which can have a value of east or west, and a field for the y-direction, which can have a value of north or south.
- The *route bit vector* maintains a record of the direction, x or y, in which the message is advanced at each processor. The bit indicated by the bit position portion of the message is set in the route bit vector to "1" if the message is advanced in the x-direction and "0" if the message is advanced in the y-direction.
- The *bit position* identifies the current position in the route bit vector; it is also a measure of how far, in global cells, a message is from its origin.
- The *route cost* keeps a running total of the cost incurred by the message at each boundary crossing. The cost of crossing a cell boundary during global expansion is computed as an inverse exponential function of the edge capacity of the boundary.

A message also has a *type* associated with it. The two types of messages used in the global routing phase are expansion and backtrace.

The processor responsible for the cell in which the source point of the net resides initiates the global expansion phase by initializing the fields of a global expansion message and sending the message to its neighboring processors contained within the minimum global bounding rectangle. On the reception of an expansion message, a processor compares the source portion of the message with its processor location to determine if it is the target processor. If the processor is not the intended destination of the message, it adjusts the fields of the message before advancing the expansion message to its neighbors within the global bounding rectangle. If a processor receives an expansion message for a net whose source does not have either an x or y coordinate in common with the processor's location it waits for another message for that net to arrive from the other direction, x or y, before advancing the message with least cost. Once the target processor has been reached by an expansion message from both the x and y directions, the backtrace phase is initiated for the least cost message.

The global backtrace phase consists of retracing the path taken by the least cost expansion message from the source to the target processor. To accomplish this, the direction portion of the message is replaced with the direction of the source processor in relation to the the target processor. The route bit vector, scanned in the opposite direction to that used during expansion, is used to determine in which direction, x or y, to move at each processor. The backtrace message traverses the path from the target processor to the source processor utilizing the route bit vector, the direction and bit position to direct its course. At each cell boundary along the backtrace, path crossings used by the net are claimed by reducing the value of the edge capacity of the boundary by one.

The global expansion phase generates M_E messages given by,

$$M_E = 2mn - m - n \qquad (11.2)$$

The backtrace phase generates M_B messages given by,

$$M_B = m + n - 2 \qquad (11.3)$$

The total number of messages, M_G, generated for the global routing of a two point net is, thus

$$M_G = M_E + M_B = 2(mn - 1) \qquad (11.4)$$

Besides the transmission time, each message has associated with it a certain amount of computation time which causes its transmission and arises from its reception. If we assume that any computation time that does not arise from the transmitting and receiving of messages is negligible in comparison with computation time that does, the number of messages generated is a relative measure of the time spent during the global routing phase.

Each processor is responsible for the global routing of nets whose source points are contained within the processor's cell. A processor initiates the global routing of a net until all nets for which it is responsible have been routed. This happens simultaneously in all processors. Although significant speed-up is achieved by this parallel global routing scheme, it is possible that during the backtrace a message may arrive at a particular processor to discover that all the available path crossings have been claimed by other nets since the expansion phase. If this situation occurs a completely a new global routing could be found from the cell where the clock occurred to the source cell. Intelligent assignment of boundary crossing costs can minimize the occurrence of this problem. In this algorithm there is no facility to recover from blocked boundary crossings as this problem did not occur while running the benchmark used to test this routing algorithm.

After a processor has completed the global routing of all the nets with source pins within its cell, it reports to the host. When all the processors have completed the global routing phase the host initiates the next phase of the routing algorithm.

11.2.5.4 CROSSING PLACEMENT

In order to decompose the routing problem into a set of independent routing problems that can be solved in parallel, it is necessary for a processor and its four neighbors to decide on the placement of nets that cross their common boundaries. At the completion of the global routing phase each processor contains a list of nets with portions within its cell, including nets that are completely contained within its cell, and the borders across which these nets pass. A *strand* is defined, as in [59], to be a connected portion of a net within a cell together with the boundary crossings with which this portion connects. These strands represent the input to the crossing placement algorithm.

Initially, all crossings have an undefined value outside the allowed range of crossing values. The crossing placement routine uses an iterative refinement method, in which each processor calculates the position of a crossing, on either its southern or eastern borders, based on a weighted average of positions of the crossings as projected on the crossing border of strands to which the crossing is connected. The closer a crossing is to the one being placed, the more weight it is given. Crossing on an opposite border to the crossing being placed are given much more weight, and source and target points, even more weight.

A crossing can become defined if a strand connects this crossing to one or more already defined crossings. Initially, only source and target points are defined. This scheme has the effect of giving the source and target end points even more weight in determining the crossing placement of strands connected to them.

Each iteration of the crossing placement algorithm consists of two steps. First, place all the eastern crossings, and second, place all the southern crossings. The first step starts from the processors on the western border of the mesh and proceeds toward the east, while the second step begins with processors on the northern border and proceeds toward the south. After each net has been placed, a message is sent to the appropriate neighbor processor with the new crossing position.

At the completion of each step each processor checks for strands that occupy the same crossing position. Any conflicts are resolved by moving the crossing of the strand belonging to the longer net to another position. The number of iterations the crossing placement algorithm exercises is predetermined by the user.

In practice, convergence occurs quickly, and weight given to crossing on opposite boundaries of a cell tend to cause the strands to form straight lines as one would expect. However, conflicts may cause jogs in the wiring paths.

Each iteration of the crossing placement algorithm generates three messages for every crossing placement iteration. Thus the number of messages generated by a net, N_c, during crossing placement is given by

$$M_c = 3I(m + n - 2) \tag{11.5}$$

where I is the number of crossing placement iteration.

11.2.5.5 DETAILED ROUTING

In the final step of the parallel routing algorithm each processor performs the detailed routing of its cell. This is done using a software maze router following the algorithm outlined in the introduction and using the additional idea from [10] of storing the location of the wavefront grid points in a stack. This step is executed in parallel on all processors and requires no interprocessor communication.

11.3 PARALLEL ROUTING WITH MULTIPROCESSORS

There are no general rules of classification on the parallel routing technologies. However, two ways of classification may be possible by considering which kinds of multiprocessor environments are utilized and how the tasks are assigned to each processor. For the multiprocessor environment, M. Zargham [89], J. Rose [63] and T. Cho et al. [16] use the shared memory environment through which each processor fulfills its routing task by communicating, while H. Date et al. [20] use the distributed environment in which the tasks are fulfilled by passing messages. Recently, N. Funabiki et al. introduced a neural network approach for the channel routing problem [26].

For the assignment of tasks, Zargham assigned a partitioned region of a routing area to each processor, while Rose, Cho and Funabiki assigned each routing task of a net to each processor. Thus, one will finish the routing task by merging properly for the border of the adjacent region partitioned. The other will finish the routing task by the routed results from each processor without any conflicts on a routing area.

11.3.1 REGIONAL PARALLELISM
ON A SHARED MEMORY SYSTEM

Zargham [89] introduced a parallel channel routing on a shared-memory multiprocessor environment. This section introduces the algorithms and its parallel processing.

11.3.1.1 ALGORITHM

The algorithm has three main phases as follows:

Phase 1: Dividing the channel into several regions(or sections) by selecting some columns.

Phase 2: Assigning tracks to nets of the selected columns.

Phase 3: Assigning tracks to nets of the columns in each region.

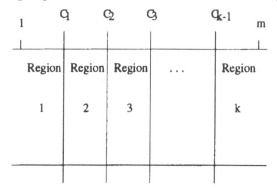

Figure 11.8 A channel, which has m columns, is divided into k regions (From *25th Des. Auto. Conf.* ©1988 IEEE. With permission.)

In phase 1, the router begins by reading in the data about nets, number of regions (r) and number of available processors (p). Given r (where $r < p$), the router will divide the channel into r regions, This is done by selecting $r - 1$ columns among all the columns in the channel. Figure 11.8 shows a channel which is divided into k regions by selecting columns $C1, C2, \cdots, Ck - 1$.. The selection of columns is made by optimizing the following two goals:

(1) Increasing number of nets in each column.

(2) Dividing the channel into equal (or near equal) regions.

That is, while channels are divided into regions, choose the columns with a density, as much as possible, near to the maximum density (the density of a column is defined to be the number of nets in that column, see [10, 57].) The selection of a column with a large number of nets will limit the number if choices for assigning tracks to its nets. Thus, the problem of assigning ideal (or near ideal) tracks to their nets becomes easier to solve.

To find suitable track assignments to the nets of a column, say column j, the following three steps are done in phase 2.

(1) Finding a preliminary partial ordering between the nets which pass through j.

(2) Finding a final ordering (with or without alternative) between the nets.

(3) Assigning tracks to the nets.

The preliminary ordering is made on the basis of a Vertical Constant Graph, denoted by VCG, that is a dependency graph specifying the order in which nets must be placed from top to bottom in a channel. For example, if there are three nets a, b, and c in column j, and there is a path from a to c, and a path from b to c, the nets will be ordered as:

level 1: $[a, b]$, and level 2: $[c]$,

which means nets a and b should be assigned to tracks above the track for c. If the length of a path between two nets is greater than a factor of maximum density, the path will be ignored at this step. This rejection eliminates the side effects of some long cycles that may exist in a VCG. However, if there is a short cycle, some of the edges will be ignored in this step to create an acyclic graph.

The final ordering and track assignment is made by grading the nets. The grading is done with a grade function that assigns a value within the range [-1, 1] to each net. The value -1 means that the net should be assigned to one of the bottom tracks of the channel. As the value approaches 1, the choice for the track will approach higher tracks of the channel.

Given the net b in column j, the grade function f is modeled as:

$$f(b) = p_1(VCG - D) + p_2(x - D) \qquad (11.6)$$

where D = distance; $p1$ and $p2$ are tunning parameters; VCG-distance is the relative distance of b from other nets in VCG; and x-distance is the relative horizontal distance of b to the nearest terminal node on the right and left of column b.

The VCG-distance gives a global view of the net's position with respect to the other nets in the channel, and is computed as:

VCG-distance = (maximum path size from every node to b in VCG)

$\qquad$ − (maximum path size from b to every node in VCG).

The x-distance gives relative horizontal distance to the nearest terminal node on the right and left of column j, and is evaluated according to their position in VCG and distances from their terminals.

$$x_D = \frac{d}{1 + |x_{br} - x_j|} + \frac{d}{1 + |x_{bl} - x_j|} \qquad (11.7)$$

where D = distance; x_{br} is the x-coordinate of the nearest terminal of net b on the right side of column j (including column j); x_{bl} is the x-coordinate of the nearest terminal of net b on the left side of column j (including column j); and d is an integer which represents the position of terminal, and is defined as:

- $d = 1$, if the nearest terminal of b is connected to the top of channel.
- $d = -1$, if the nearest terminal of b is connected to the bottom of channel.

Thus, if the x-distance approaches -1, it is better to assign a track near to the bottom of channel. Otherwise, if it approaches 1, assign a track near to the top of channel.

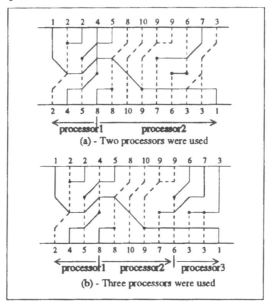

Figure 11.9 Brustein's difficult channel (From *25th Des. Auto. Conf.* ©1988 IEEE. With permission.)

In phase 3, tracks are assigned first to each column of region from left (right) to the right (left). Given tracks that are already assigned to column $j + k$ (for $0 < k < i - 1$), the column $j + k + 1$ can be routed by considering the recent assignment of $j + k$ and i, and the grade function f expressed as

$$f = p_1(VCG - distance) + p_2(x - distance) + pcw_3(y - distance) \quad (11.8)$$

where p_1, p_2, and p_3 are tuning factors.

For a given net b, which is already assigned to a track t in column $j + k$, the y-distance is the vertical distance between t and b's terminal. This distance, as x-distance, helps in deciding which direction (up, down, or straight) the net *b* should be routed. Based on the number of *b*'s terminals, the *x*-distance and *y*-distance are calculated differently.

To clarify some of the routing techniques of the algorithm, one examples is given in Figure 11.9, which is a well-known difficult routing problem. Figure 11.9a is a solution with two processors, while Figure 11.9b is a solution with three processors.

11.3.1.2 PARALLEL PROCESSING

One of the main issues that must be solved is how to assign a new work to a processor. To discuss this problem, a search tree for one of the solutions of Brustein's channel are presented as shown in Figure 11.10. Figure 11.10 shows an AND/OR search tree for the solution which was presented in Figure 11.9. In this figure, except node 1, a column is assigned to each node. The nodes 1 and 2 are called AND nodes, and the other nodes are called OR nodes. In this tree, we say a leaf node (or terminal node) succeeds if there is a track assignment for the nets in the column which has been assigned to this node. An OR node, which presents several alternatives, succeeds if all the successor nodes on only one of its branch succeed (note: an OR node may have only one branch or more than one branch).

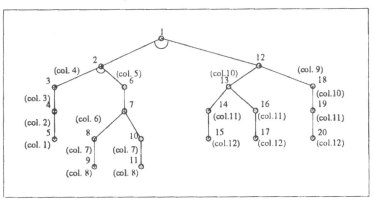

Figure 11.10 Search tree for one of the solutions of Brustein's channel (From *25th Des. Auto. Conf.* ©1988 IEEE. With permission.)

The successor nodes for a node, say n, are the nodes of the subtree which is rooted by n. An AND node succeeds if there is at least one track assignment to every column which is assigned to its successors. Thus, the channel pr oblem has a solution, if all the AND nodes succeed.

In the beginning, all of the processes are in a global pool. Initially, one processor, say processor 1, will leave the pool and begin dividing the channel into three regions. As the result, processor 1 will produce an open AND node 1, and two other nodes, 2 and 12. By open node we mean a node that has child which is not yet claimed by a processor.

At this time, processor 1 will move down to node 2 and another processor, say processor 2, will go to node 12. Processor 1 will assign a unique track assignment to column 4, and makes node 12 an open AND node by producing two nodes, 3 and 6. Notice that node 3 is assigned to the column on the left side of column 4 (that is column 3) and node 6 is assigned to the right side column. In parallel with processor 1, processor 2 will assign two alternative track assignments to column 9, and makes node 12 an open OR node by producing two nodes, 12 and 18. The best alternative will be given to node 12 (the leftmost branch, which is also called favorite branch), and the second alternative will be given to node 18. At this time, there are two open nodes, 2 and 12.

Each processor at that node will move to the node on the leftmost branch. That is, processor 1 will go to node 3 and processor 2 to node 13. However, there is still one more processor in the pool, say processor 3. Processor 3 can go either to node 6 or node 18. We move processor 3 to node 6. This is done in case we are looking for only a single answer; the execution time will be reduced if we try to assign track to the columns which are not considered by any processor yet. When more than one answer is desired, the choice may not make so much difference in this step. Once processor 3 is assigned to node 6, node 2 becomes closed, which means every child is processed. The mechanism for assigning a new work to a processor could be changed as a result of requesting one or all of the solutions to the problem. In the case when all answers are desired, it is better to give a processor a great amount of work (a large subtree) to do, thus preventing its quick return to the global pool.

This method prevents the overhead associated with switching between unrelated works (by unrelated works we mean the works which are on different subtree).

The algorithm works well for implementation on any shared memory multiprocessor environment, and does not impose restrictions on the number of processors. The increased of number of processors will reduce the computation time. The algorithm is general purpose in nature and could be used in a variety of routing problems.

11.3.2 GLOBAL ROUTING
ON A SHARED MEMORY SYSTEM

Considering the n-nets routing problem with n processors, the routing task of each net can be assigned to each processor. Thus, each processor chooses the proper path of a net which has no conflicts to other nets by sharing the information from other processors. J. Rose [64] introduced the LocusRoute algorithm which uses this idea for placement and global routing in the shared memory environment. He exploited three orthogonal parallel decompositions of the router: *wire-by-wire parallelism* which is the same as net parallelism, *segment-based parallelism* which routes each two-point segment produced by the minimum spanning tree decomposition in parallel and *route-based parallelism* which evaluates, in parallel, the potential routes of a segment for each permutation. In this section, the LocusRoute algorithm with uniprocessor is described first. Then in the *wire-by-wire* parallel approach, each processor routes a different wire at the same time, using the algorithm described.

11.3.2.1 PROBLEM DEFINITION

Global routing for standard cells decides the following for each wire in the circuit.

(1) For each group of electrically equivalent pins (pin clusters in the terminology of [69]) it determines which of those pins are actually to be connected.

(2) If there is no path between channels when one is required, it must decide either which built-in feed-through to use or where to insert a feed-through cell.

(3) It decides which part or parts of a channel a wire should occupy.

(4) It must determine the channel to use in routing from a pad into the core cells

The objective of a global router is to minimize the sum of the channel densities of all the channels (hereafter called *the total number of tracks*). In this discussion of global routing there will be no differentiation between feed-through cells and built-in feed-throughs — they are referred to joinly as *vertical hops*. The decision to insert a feed-through cell or use a built-in feed-through is deferred to a post-processing step. Whenever comparisons are made to other global routers, the post-processing step is invoked to obtain accurate track counts.

11.3.2.2 LOCUSROUTE ALGORITHM

In the LocusRoute algorithm, the following five steps are executed for each wire. The details of each step are contained later.

(1) *Segment decomposition*: Each multipoint wire is divided into a set of two-point segments (where a "point" is a pin cluster), using a minimum spanning tree algorithm. By considering only two-point wires we introduce a suboptimality, but some of the negative effect is mitigated by a local optimization performed in the reconstruction: step (4) below.

(2) *Permutation decomposition*: Since each pin cluster can consist of two physical pins (one on the top of a cell and one on the bottom), there are four possible ways to connect two pin clusters. Each of these possible routes is called a permutation. This step decomposes the segment into permutations and uses a simple heuristic (described below) to see if all the permutations should be evaluated.

(3) *Route generation and evaluation*: A low-cost path is found for each permutation by evaluating a subset of the two-bend routes between each pin pair. The permutation with the best cost is selected as the route for that segment.

(4) *Reconstruct*: This step joins all the segments back together, performs a local optimization, and assigns unique numbers to distinct segments of the same wire in each channel. This is so that a channel router can distinguish between two segments and will not inadvertently join them together.

(5) *Record*: The presence of the newly routed wire is recorded so that later wires can take it into account.

11.3.2.3 WIRE-BY-WIRE PARALLELISM

In the wire-by-wire parallel approach, each processor routes a different wire at the same time, using the algorithm. The cost array is stored in shared memory and all processors access and change this structure simultaneously. The general flow of this parallel decomposition is shown in Figure 11.11.

The master processor initializes the cost array, and sets up each individual wire as a task on one central task queue. All processors then remove wire tasks from the queue and execute the LocusRoute algorithm by reading the shared cost array to evaluate the routes for each wire and then updating the cost array with the best route that is found.

This wire-by-wire approach does not exactly reproduce the serial LocusRoute algorithm. In the sequential algorithm there is a *data dependency* which dictates that wires be routed sequentially: after the wire is routed, its presence is recorded in the cost array so that subsequent wires may make the new path into account. In the wire-by-wire parallel approach, this data dependency in relaxed by allowing several wires to be routed simultaneously. This kind of approach is called *chaotic* parallelism — where data dependencies are relaxed under the assumption that the data used will only be marginally different from the correct data, and thus will still lead the system toward a good solution.

In a chaotic approach, iteration is typically used so that the system converges to the same answer as the sequential problem. For example, in the parallel simulated annealing work for placement of standard cells

[2, 8, 9, 61, 62] the iteration of simulated annealing overcomes the error due to the chaotic parallel approach. While the LocusRoute does iterate over the solution, there is no guarantee that this kind of iteration will converge to the same answer as the sequential algorithm. Hence, there is a degradation in the quality of the final answer. We are willing, as discussed earlier, to trade some amount of quality for increased speed, for the end purpose of combined placement and routing.

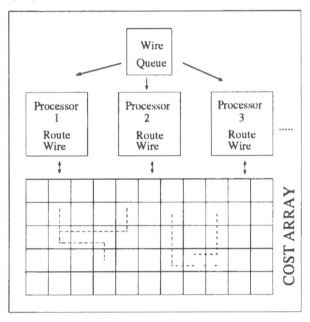

Figure 11.11 Single bus architectures (From *IEEE Trans. CAD*. Vol. 9, No. 10. ©1990 IEEE. With permission.)

11.3.3 DETAILED ROUTER
ON A SHARED MEMORY SYSTEM

While Rose [63] exploited a global router on a shared memory system, Cho et al. proposed a detailed router, PARALLEX, in the same environment [14 - 17]. In the detailed router, every conflict among nets should be removed in parallel. Thus, the implementation of the detailed router is more difficult than that of the global router. The router PARALLEX gives a general idea of how to utilize an MIMD system with shared memory environment.

11.3.3.1 REPRESENTATION OF A ROUTED NET

The representation scheme of a routed net is shown by two types of segments: *horizontal-wire-segment* and *vertical-wire-segment*. Each segment has eight possible directions in which other segments at both ends can be connected. High North (HN), High East (HE), High South (HS) and High West (HW) are for one end (called high end) of the segment, and Low North (LN), Low East (LE), Low South (LS) and Low West (LW) are for the other end (called low end) of the segment. A segment is represented by five tuples: net ID, Is_horizontal_or_vertical, row (column) index for horizontal (vertical) segment, two column (row) indices of a low end and a high end of a horizontal (vertical) segment.

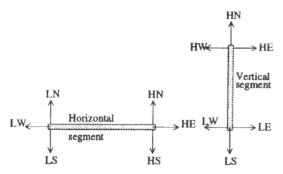

Figure 11.12 Horizontal and vertical segment

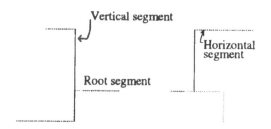

Figure 11.13 A net segment tree with root segment

Connections between segments are made by a pair of pointers as shown in Figure 11.12. Whenever two segments are connected, there implicitly is a via or a contact cut. A segment represents an interconnection between two vias.

11.3.3.2 DATA STRUCTURES

For managing the horizontal/vertical segments and easily detecting conflicts, segments are stored in three data structures.

1. *Net Segment Tree* (NST): An NST stores information of the vertical and/or horizontal segments of a net in a tree structure as shown in

Figure 11.13. Those segments are found by the Rectilinear Steiner Tree [27, 51, 56]. An NST shows physical connections of a net. Furthermore, two adjacent segments in an NST are doubly linked, and an NST can be reached from any segment in the tree.

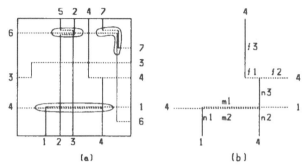

Figure 11.14 Initial routing and types of segments
(a) Initial routing phase of the pedagogical problems
(b) Types of segments in a net segment tree

Thus, this net segment tree structure is suitable for manipulating the changes obtained from the path-finding procedure by using MOVE, FIND, DELETE and MERGE operations.

2. *Horizontal Segment Tree* (HST): An HST stores information on the horizontal segments of all of the nets. For each row, horizontal segments located in the row are stored in a HST and sorted by the non-decreasing order of the left end. If the left ends of two or more segments are equal, they are sorted by the non-decreasing order of the right end.

3. *Vertical Segment Tree* (VST): Similarly, a VST stores information on the vertical segments of all of the nets.

A set of NSTs for all nets is global data and globally accessible and maintained by all the processes, while an HST and a VST are private data for each process. A set of NSTs of all nets forms a switchbox. Also, an HST and a VST form a switchbox. Thus a set of NSTs of all nets can be converted to an HST and a VST, and *vice versa*. Generally, NSTs will be used to hold the nets' information, while the HST/VST will be used to determine the relationships among the segments of nets. Therefore, the existence of any conflicts is detected by traversing the HST and the VST. If we trace by each net, it means that we are tracing each NST. If we trace by each row or column, then we are tracing an HST or a VST.

11.3.3.3 CONFLICT GROUP

There are three kinds of conflicts among segments in a row or a column; a *grid-conflict*, a *segment-conflict* and a *via-conflict*. A *grid-*

conflict exists if two or more horizontal (or vertical) segments, belonging to different nets, meet in a grid unit of a row (or a column). A *segment-conflict* exists between two grid-conflicts if a segment is involved in both grid-conflicts. A *via-conflict* exists if there are two segment-conflicts, or grid-conflicts. One or more conflicts related to each other by a *grid* or *segment* or *via-conflict* are considered a *conflict group*, which implies that the solution of a conflict influences the solution of another conflict in a group. The conflict group is constructed with a set of segments involved in the conflicts when an HST and a VST are built from NSTs. If we look at Figure 11.14a, we observe two conflict groups, one group by conflicts 1, 2 and 3, the other by conflict 4 itself.

11.3.3.4 FIXED SEGMENTS AND REROUTABLE SEGMENTS

For the convenience of explanation, we classify segments into three types: the *main-conflict segment*, the *neighbor segment* and the *fixed segment*.

A *main-conflict segment* is the segment directly involved in the conflict, as shown in Figure 11.14b, where net 1 and 4 are extracted from Figure 11.14a. *Neighbor segments* are segments connected at both ends of the main-conflict segment, which may have a maximum of 6 *neighbor segments*. A *main-conflict segment* and *neighbor segments* will be rerouted in the path-finding procedure. All of the segments, except main-conflict segments and neighbor segments are *fixed segments*, including segments in nets not related to any conflicts at all. Also, segments in other conflict groups are considered to be *fixed segments* for the current conflict group. The *fixed segment* is regarded as an obstacle in the path-finding procedure.

11.3.3.5 SEARCH LISTS

A search list is an order of segments, which will be used for searching the paths for the conflict group. If there are n segments in a conflict group, n search lists can be constructed. The order of the first search list is identical to the order of the conflict group, and the second search list can be constructed by placing the last segment of the first search list in the first place of the second search list. The next search list can be constructed by placing the last segment of the current search list in the first place of the current search list. Similarly, the remaining $n - 3$ search lists can be constructed. These n search lists are sufficient to find the conflict-free routes for a conflict group. This is possible with the aid of the maze routing procedure described later in the next section.

Each search list will be assigned to the process which bears the same

ID of the first segment in the search list. The order in each search list is used to find the local solutions in a conflict area by keeping the first segment path (optimal by itself) unchanged, while changing the paths of the other segments, one-by-one, according to the order in the search list.

11.3.3.6 ARCHITECTURE OF PARALLEX

PARALLEX was developed and was experimentally tested on a Sequent Symmetry Computer System which has 26 processors. However, PARALLEX utilized processes instead of utilizing processors. The reasons are that resource scheduling is done automatically by the operating system and processes are available to the general users while processors are privileged to superusers. Thus, we may not assume that a process is bound to a processor with no other processes running in it, though more processors were installed than the number of processes needed. In fact, if the processes are used for the tasks, then this implies that PARALLEX is adaptable to larger problems with more nets than number of processors.

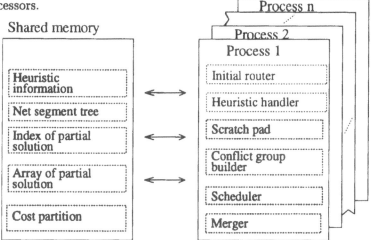

Figure 11.15 Architecture of PARALLEX

Each process shares the information of net segment tree, partial solution index and partial solutions in the shared memory as shown in Figure 11.15. Each Net Segment Tree (NST) partition of memory is filled with the result of the rectilinear Steiner tree from each process. Once NSTs are completed, the search lists of each process are found by using the conflict group finder. Using these search lists, the partial solution finder finds partial solutions and records the results to the partial solution partition of the shared memory, while it is posting the index of the partial solution to the index partition. Partial solutions obtained from each search list, PS_{ij} (i is an ID of the conflict group and j is the process ID) are stored in a two-dimensional array of PS_{ij} to the partial solution partition.

The Scheduler of each process is constantly checking the index partition for the solution of other conflict groups and makes a queue, by indices of the partial solutions with the order of queue, so that the Merger is able to proceed with the merging task. The task of the Merger is to construct an HST and a VST and if there is no conflict, then an HST and a VST is a solution of a switchbox. If there is no solution, then the Merger uses the next index of partial solutions from the queue. If the Merger finds a solution, it records the cost of the solution to the cost partition.

If only one solution (single solution) is required, all the active processes stop their tasks in case any cost is reported to the cost partition. If an optimal solution is required within an allowed cost, each process keeps comparing its cost to the cost partition and replaces the value of cost partition whenever it is lower.

11.3.3.7 OVERVIEW OF ALGORITHMS

A procedure of PARALLEX is described as follows. The data structure of a segment and main variables of PARALLEX are shown as follows.

```
typedef struct SEGMENT {
        int net_id, row_col_number, hor_or_vertical;
        int low_end, high_end;
        struct SEGMENT *HN, *HE, *HW, *HS;
        struct SEGMENT *LN, *LE, *LW, *LS;
}
shared int NumNet;
shared SEGMENT *NetSegTree;
```

The structure of a segment, SEGMENT, includes five tuples of information to identify each segment and 8 pointers to connect other segments. NetSegTree is a root segment structured as SEGMENT for the net segment tree, and NumNet is an integer to create processes.

The child processes with the same number of nets are created by the *for* loop. The system call *fork()* is to create each process and it returns zero if a process is successfully created. Each process finds a rectilinear Steiner tree by the procedure, Find RST( ). The rectilinear Steiner tree is converted into an HST, a VST by HST_VST( ) and a NetSegTree by ConstructNST( ), respectively. Conflict groups, are constructed by the procedure Conflict group Builder( ). According to information of ConfGrp and NetSegTree, partial solutions (PS) are found by PartSolSearcher( ) described in the next section. If there is a solution after the merging procedure, Merger( ), ExistSol becomes 1 and releases the current process.

```
PARPRALLEX( ) {
    int procs_id; Process ID assigned.
    int pid;      If pid == 0, process is created.
    int ExistSol; If solution exists ExistSol = 1.
    NumNet = ReadNet( ); Read net information.
    InitRouter( );          Extend inward 1 grid unit.
    for (procs_id = 1;procs_id ≤ NumNet ; ++procs_id) {
        if ((pid = fork( )) ≤ 0) { Process creation.
            if (pid<0) { Child is not created.
                exit(1);   Stop execution.
            }
            if (pid==0) { Child process is created.
                FindRST(procs_id,Rst); Rst is a root of RST.
                HST_VST( );          Build HST and VST.
                ConstructNST(procs_id,NetSegTree);
                ConfGrpBuilder(ConfGrp);
                PS is a root of partial solutions.
                PartSolSearcher(procs_id,NetSegTree,PS);
                ExistSol = Merger(procs_id,PS);
                exit(0);         Release child process.
            }
        }
    }
    exit(0); Release the parent process.
}
```

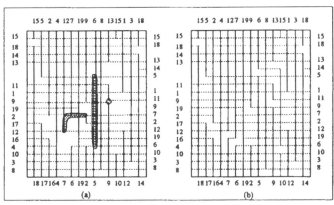

Figure 11.16 Conflicts shown on HST/VST and
solution of Modified Dense Switchbox.
(a) Three conflict group. (b) Final solution after merging

11.3.3.8 EXPERIMENTAL RESULTS

Three experiments are performed by running PARALLEX on a Sequent Symmetry parallel computer. One experiment among them is a variant of the Modified Dense Switchbox [18] as shown in Figure 11.16.

Modified Dense Switchbox: With respect to the experiment of the Modified Dense Problem, the results are illustrated in Table 11.1. The increase in the via weight, **V**, is slower than that of slack-cost, which needs more time to arrive at the solution. The time for Process 12 was not available for the single solution finding procedure.

Table 11.1 Results of Modified Dense Switchbox

Type of router	Process ID	A-B-V 4-4-0	A-B-V 6-9-1	A-B-V 7-9-1
	1	1.09	1.04	1.11
	2	na	na	na
	3	.89	.89	.80
	4	.87	.69	.97
	5	1.94	1.65	1.62
	6	1.20	1.24	1.20
	7	na	na	na
	8	.88	.90	.68
	9	1.43	1.47	1.53
PARALLEX	10	.64	.67	.83
	11	1.87	2.08	2.18
	12	na	na	na
	13	.59	.59	.74
	14	.68	.53	.70
	15	.54	.54	.66
	16	.50	.47	.65
	17	.60	.56	.60
	18	.57	.39	.57
	19	.56	.46	.55
	Parent	1.94	2.08	2.18
Sequential		17.80	20.68	25.22

11.3.4 A PARALLEL LOOKAHEAD LINE-SEARCH ROUTER IN A DISTRIBUTED MEMORY SYSTEM

Date and Taki [20] examined a parallel lookahead line search router in a distributed memory system with 256 processors. In its nature of message passing, two problems are inherent; one is memory overflow for communications paths between processors due to increased number of objects in the large scale data, the other is the degradation in wiring rate due to net conflicts among concurrently connected nets. For the first problem, a process called a *distributed process*, each processor is assigned. The distributor process reduces the number of message paths between processors. For the second problem, the rip-up and reroute operation to retain the routing order is adopted.

11.3.4.1 MEMORY OVERFLOW FOR COMMUNICATION PATHS

For implementing the concurrent program using KL1(a concurrent logic language), two kinds of memory are necessary. One is for representing processes. The other is for communication paths among processes. In the routing program, each process corresponding to the column or row is mapped to a *master line process* and each line segment on a column or a row is mapped to a *line process*. A master line process manages line processes on a column or row and passes messages between the line processes and crossing line processes. So each master line process must communicate with all the other mater line processes orthogonal to it. Therefore, the number of communication paths per master line process increases in the large-scale data. For efficient parallel execution on a distributed memory machine like PIM/m, processor mapping is important. In this router, a master line process and line processes on it are grouped. Then, each group is assigned to a processor randomly. This strategy mainly focuses on load balancing. In this case, the communication between master line processes may be inter-processor communication. When the size of the routing grid becomes larger, many communication paths arise between the processors. The experimental results shows that the memory overflow for representing the communication paths between processors occurs for 750×750 grid size data on PIM/m. This limitation is too severe for the use in practical amounts of data.

11.3.4.2 DEGRADATION OF WIRING RATE

Considering the routing of multiple nets, different nets may conflict on the same grid line segment. In this situation, the first message to arrive (corresponding to net A) occupies the segment. The second message to arrive (net B) fails to complete a route and backtracks. However, net A may backtrack later and may release the line segment. In this case, net B does not visit the line segment any more and the line segment may be left unused. This causes both lower quality routes (longer paths)and a lower wiring rate (more unconnected nets) which degrades for data with dense terminals.

Thus distributor processes are introduced to reduce the number of communication paths between processors. A distributor process is placed in each processor. When the distributor processes communicate directly with each other, bottlenecks occur in each distributor process.

11.3.4.3 RIP-UP AND REROUTE

In order to retain the sequential routing order, the automatic operation of rip-up and reroute is added to the basic algorithm. In the basic algorithm, two types of parallelism are realized. One is parallelism within a single lookahead operation and the order is concurrent routing of multiple nets. Assuming that the routing order of nets is given in a sequential routing program, let the order of net_i be i, $(i = 1, \cdots, N)$. Then each line process is in one of three states; *free*, *occupied*, or *concrete*. The *free* status means that the line segment is not occupied by a net. The *occupied* status means that the line segment is occupied by some nets permitting the multiple overlapping of routing paths. The *concrete* status means that the line segment is occupied by a net and is treated as a routing segment inhibited by other nets. The lookahead algorithm for a uniprocessor stated below is almost identical to the conventional line search algorithm except a lookahead scheme included. In lookahead operations, the algorithm searches the line segment whose expected point is defined as the closest location to a goal on a line segment. Its routing segment is decided from that search.

11.3.4.4 AN ALGORITHM WITH A PROCESSOR

An algorithm with a processor can be expressed as follows:

(1) Start (connect [S] to [T])
(2) Operate Lookahead scheme [42].
(3) If there is an expected point to be connected, go to step 4, else go to step 6.
(4) If the expected point reaches to [T], go to step 8.
(5) If the expected point does not reach to [T], go to step 2.
(6) Backtrack and if backtracking point is [S] then go to step 7, else go to step 2.
(7) Report a failure of routing.
(8) Report a success of routing.

In parallel operation, each processor will execute the sequential algorithm in parallel. In the reconfirm operation, the status of all line processes on the routing path changes to *concrete*. If a line process is occupied by other nets, the *cancel* message is passed to them. After completing the reconfirm operation, a message is passed to the next net to keep the routing order. The net receiving the *cancel* message starts the rip-up and reroute.

11.3.5 NEURAL NETWORK APPROACH

N. Funabiki et al. [26] proposed a parallel channel routing problems on a neural network environment. T. Cho et al. reported a distributed genetic approach for channel routing problems recently [14, 15]. These approaches utilize a general purpose parallel machine. The genetic approach is considered better for finding global solutions compared to the neural network model. In this chapter only neural network will be described.

The proposed parallel algorithm is based on a three-dimensional artificial neural network model which is composed of a large number of massively interconnected simple processing elements. The processing elements are called neurons because they perform the function of simplified biological neurons. The artificial neural network model for solving combinational optimization problems was introduced by Hopfield and Tank [36]. However, they use not only the sigmoid neuron model, which is slow for the convergence, but also a decay term which is proven to disturb the convergence. In order to speed up the convergence, the McCulloch-Pitts neuron model [49] has been used without the decay term and applied to several NP-complete and optimization problems [75,76,77,78,79]. The output V_{ijk} of the ijk-th processing element based on the modified McCulloch-Pitts neuron model [77]

$$
\begin{aligned}
V_{ijk} &= 1 \quad \text{if} \quad U_{ijk} > 0 \quad \text{and} \quad U_{ijk} = \max\{U_{ijk}\} \\
&\qquad \text{for} \quad q = 1, \cdots, m \quad \text{and} \quad r = 1, 2 \\
&= 0 \quad \text{otherwise}
\end{aligned}
\tag{11.9}
$$

where U_{ijk} is the input of the ijk-th processing element and m is the number of tracks of the channel. The change of the input U_{ijk} is given by the partial derivatives of the computational energy E with respect to the output V_{ijk}, where E is a $2nm$-variable function; $E(V_{111}, V_{112} \cdots V_{nm2})$. Note that n is the number of given nets. The equation is called a motion equation, given by

$$
\frac{dU_{ijk}}{dt} = -\frac{\partial E(V_{111}, V_{112}, \cdots, V_{nm2})}{\partial V_{ijk}}
\tag{11.10}
$$

The motion equation also presents the interconnections between the processing elements. Whatever computational energy function E is given, the motion equation forces it to monotonically decrease. The following proof shows that the motion equation forces the state of the system to converge to the local minimum where the energy function E is usually nonconvex. The motion equation performs the gradient method to minimize the energy function.

Proof: Consider the derivatives of the computational energy function E with respect to time t:

$$
\frac{dE}{dt} = \sum_i \sum_j \sum_k \frac{dV_{ijk}}{dt} \frac{\partial E}{\partial V_{ijk}}
$$

$$
= \sum_i \sum_j \sum_k \frac{dV_{ijk}}{dt} \left(-\frac{dU_{ijk}}{dt} \right)
$$

(Because of the motion equations, $\dfrac{\partial E}{\partial V_{ijk}} = \left(-\dfrac{dU_{ijk}}{dt} \right)$)

$$
= -\sum_i \sum_j \sum_k \left(\frac{dU_{ijk}}{dt} \frac{dV_{ijk}}{dU_{ijk}} \right) \left(\frac{dU_{ijk}}{dt} \right)
$$

$$
= -\sum_i \sum_j \sum_k \left(\frac{dV_{ijk}}{dU_{ijk}} \right) \left(\frac{dU_{ijk}}{dt} \right)^2 \leq 0 \qquad (11.11)
$$

As long as the input/output function of the processing elements obeys the nondecreasing function, dV_{ijk}/dU_{ijk} must be positive or zero so that dE/dt is negative or zero. Therefore, the state of the system is always guaranteed to converge to the local minimum [78].   □

11.3.5.1 SYSTEM REPRESENTATION

Figure 11.17a shows a channel routing problem in [88] where ten nets are given to be routed in a four-layer channel which has three tracks. The ten nets are (T2,T5), (B1,B6), (B2,B4), (T3,T9), (B3,T4,B5), (T6,B7), (T7,B11), (B8,B10), (B9,T10,B12), and (T11,T12), where T_i indicates the top terminal at the i-th column and B_j indicates the bottom terminal at the j-th column. For example, net 1 (T2,T5) has two terminals to be interconnected; one is the top terminal at the second column and the other is the top terminal at the fifth column. Figure 11.17b shows the system representation for this problem.

Only one horizontal segment is used for one net, and the vertical segments of the net are automatically assigned on the layer corresponding to the layer on which the horizontal segment of the net is assigned. The channel routing problem can be simplified into the layer-track problem.

This involves finding the layer number and track number for embedding the horizontal segment of the given net without violating the overlapping conditions. Six ($=3 \times 2$) processing elements are used to indicate layer and track on which a given net should be embedded. Because three tracks of two layers for the horizontal segments are available for the ten net and three track problem, a total of 60 ($=10 \times 3 \times 2$) processing elements are used in this problem. V_{ijk} represents the output of the ijk-th processing element, which corresponds to the j-th net and the j-th track on the k layer for $i = 1, \cdots, 10, j = 1, 2, 3$, and $k = 1, 2$.

Generally, $n \times m \times 2$ processing elements are used to represent n nets, m tracks, and two horizontal segment layers for the four-layer channel routing problems where $m \times 2$ processing elements describe the track number and the layer number for a single net.

The output of one and only one processing element among the $m \times 2$ processing elements should be nonzero to locate the net on one of the m tracks of two layer. The nonzero output means that the net should be embedded on the corresponding track and layer.

Figure 11.17b shows one of the solutions for this problems, where the black squares indicate the nonzero output of the processing element. Figure 11.17b shows that the net 1 (T2,T5) is assigned on the second track of the first two-layer channel and the net 2 (B1,B6) is assigned on the first track of the second two-layer channel, and so on. Note that the two-layer channel means a pair of two layers for the horizontal segments and vertical segment. Figure 11.17c shows the routing solution corresponding to Figure 11.17b.

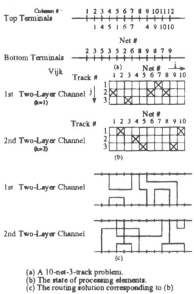

(a) A 10-net-3-track problem.
(b) The state of processing elements.
(c) The routing solution corresponding to (b).

Figure 11.17 System representation for a 10-net-3-track problem (From *IEEE Trans. on CAD*, Vol. 11, No. 4. ©1992 IEEE. With permission.)

Each net must satisfy the separation condition; in other words, no two different nets must violate the overlapping conditions. The overlapping conditions for the horizontal segments of the nets must meet the following condition, where head indicates the leftmost column number of the i-th net. Overlapping condition: $head_i \leq head_p \leq head_i, head_p \leq head_i \leq head_p$.

The horizontal overlapping conditions for the i-th net j-th track k-th layer processing element are given by

$$\sum_{\substack{p=1 \\ p\neq i \\ head_i \leq head_p \leq tail_i}}^{n} V_{pjk} + \sum_{\substack{p=1 \\ p\neq i \\ head_p \leq head_i \leq tail_p}}^{n} V_{pjk} \qquad (11.12)$$

This horizontal condition is nonzero if the horizontal segments of the other nets overlap the horizontal segment of the i-th net on the j-th track of the k-th layer. The overlapping conditions for the vertical segments of the nets can be determined in a similar manner as in horizontal segments, where the i-th net and the p-th net have terminals on the opposite sides of the same column. The vertical overlapping conditions for the i-th net j-th track k-th layer processing element are given by

$$\sum_{\substack{p=1 \\ p\neq i}}^{n} T_{ip} \sum_{q=1}^{j} V_{pqk} + \sum_{\substack{p=1 \\ p\neq i}}^{n} B_{ip} \sum_{q=j}^{m} V_{pqk} \qquad (11.13)$$

where T_{ip} is 1 if the p-th net has a bottom terminal at the column at which the i-th net has a top terminal and is 0 otherwise. B_{ip} is 1 if the p-th net has a top terminal at the column at which the i-th net has a bottom terminal and is 0 otherwise. The vertical condition is nonzero if the vertical segments of the other nets overlap the vertical segments of the i-th net. The motion equation of the i-th net j-th track k-th layer processing element for the n-net-m-track problem is given by

$$
\begin{aligned}
\frac{dU_{ijk}}{dt} = \quad &- \quad A\left(\sum_{q=1}^{m}\sum_{r=1}^{2} V_{Iqr} - 1\right) \\
&- \quad B\left(\sum_{\substack{p=1 \\ p\neq i \\ head_i \leq head_p \leq tail_i}}^{n} V_{pjk} + \sum_{\substack{p=1 \\ p\neq i \\ head_p \leq head_i \leq tail_p}}^{n} V_{pjk}\right) \\
&- \quad B\left(\sum_{\substack{p=1 \\ p\neq i}}^{n} T_{ip}\sum_{q=1}^{j} V_{pqk} + \sum_{\substack{p=1 \\ p\neq i}}^{n} B_{ip}\sum_{q=j}^{m} V_{pqk}\right) \\
&+ \quad Ch\left(\sum_{q=1}^{m}\sum_{r=1}^{2} V_{iqr}\right) \qquad (11.14)
\end{aligned}
$$

The first term (A term) forces one and only one output among the $2m$ processing elements to be nonzero where the i-th net is assigned. The second and third terms (B terms) exert the inhibitory forces. The B terms discourage the output of the ijk-th processing element from being nonzero if the other nets overlap with the i-th net. The last term (C term) provides the hill climbing which allows the state of the system to escape from the local minimum and to converge to the global minimum. The C term encourages the ijk-th processing element to be nonzero if the output of all the processing elements for the i-th net is zero. The function $h(x)$ is 1 if $x=0$ and is 0 otherwise. A,B, and C are constant coefficients.

11.3.5.2 PARALLEL ALGORITHM

The following procedure describes the proposed parallel algorithm based on the first-order Euler method for the channel routing problem where n nets and m tracks on the four-layer channel are given.

1. Set $t = 0$, $A = 1$, $B = 1$, $C = 10$, $U_{min} = -20$ and $T\_max = 500$.
2. The initial values of $U_{ijk}(t)$ for $i = 1, \cdots, n$, $j = 1, \cdots, m$ and $k = 1, 2$ are uniformly randomized between 9 and U_{min}, and 0 is assigned as the initial values of $V_{ijk}(t)$ for $i = 1, \cdots, n$, $j = 1, \cdots, m$, and $k = 1, 2$.
3. Use the motion equation to compute $\Delta U_{ijk}(t)$ for $i = 1, \cdots, n$, $j = 1, \cdots, m$, and $k = 1, 2$. If $(t \bmod 10) \leq 5$, then

$$
\begin{aligned}
\Delta U_{ijk}(t) = \quad - \quad & A\left(\sum_{q=1}^{m}\sum_{r=1}^{2} V_{iqr}(t) - 1\right) \\
- \quad & B \cdot V_{B_1}(t) \cdot V_{ijk}(t) \\
- \quad & B \cdot V_{B_2}(t) \cdot V_{ijk}(t) \\
+ \quad & Ch\left(\sum_{q=1}^{m}\sum_{r=1}^{2} V_{iqr}(t)\right)
\end{aligned}
\tag{11.15}
$$

where

$$
V_{B_1}(t) = \sum_{\substack{p=1 \\ p\neq i \\ head_i \leq head_p \leq tail_i}}^{n} V_{pjk}(t) + \sum_{\substack{p=1 \\ p\neq i \\ head_p \leq head_i \leq tail_p}}^{n} V_{pjk}(t)
$$

and

$$
V_{B_2}(t) = \sum_{\substack{p=1 \\ p\neq i}}^{n} T_{ip} \sum_{q=1}^{j} V_{pqk}(t) + \sum_{\substack{p=1 \\ p\neq i}}^{n} B_{ip} \sum_{q=j}^{m} V_{pqk}(t)
$$

Otherwise,

$$
\Delta U_{ijk}(t) = \quad - \quad A\left(\sum_{q=1}^{m}\sum_{r=1}^{2} V_{iqr}(t) - 1\right)
$$

$$
\begin{aligned}
&- \quad B \cdot V_{B_1}(t) \\
&- \quad B \cdot V_{B_2}(t) \\
&+ \quad Ch\left(\sum_{q=1}^{m}\sum_{r=1}^{2} V_{iqr}(t)\right)
\end{aligned}
\tag{11.16}
$$

4. Compute $U_{ijk}(t+1)$ for $i = 1, \cdots, n$, $j = 1, \cdots, m$, and $k = 1, 2$ based on the first-order Euler method:

$$
U_{ijk}(t+1) = U_{ijk}(t) + \Delta U_{ijk}(t) \tag{11.17}
$$

5. $\qquad$ If $U_{ijk}(t+1) > U_{max}$ then $U_{ijk}(t+1) = U_{max}$ $\qquad$ (11.18)

$$
\begin{aligned}
&\text{If } U_{ijk}(t+1) < U_{min} \text{ then } U_{ijk}(t+1) = U_{min} \\
&\text{for } i = 1, \cdots, n, \quad j = 1, \cdots, m, \quad \text{and} \quad k = 1, 2
\end{aligned}
\tag{11.19}
$$

6. Evaluate the values of $V_{ijk}(t+1)$ for $i = 1, \cdots, n, j = 1, \cdots, m$, and $k = 1, 2$

$$
\begin{aligned}
V_{ijk}(t+1) \quad &= \quad 1 \text{ if } U_{ijk}(t+1) > 0 \text{ and } U_{ijk}(t+1) \\
&= \quad max\{U_{iqr(t+1)}\} \\
&\qquad \text{for } q = 1, \cdots, m, \quad \text{and} \quad r = 1, 2 \\
&= \quad 0 \text{ otherwise}
\end{aligned}
\tag{11.20}
$$

7. If $V_{ijk}(t) = 1$ and $\Delta U_{ijk}(t) = 0$ for $i = 1, \cdots, n, \exists j \in \{1, \cdots, m\}$, and $\exists k \in \{1, 2\}$ (all nets are embedded without conflicts) or $t = T_{max}$, where T_{max} is the maximum number of iteration steps, then terminate this procedure; otherwise increment t by 1 and go to step 2. The modified motion equations in step 2 and the range limitation of the input $U_{ijk}(t+1)$ in step 4 improve the convergence frequency to the global minimum.

The given nets are divided into two groups to perform this algorithm in two phases. The first group consists of the longer nets, where the net width between the leftmost column and the rightmost column is more than 30% of the total channel width, and the other group consists of the remaining, shorter nets. It is necessary to fix the locations of the nets in the first group in the beginning because it is difficult or impossible to find the locations for the longer nets in the channel if many nets are embedded. The strategy is that in the first phase we assign the first group nets by applying the algorithm only to them and in the second phase we assign the second group nets by applying the algorithm for all the nets where the outputs of the processing elements corresponding to the first group nets are fixed.

The state of $n \times m \times 2$ processing elements for the n-net-m-track problem can be updated synchronously or asynchronously. In this paper, the synchronous parallel system is simulated on a sequential machine.

The synchronous parallel system can be performed on maximally $n \times m \times 2$ processors while the performance is improved as more processor are used. Figure 11.18 shows solutions from the algorithm given with different initial value of $U_{ijk}(t)$.

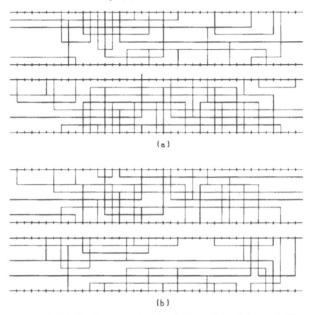

(a)

(b)

Figure 11.18 Solution for example 1 (a) Solution 1 (b) Solution 2 (From *IEEE Trans. on CAD*, Vol. 11, No. 4. ©1992 IEEE. With permission.)

11.4 CONCLUSION

This chapter provided a broad treatment of parallel processing from the standpoint of hardware and routing algorithms in a VLSI design. While much has been reviewed and significant results have been described, opportunities for intensive research and development remain. A variety of multiprocessing system environments are improved day by day.

For instance, massively parallel processing systems with tens of thousands of processors are on the verge of being commercially available products according to the development of VLSI technology. Low-cost high-performance microprocessors will allow the researchers to study more efficient parallel architecture, which in turn enables the developer of routing algorithm to explore faster routers.

As N. Funabiki and Y. Takefuji [26] experimented the routing algorithm with neural network system, genetic methods by Cohoon and Cho [14, 15, 17, 50] may be a good candidate for future studies. Genetic algorithms are said to be better in finding global solutions than the neural network system.

REFERENCES

1. H. G. Adshed, "Towards VLSI Complexity: The DA Algorithm Scaling Problem: Can Special DA Hardware," *Proc. 19th Des. Auto. Conf.*, June 1982, pp. 339–344.

2. P. Banerjee and M. Jones, "A Parallel Simulated Annealing Algorithm for Standard Cell Placement on a Hypercube Computer," *Proc. ICCAD 86*, Nov. 1986, pp. 34–37.

3. T. Blank, "A Survey of Hardware Acceleraors Used in Computer-Aided Design," *IEEE Design and Test*, Aug. 1984, pp. 21–39.

4. T. Blank, M. Stefik, and W. van Cleemput, "A Parallel Bit Map Processor Architecture for Da Algorithms," *Proc. 18th Des. Auto. Conf.*, 1981, pp. 837–845.

5. M.A Breuer and K.Shamsa, "A Hardware Router," *J Digital Syst.*, vol. 4(no. 4), 1980.

6. R. J. Brouwer and P. Banerjee, "Fhigure: A Parallel Hierarchical Global Router," *Proc. 27th Des. Auto. Conf.*, 1990, pp. 650–653.

7. David A. Carlson, "Performing Tree and Prefix Computations on Modified Mesh-Connected Parallel Computers," *Int. Conf. Parallel Processing*, 1985, pp. 715–718.

8. A. Casotto, F. Romeo, and A. Sangiovanni-Vincentelli, "A Parallel Simulated Annealing Algorithm for the Placement of Macro-Cells," *IEEE Trans. CAD*, vol. 6, Sept. 1987, pp. 838–847.

9. A. Casotto, F. Romeoo, and A. Sangiovanni-Vincentelli, "A Parallel Simulated Annealing Algorithm for the Placement of macro-cells," *Proc. IC-CAD 86*, Nov. 1986, pp. 30–33.

10. W. S. Chan, "A Channel Routing Algorithm," *Proc. Third Caltech Conf. on VLSI, Bell Lab*, 1983, pp. 117–140.

11. Shing Chong Chang and Joseph JaJa, "Parallel Algorithm for Channel Routing in the Knock Model," *Int. Conf. Parallel Processing*, 1988, pp. 18–25.

12. T. Cheung and J.E. Smith, "An Analysis of thr Cray X-mp Memory System," *Proce. ICCP*, pp. 499–505.

13. E. Chiricozzi and A. D'amico, *Parallel Processing and Applications*, North-Holland, Amsterdam, New York, 1988.

14. Tae Won Cho, Hee Il Ahn, and Seung Kee Han, "Genrouter: A Genetic Algorithm for Channel Routing Problems," *IEEE TenCon95 on Microelectronics and VLSI*, Nov. 1995.

15. Tae Won Cho, Hee Il Ahn, and Chul Eui Hong, "A Distributed Genetic Approach for Channel Routing Problems," *4th ICVC*, Oct. 1995.

16. Tae Won Cho, Sam S. Pyo, and J. Robert Heath, "A Parallel Approach to Swichbox Routing," *IEEE Trans. on CAD*, vol. CAD-13, no. 6, 1992, pp. 684–693.

17. James P. Cohoon. "Genetic Placement", *IEEE Trans. on CAD*, CAD-6(6), Nov. 1987, pp. 956–964.

18. James P. Cohoon and P. Heck, "Beaver: A Computational Geometry Based Tool for Switchbox Routing," *IEEE Trans. on CAD*, CAD-7, 1988.

19. H. Date, Y. Matumoto, K. Kimura, K. Taki, H. Kato, and M. Hoshi, "LSI-CAD Programs on Parallel Inference Machine," *Proc. Int. Conf. on FGCS, ICOT, Tokyo*, 1992, pp. 237–247.

20. Hiroshi Date and Kazuo Taki, "A Parallel Lookahead Line Search Router with Automatic Ripup-and-Reroute," *Proc. IEEE Euro ASIC*, 1993, pp. 117–121.

21. H. Dietz and D. Klappholz, "Refining a Conventional Language for Race-Free Specification of Parallel Algorithms," *Proc. ICPP*, 1984, pp. 380–382.

22. H. Dietz and D. Klappholz, "Refined c: A Seqquential Language for Parallel Proogramming," *Proc. ICPP*, Aug. 1985.

23. W. A. Dees et al, "Automated Rip-Up and Reroute Techniques," *Proc. 19th ACM/IEEE Des. Auto. Conf.*, June 1982, pp. 432–439.

24. A.K. Ezzat and R. Agrawal, "Making Oneself Known in a Distributed World," *Proc. ICPP*, Aug. 1985, 139–142.

25. S. Fernbach, "Parallelism in Computing," *Proc. Inter. Conf. Paral. Proc. (ICCP)*, Aug. 1981, pp. 1–4.

26. Nobuo Funabiki and Yoshiyasu Takefuji, "A Parallel Algorithm for Channel Routing Problems," *IEEE Trans. CAD*, vol. 11, no. 4, 1992, pp. 464–474.

27. M. Hanan, "On Steiner's Program with Rectilinear Distance," *SIAM J. Appl. Math.*, vol. 14, Mar. 1966, pp. 255–265.

28. J. M. Hancock and S. DasGupta, "Tutorial on Parallel Processing for Design Automation Applications," *23rd Des. Auto. Conf.*, 1986, pp. 69–77.

29. P. B. Hansen, "The Programming Language Concurrent Pascal," *IEEE Trans. Soft. Eng.*, vol. SE-1, June 1975, pp. 199–207.

30. J. P. Hayes, T.N. Mudge, Q.F. Stout, S. Colley, and J. Palmer, "Aritecture of a Hypercube Supercomputer," *Int. Conf. Parallel Processing*, Aug. 1986, pp. 653–660.

31. S. Heiss, "A Path Connection Algorithm for Multi-layer Board," *Proc. 5th Des. Auto. Conf.*, 1986, pp. 1–14.

32. W. Heyns, W. Sanse, and H. Beke, "A Line-Expansion Algorithm for the General Routing Problem with a Guaranteed Solution," *Proc. 17th Des. Auto. Conf.*, 1980, pp. 243–249.

33. C. A. R. Hoarse, "Communicating Sequential Processes," *Com. Acm*, Aug. 1978, pp. 666–667.

34. S. J. Hong and R. Nair, "Wire Routing Machines-New Tools for VLSI Physical Design," *Proc. IEEE*, vol. 71, no. 1, Jan. 1983, pp. 57–65.

35. S. J. Hong, R.Nair, and E.Shapiro, "A Pysical Design Machine," *VLSI 81 Conf. Proc.*, ed. by J. P. Gray, Academic Press, New York, 1981.

36. J. J. Hopfield and D.W. Tank, "Neural Computation of Decisions in Optimization Problems," *Biol. Cybern*, vol. 52, 1985, pp. 141–152.

37. A. Iosupovici, "Design of an Iterative Array Maze Router," *Proc. IEEE Int. Conf. on Circuits and Computers*, Oct. 1980, pp. 908–911.

38. A. Iosupovici, "A Class of Array Architectures for Hardware Grid Routers," *IEEE Trans. CAD*, vol. CAD-5, no. 2, April 1986, pp. 245–255.

39. A. Iosupovici and A. Vachidsafa, "Parallel Routing on a Hardware Array Router," *Int. Conf. of CAD*, 1985, pp. 136–138.

40. D. S. Johnson, "The NP-Completeness Column: An Ongoing Guide," *J. Algorithms*, Dec. 1982, pp. 381–395.

41. Masayoshi Tachibana Kei Suzuki, Yusuke Matsunaga and Tatsuo Ohtsuki, "A Hardware Maze Router with Application to Interactive Rip-Up and Reroute," *IEEE Trans. CAD*, vol. CAD-5, no. 4, 1986, pp. 466–476.

42. H. Kitazawa, "A Line Search Algorithm with High Wireability for Custom VLSI Design," *Proc. IS-CAS'85*, 1985, pp. 1035–1038.

43. K.Suzuki, Y.Matsunaga, M.Tachibana, and T.Ohtsuki, "A Hardware Maze Router with Application to Interactive Rip-up and Reroute," *IEEE Trans. CAD.*, vol. CAD-5, no. 4, Oct. 1986, pp. 466–476.

44. C. Y. Lee, "An Algorithm for Path Connections and Its Applications," *IRE Trans. Electronic Computers*, vol. EC-10, 1961, pp. 346–365.

45. C. E. Leiserson, "Fat-trees: Universal Networks for Hardware-Efficient Supercomputing," *Proc. ICPP*, Aug. 1985, pp. 393–402.

46. S. P. Levitan, "Evaluation Criteria for Communication Structures in Parallel Architectures," *Proc. ICPP*, Aug. 1985, pp. 147–154.

47. R. Maenner, "Hardware Task Processor Scheduling in a Polyprocessor Environment," *IEEE Tran. Computer*, vol. C-33, July. 1984.

48. C. Maples, "Pyramids, Crossbars and Thousands of Processors," *Proc. ICCP, St. Charles, ILL*, Aug. 1985, pp. 681–688.

49. W.S. McCulloch and W.H. Pitts, "A Logical Calculus of Ideas Immananent in Nervous Activity," *Bull. Math. Biophys.*, vol. 5, 1943, pp. 115.

50. Edward J. Mello and Robert H. Fujii, "Darwin: A Timing Driven Placement Algorithm Using the Genetic Optimazation Approach," *ICVC'93 TAEJON*, 2, 1993, pp. 219–222.

51. Jan ming Ho, Gopalakrishnan Vijayan, and C. K. Wong, "A New Approach to the Rectilinear Steiner Tree Problem," *Proc. 26th ACM/IEEE Design Automation Conf.*, 1989, pp. 161–166.

52. A. Norton and G.F. Pfister, "A Methodology for Predicting Multiprocessor Performance," *Proc. ICPP*, Aug. 1985, pp. 772–781.

53. O.A. Olukotun and T.N. Mudge, "A Preliminary Investigation into Parallel Routing on a Hypercube Computer," *24th Des. Auto. Conf.*, 1987, pp. 814–820.

54. Mikael Palczewski, "Plane Parallel A* Maze Router and Its Application to FPGAS," *29th Des. Auto. Conf.*, 1992, pp. 691–697.

55. G.F. Pfister, W.C. Brantley, D.A. George, S.L. Harvey, W.J. Kleinfelder, K.P. Mcauliffe, E.A. Melton, V.A. Norton, and J. Weiss, "The IBM Research Parallel Processor Prototype (rp3): Introduction on Architecture," *Proc. ICPP*, 1985, pp. 764–771.

56. R. C. Prim, "Shortest Connection Networks and Some Generalizations," *Bell System Tech. J.*, vol. 36, 1957, pp. 1389–1401.

57. R. Y. Printer, "River Routing: Methodology and Analysis," *Proc. Third Caltech Conf. on VLSI, Bell Lab*, 1983, pp. 141–163.

58. S. Renben and P. C. Patton, "BCA: A Bus Connected Architecture," *Proc. ICCP*, St. Charles, ILL, Aug. 1985, pp. 79–88.

59. R. L. Rivest, "The PI (Placement and Interconnect) System," *19th Des. Auto. Conf.*, 1982, pp. 418–424.

60. R.Nair, S.J.Hong, S.Liles, and R.Villani, "Global Wiring on a Wire Wouting Machine," *19th Des. Auto. Conf.*, Nov. 1982.

61. J. S. Rose, D. R. Blythe, W. M. Snelgrove, and Z. G. Vranesic, "Fast, High Quality VLSI Placement on an MIMD Multiprocessor," *Proc. IC-CAD 86*, Nov. 1986, pp. 42–45.

62. J. S. Rose, W. M. Snelgrove, and Z. G. Vranesic, "Parallel Standard Cell Placement Algorithms with Quality Equivalent to Simulated Annealing," *IEEE Trans. CAD*, vol. 7, March 1988, pp. 387–396.

63. Jonathan Rose, "LocusRoute: A Parallel Global Routing for Standard Cells," *25th Des. Auto. Conf.*, 1988, pp. 189–195.

64. Jonathan Rose, "Parallel Global Routing for Standard Cells," *IEEE Trans. CAD*, vol. 9, no. 10, Oct. 1990, pp. 1085–1095.

65. R. M. Russell, "The CRAY-1 Computer Systems," *Comm. ACM*, Jan. 1978, pp. 63–72.

66. R. A. Rutenbar, *A Class of Cellular Computer Architectures To Support Design Automation*, Ph.d. thesis, Dept of CICE, University of Michigan, Ann Arbor, MI, 1984.

67. R. A. Rutenbar, T. N. Mudge, and D. E. Atkins, "A Class of Cellular Architectures to Support Physical Design Automation," *IEEE Tran. CAD of IC's and Systems*, vol. 3, no. 4, Oct. 1984, pp. 264–278.

68. Thomas Ryan, "An ISMA Lee router Accelerator," *IEEE Design and Test of Computers*, Oct. 1987, pp. 38–45.

69. C. Sechen and A. Sangiovanni-Vincentelli, "The Timeberwolf Placement and Routing Package," *IEEE J. Solid-State Circuits*, vol. SC-20, Apr. 1985, pp. 510–522.

70. M. Seto, K. Kubota, and T. ohtsuki, "A Hardware Implementation of Gridless Routing Based on Content Addressable Memory," *Proc. Des. Auto. Conf.*, 1990, pp. 646–649.

71. J. Soukup, "Fast Maze Routor," *Proc. 15th Des. Auto. Conf.*, 1978, pp. 100–101.

72. R. A. Stokes, "Burroughs Scientific Processor," *High Speed Computer and Algorithm Organization*, 1977, pp. 85–89.

73. R. J. Swan, S. H. Fuller, and D. P. Siewiorek, "CM*-A Modular, Multi-Microprocessor," *Proc. of National Compu. Conf. (NCC)*, 1977, pp. 645–655.

74. M. Tachibana, S. Nakajima, and T. Ohtsuki, "An Architecture of Parallel Routing Processor," *Japanese TG CAD83-75 IECE*, Japan, 1983, pp. 25–30.

75. Y. Takefuji, L. L. Chen, K.C. Lee, and J. Huffman, "Parallel Algorithms for Finding a Near-maximum Independent Set of a Circle Graph," *IEEE Trans. Neural Networks*, vol. 1, Sept. 1990, pp. 263-267.

76. Y. Takefuji and K. C. Lee, "A Near-Optimum Parallel Planarization Algorithm," *Science*, vol. 245, Sept. 1989, pp. 1221-1223.

77. Y. Takefuji and K. C. Lee, "A Parallel Algorithm for Tilting Problems," *IEEE*, vol. 1, Mar. 1990, pp. 143-145.

78. Y. Takefuji and K. C. Lee, "A Super Parallel Sorting Algorithm Based on Neural Networks," *IEEE Trans. Circuits. Syst.*, vol. 37, Nov. 1990, pp. 1425-1429.

79. Y. Takefuji and K. C. Lee, "An Artificial Hysteresis Binary Neuron: A Model Suppressing the Oscillatory Behaviours of Neural Dynamics," *Biol. Cybern.*, vol. 64, 1991, pp. 353-356.

80. K. Taki, "Parallel Inference Machine PIM," *Proc. Int. Conf. FGCS, ICOT*, Tokyo, 1992, pp. 50-72.

81. Masayoshi Tachibana Tatsuo Ohtsukki and Kei suzuki, "A Hardware Maze Router with Rip-up and Reroute Support," *Int. Conf. CAD*, 1985, pp. 220-222.

82. Takumi Watanabe, Hitoshi Kitazawa, and Yoshi Sugiyama, "A Parallel Adaptable Routing Algorithm and Its Implemention on a Two-Dimensional Array Processor," *IEEE Trans. CAD*, vol. CAD-6, no. 2, March 1987, pp. 241-250.

83. Takumi Watanabe and Yoshi Sugiyama, "A New Routing Algorithm and Its Hardware Implementation," *Des. Auto. Conf.*, 1986, pp. 574-580.

84. Michael Weiss, Robert Morgan, and Zhixi Fang, "Dynamic Scheduling and Memory Management for Parallel Programs," *Int. Conf. Paral. Proces.*, 1988, pp. 161-165.

85. Linda F. Wilson and Mario J.Gonzalez, "Design Guidelines for Parallel Algorithms Using Continuous Job Profiles," *5th Int. Paral. Proces. Sym.*, 1991, pp. 30-36.

86. Youngju Won and Sartaj Sahni, "Maze Routing on a Hypercube Multiprocesor Computer," *Int. Conf. Paral. Proces.*, 1987, pp. 630-637.

87. Youngju Won, Sartaj Sahni, and Yacoub El-Ziq, "A hardware accelerator for maze routing," *24th Des. Auto. Conf.*, 1987, pp. 800-806.

88. T. Yoshimura, "Efficient algorithm for channel routing," *IEEE trans. CAD*, CAD-1, Jan. 1982, pp. 25-35.

89. Mehdi R. Zargham, "Parallel Channel Routing," *25th Des. Auto. Conf.*, 1988, pp. 128-133.

APPENDIX A: SYMBOLS

ADJNET(j): The set of adjustable overlap nets whose adjustable regions cover the column j

$A(n)$: The sensitivity of the voltage on net n to the voltage fluctuations in the adjacent nets

B(V,E): An intersection position graph

$C_{A,B}^{(b)}$: The bound of the coupling due to crossover between nets A and B

$C_{A,B}^{(max)}$: The maximum coupling due to crossover between nets A and B

$C_{A,B}^{(min)}$: The minimum coupling due to crossover between nets A and B

C_{cr}: The unit-length capacitance between two crossing wires

CD: Dynamic ordering in terms of the number of 1-segment choices available for each net

$CD(i,j)$: The density of column j in channel i

C_h: The coupling contribution due to horizontal parallel segments

$C_h^{(b)}$: The bound of the coupling contribution due to horizontal parallel segments

Chnu: The number of channels

C_i: An i-edge-set (nondegenerated) clique

C_{id}: An i-edge-set degenerated clique

C_{in}: An i-edge-set NN clique

C_q: A cluster in wiring

CS: Static ordering in terms of the number of 1-segment choices available for each net

C_u: The capacitance between a unit-length wire segment and the substrate

C_u': The unit-length capacitance between two wire segments

C_v: The coupling contribution due to vertical parallel segments

$C_v^{(b)}$: The bound of coupling contribution due to vertical parallel segments

c_w: The allowed wire density of a grid edge in a topography map

C_y: An edge-set cycle; a connected graph in that every edge set is exactly adjacent to two edge sets

$D(n)$: A function whose value is either 0 or 1

$d_q^{(j)}$: A weight vector of each state j of cluster c_q

$d_{q,i}^{(j)}$: The ith component of weight vector $d_q^{(j)}$, which is the number of units of wire W_i embedded in the poor layer L_2 if cluster c_q is in state j, where $j = 0$ or 1 for two-layer wiring

$d_r(p_i, p_j)$: The rectilinear distance between p_i and p_j

$d_s(p_i, p_j)$: Slant distance selecting a minimum distance from the horizontal distance and the vertical distance between p_i and p_j

d_w: The wire capacity of the vicinity of a grid edge in a topography map

$E(G)$: A set of all edges in graph G

e_{pq}: A shortest rectilinear edge from p to q, $e_{pq} \in E_{pq}$

E_{pq}: An edge-set that contains a set of all shortest edges between points p and q

E_{pq}^{max} (or E_{pq}^{min}): A fixed edge-set in E_{pq} when the distance between p and q is maximized (or minimized)

E_T (or $E(T)$): A set of all edge-sets in an edge-set tree T

F_b: The cost contribution in case a bend is needed to add an edge to the wire

$F(e)$: The cost function of analog area routing for edge e

F_j: A cluster of incidence intersection column j

F_v: The cost contribution in case a via is needed for adding an edge to a wire

$G_c(V, E)$: A minimum intersection graph

G_d: A dual graph

$G = (E, V)$: A graph formed by a set of edges, $E = E(G)$, and a set of vertices, $V = V(G)$

$G_L(V, E)$: A minimum chain graph

$group_t$: A set of segments belonging to track t

H_x: The number of horizontal antifuses

I: An index measuring the total contributions of an algorithm

I_i: An index measuring the contributions of the ith subalgorithm

$I(v_i)$: A set of free grid points which can be reached by finding a path starting from v_i

$I_x BC$: The blocking connections related to the source or the target island

$I_x BP$: The blocking points of island I_x

$\mathbf{K}(\mathbf{p})$: An array of all performance functions for parasitics

$\mathbf{K}_i(p)$: An array of a performance function for the ith parasitic

LE: Left-edge first in ordering nets in terms of left edges of nets

$left_j$: The leftmost x-coordinate of segment j

$LefnetN$: The left-hand side net of a *span*

LF: Longest first in ordering nets in terms of length

$l_{R_{pq}}$: The perimeter of R_{pq}; $l_{R_{pq}} = 2l_{E_{pq}}$

L_{total}: The total routing length in a topography map

L_{track}: The total track length in a topography map

l_x: The length of $x \in \{e_{pq}, E_{pq}, C_i, C_{id}, C_y, t, T, T_{NR}\}$

$max(a, ..., z)$: A function that selects the maximum one in $\{a, ..., z\}$

m_c: A figure of merit for congestion

$\mathbf{M}_i$: An initial matrix

$mid(a, b, c)$: A function that selects one with the mid-value in $\{a, b, c\}$

$min(a, ..., z)$: A function that selects the minimum one in $\{a, ..., z\}$

$\mathbf{M}_k$: An assignment matrix

M_n: A net matrix

M_p: Domain of p, a set of points enclosed by a rectangle

$MRS(s,t)$: A minimal rip-up set of the failed connection from points s to t

M_s: A segment matrix

NC_k^A: The total number of *1*s in column k of a *(0,1)* matrix A

N_{layer}: The number of layers

$n_p(p_i, p_j)$: A node priority that indicates $min(i,j)$

N_x: The minimum cardinality of a set of rows of a segment matrix

$O(n^x)$: The order of the xth power of size n, that usually indicates the time complexity of a deterministic problem with the size of n

$OR(a,b)$: Logical function, a or b

$\mathcal{P}$: A set of all parasitics

$\mathbf{p}^{(b)}$: An array of bounds for all parasitics

$p_i^{(0)}$: The normal value of the ith parasitic, p_i

$\mathbf{P}_k$: An active track matrix

$p^{(\mathbf{max})}$: An array of the lower bounds for all parasitics

PS_{ij}: A two-dimensional array for searching a partial solution of the ith conflict group in the jth process

$right_j$: The rightmost x-coordinate of segment j

$RignetN$: A right-hand side net of a $span$

R_{pq}: A rectangle with p and q as vertices

$RS(s,t)$: A rip-up set of the failed connection from points s to t

$RS_{min}(s,t)$: A rip-up set of the failed connection from points s to t with the minimum-cost constraint

R_x: The xth row of a net matrix

S: A sensitivity array for constraint generation

$S(n)$: The sensitivity of a digital net n to the noise coming from the surrounding nets

$span$: A region between two adjacent pins

S_x: A set of rows of a segment matrix

T: An edge-set tree that contains a set of rectilinear trees

t_b: A binary tree

$T(C_1, C_3, ..., C_i)$: An edge-set tree tested by cliques C_1, C_2, ..., and C_m

$T(C_{1n}, C_{3n}, ..., C_{in})$: An edge-set tree tested by cliques C_{1n}, C_{3n}, ..., and C_{mn}

T_{NR}: A no-redundancy edge-set tree

$Tranu(i)$: The number of tracks in the ith channel

$VCD(j)$: The vertical density at the column j

V_{IH}: Low to high transition

V_n: A set of all given nodes for a graph

$VOF(i,j)$: The overflows at the jth column of channel i

$V_s(T)$: A set of all Steiner points in edge-set tree T

$V_T(V(T))$: A set of all nodes and all Steiner points, $V_t = V_T = V_n \cup V_s(T)$ if $t \in T$

$VVOF(j)$: The overflows at the jth column

W: A wire layout

$w(e(p_i, p_j))$: The weight of edge $e(p_i, p_j)$

W_i: The wire layout of the ith net

w_{pq}: A wire connecting points p and q in a topography map

w_{p_j}: A weight of parasitic p_j

(x_r, y_r): A point r with x_r and y_r as horizontal and vertical coordinates

$\{x, y\}$: An edge with terminals (x_x, y_x) and (x_y, y_y)

$|X|$: The total number of elements in set X

$\alpha(P)$: A Path vector

$\Delta \mathbf{K}(\mathbf{p})$: An array of the corresponding degradations of $\mathbf{K}$ due to variations $\mathbf{p}$

$\overline{\Delta \mathbf{K}(\mathbf{p})}$: An array of the (user-provided) constraints on degradation of $\mathbf{K}$ due to $\mathbf{p}$

APPENDIX B: ACRONYMS

ASIC: Application-specific integrated circuit

AWE: Asymptotic waveform evaluation

BG: Blocked-gulf detection (procedure)

BN: Bottemneck (procedure)

BnS: Batched n-Steinerization

CA: Cell assignment (technique)

CMOS: Complementary metal-oxide-semiconductor

CRABAR: Channel routing algorithm based on assigning resources

CSP: Candidate Steiner point

CVM: Constrained via minimization problem

CVMPP: Constrained via minimization with pin preassignment

DIP: Dual in-line package

DMA: Data memory allocation

FPGA: Field-programmable gate array

GnS: Greedy n-Steinerization

HE: High east

HN: High north

HS: High south

HST: Horizontal segment tree

HW: High west

ILP: An integer linear programming problem

LE: Low east

LM: Lee-Moore (algorithm)

LN: Low north

L-RST: L shaped RST

LS: Low south

LSP: Local Steiner point

LW: Low west

MCM: Multichip module

MRS: Minimum rip-up set

MRST: Minimum rectilinear Steiner tree

MSPT (RMST): Minimum rectilinear spanning tree

MWIS: Maximum-weight independent set

n-CVM: N-layer constrained via minimization problem

NST: Net segment tree

OARST: Obstacle-avoiding rectilinear Steiner tree

OASP: Obstacle-avoiding shortest path

ODD: Odd faces of a wire layout

PDW$_2$: Performance driven 2-layer wiring

PDW$(=)_2$: Performance driven 2-layer wiring with the same conductivity in two layers

PDW$(\neq)_2$: Performance driven 2-layer wiring with different conductivities in two layers

PLM: Parallel Lee-Moore (algorithm)

PWB/PCB: Printed wiring/circuit board

RAM: Random access memories

RST: Rectilinear Steiner point

SAM: Synchronous active memory

spp: Shortest-paths preserving graph

SSN: Simultaneous switching noise

VCG: Vertical constant graph

VDB: Vertex-deletion graph bipartization problem

VLSI: Very large-scale integration

VST: Vertical segment tree

WRM: Wire routing machine

1-D: One-dimensional

2-D: Two-dimensional

APPENDIX C: GLOSSARY

Adjacent edges (edge sets): Two edges (edge sets) that intersect at one or more points.

Adjacent points (vertices): Two points (vertices) that are directly connected by an edge or an edge set.

Algorithm expansion: An approach that expands an inefficient algorithm for an exact solution into a series of subalgorithms for finding a subseries of efficient subalgorithms that converge the major contribution of the exact solution.

Antifuse: A programmable connection for connecting two broken wire segments separated by a "fuse," that is used in an algorithm for segmented channel routing.

Backward search: A search process searching a path connecting two given terminals from the progressing information provided by the forward search.

Basic cell block: A block in a cell row for accommodating a basic cell with equal height in channel routing.

Bend: An angle in an edge, especially a right angle in a rectilinear edge.

Bent edge: A rectilinear edge with one or more bends.

Bent-edge set: A set of all bent edges connecting two vertices.

Blocking connections: A set of connections related to the source or the target island.

Blocking points: A set of points on the border of an island.

Bounded length: The length difference of two paths (or wire segments) bounded by the requirement for routing in high-speed analog circuits.

Bounded length problem: A routing problem to find paths satisfying the bounded length requirement of high-speed analog circuits.

Breaking rule: A rule for removing a set of edge sets from an edge-set tree.

Capacity of the main channel: The number of wiring tracks in the main channel.

Channel density: The maximum tracks occuring in a channel

Channel routing: A routing process assigning a set of connections to tracks within a rectangular region (channel).

Clearance: The insulation area between two neighboring parallel wires in a printed wiring/circuit board.

Clique: An NR edge-set subtree whose inner vertices are Steiner points only, whose length is minimal and constant, independent of movements of Steiner points, and whose total number of edge sets is the minimum for maintaining the maximum number of points in each moving domain.

Cluster: A maximal set of mutually crossing and overlapping critical wire segments.

Combination: Taking r elements at a time from n $(n \geq r)$ distinct elements, without considering the order.

Compaction: A routing process that moves wires to their vicinity to increase the density of that area, such that routing area can be reduced or the density of a layer or a board can be equalized.

Compound clique: A clique evolved from a set of adjacent LD cliques and their adjacent degenerated edge sets.

Connected graph: A graph in which each vertex is connected to any other vertices with one or more edges.

Connecting rule: A rule for connecting a set of edge-set trees for subsets of nodes into an edge-set tree for all nodes.

Connection order: The order for connecting terminals one after another to form a net.

Constraint-driven routing: A routing approach that tries to enforce constraints directly, using a simplified quantitative model of the functional relation between performance specifications and parasitics to determine a pattern for the router's behavior.

Contact region: A channel routing region where contacts are located.

Contact region routing: A routing process providing a good starting point for the main-channel routing.

Critical area: The region between two embedded islands.

Crosstalk: Undesirable electrical interference caused by the coupling of energy between signal paths in a high-speed circuit.

Crossunder: An interconnection of nets on a polysilicon bar.

Cycle: A sub-graph in which every vertex connects exactly two edges.

Degenerated clique: An interfacing edge-set subtree among cliques and given nodes in a no-redundancy (NR) edge-set tree.

Degree of a vertex (node, or point): The number of edges connected to a vertex in a graph.

Density topography: A topography whose local topographical height is indicated by the wire density of a grid edge in a gridded layer where all nets are placed.

Depression: A routing process that moves wires to their vicinity to decrease the density of that area, such that the density of a layer or a board can be equalized.

Descendant edge set: An edge set in an edge-set tree for a subset of nodes, which is generated by removing a set of edge sets from an edge-set tree for all nodes.

Detailed routing: The final determination of interconnections on a substrate by searching a path connecting a set of terminals (pins or pads) in a workspace, where the substrate associated with the routed paths is simulated.

d-stretching: To stretch the wire layout in d dimensions for inserting new horizontal and/or vertical grid tracks, where $d = 1$ or 2.

Dual graph: A pair of graphs mapping one's positive to another's negative.

Dynamic Steiner point: A Steiner point that can be moved without changing the length of an edge-set tree.

Edge: A connection of two vertices with specific measures of length and shape.

Edge set: A set of all minimum rectilinear edges that connects two vertices (nodes or Steiner points) in an edge-set tree.

Edge-set cycle: A graph in which each vertex is an intersection of two edge sets.

Edge set replacement: Replacing a set of edge sets in an edge-set tree with another set of edge sets to form a new edge-set tree.

Edge-set subtree: A subgraph of an edge-set tree.

Edge-set tree: A graph generated by extending every edge and every Steiner point in a rectilinear tree into an edge set and a dynamic Steiner point, respectively.

Edge-set tree tested by i-edge-set clique: An edge-set tree whose length cannot be reduced by replacing an arbitrary i-edge-set clique or an i-edge-set subtree whose inner vertices are not nodes with new edge sets.

Edge-set tree tested by *i*-edge-set NN clique: An edge-set tree whose length cannot be reduced by replacing an arbitrary *i*-edge-set NN clique with new edge sets.

Electro-static shield: An interconnection structure able to decouple effectively critical net pairs.

End vertex: A vertex adjacent to only one other vertex in an edge-set subtree.

Escape graph: A graph that is guaranteed to contain the shortest paths between all pairs of terminals, to which shortest-path algorithms can then be applied.

Essential internal LSP: A local Steiner point of an internal node whose adjacent edges generate one local Steiner point only.

Essential leaf LSP: A local Steiner point of a leaf node whose parent node and other neighbors of the parent node form a unique local Steiner point.

Euclidean distance: The distance of two vertices, measured by the length of a straight line-segment with the vertices as terminals.

Evolved clique: A clique formed by a stem and its adjacent straight-edge sets, in which the main vertices are, or are evolved (changed) from, the end nodes of an LD clique.

Extended edge-set tree: An edge-set tree in which each moving edge set is extended to an extended moving edge set.

Extended parallel moving edge set: A set of parallel moving edge sets whose end vertices are located on two parallel edges.

Figure of merit for congestion: An index used to indicate the average rate of the wasted wiring area in global routing.

Fixed edge set: An edge set connecting two fixed vertices.

Fixed segment: Segments in nets not related to any conflicts.

Flexibility: A measure of how easily the tool is able to meet the constraints.

Flex point: A left son having only one right son, or a right son having only one left son in a binary tree.

Forward search: A search process for providing progressing information which starts from one or more terminals and progresses with a specific search technique, until an object is encountered.

Generalized knock-knee mode wire layout: A wire layout in which no edge in a subgraph for a net shares an edge with any other nets.

General solution of the Steiner's problem: A set of NR edge-set trees containing all MRSTs for a set of given nodes.

Global routing: A routing process estimating the minimum number of layers and assigning each net or subnet to a specific layer for completing subsequent detailed routing without incomplete wires.

Graph: A super-set of a set of edges and a set of vertices used to describe the relationship between the edges and vertices.

Grid conflict: A conflict in parallel computing, in that two or more horizontal (or vertical) segments belonging to different nets meet in a grid unit of a row (or a column).

Gridded workspace: A workspace divided into square grids.

Grid expansion: A search process that progresses from a grid to its adjacent grids in a gridded workspace.

Gridless searching: A routing technique that searches a path from the coordinates of obstacles, rather than the grids in a gridded workspace.

Hightower's router: A line-searching technique that progresses from a fast line to another fast line, which is the longest line segment nearest the previous parallel fast line.

Imagined clique: A virtual clique formed by an LD clique and its adjacent straight edge sets, that does not exist in an edge-set tree, but is imagined as what an existing clique is evolved from.

Inner vertex: A vertex adjacent to more than one other vertex in an edge-set subtree.

Island: A set of free grid points which can be reached by finding a path starting from a vertex.

k-**layer wirable:** A wire layout having *k*-layer wiring.

Knock-knee mode wire layout: A wire layout modeling the popular constraint that wires can only run horizontally and vertically.

k-**segment matching problem:** A segmented channel routing problem in which each row of the net matrix can be matched with at most k rows of the segment matrix.

Layer: A conduction sheet etched to form wires or a ground place in a printed wiring substrate.

Layer assignment (layering or wiring): A process assigning wire segments of the wire layout to one of k available layers so that the segments are electronically connected in the right way.

Layout: Arranging modules and interconnecting electronic nets on a substrate.

Layout compaction: A process of converting a symbolic layout or a sketch of a topological layout to a design-rule-correct mask layout with minimal area.

Lee-cell: A simple processing element used in an L-machine.

Lee-machine: An automation machine consisting of a control unit that communicates with a host computer and sequencing the operations of an array of simple processing elements.

Lee's (Lee-Moore's) router: A maze-searching technique for searching a shortest path, by simultaneously expanding grids to their adjacent grids until the source grid is encountered.

Linear degenerated (LD) clique: A degenerated clique whose inner vertices (Steiner points) lie on a straight line terminated at two nodes.

Linear programming problem: A programming problem characterized by linear functions of the unknowns in which the objective is linear and the constraints are linear equalities or linear inequalities.

Line searching: A search technique that searches a path connecting two terminals from a set of line segments to another set of line segments, which are extended from a terminal, or imaginary grids occupied by an existing line segments.

Local congestion elimination: A routing process consisting of both compaction and depression for equalizing the wire density of a layer or a board.

Local refinement: The process determining local Steiner points ·

L-shaped layout: A rectilinear layout that considers L-shaped edges only

Main channel: A channel composed of tracks between the top and the bottom tracks where the long contacts are located.

Main channel routing: A routing process used in an environment where channels are fixed and fixed-length crossunders are preplaced.

Main-conflict segment: A segment directly involved in the conflict.

Main vertices: The end vertices of a clique that are evolved from two end nodes of an LD clique.

Manhattan distance: A distance between two points, formally called a rectilinear distance, measured by a half-perimeter of the rectangle with the points as vertices.

Matched pair: A pair of edges or wires with the same shape and the same length.

Matched-pair problem: Finding a pair of shortest paths with the same shape and the same length in a routing workspace.

Matrix row matching problem: A channel routing problem formulated by the net and the segment matrices.

Maximized edge set: A fixed edge set having the maximum length among all fixed edge sets of a moving edge set.

Maze searching: A strategy for searching a path connecting two terminals in a gridded workspace.

Median point: A point whose x- and y-coordinates are determined by the mid-values of the x- and y-coordinates of three points, respectively.

Merging tree: A directed tree with weight, in which each node represents a net, and a directed edge from node i to node j means that these two nets can be routed on the same track and net i must be placed on the left of net j.

Minimized edge set: A fixed edge set having the minimum length among all fixed edge sets of a moving edge set.

Minimum rectangle: The rectangle with the minimum perimeter enclosing a set of points.

Module Assignment: The grouping of modules to a workspace prior to placement.

Moving domain of a Steiner point: A set of points that defines the moving range of a Steiner point in an edge-set tree.

Moving edge set: An edge set with at least one dynamic Steiner point as a vertex.

Neighbor segments: Segments that are connected to two ends of a main-conflict segment.

Net: A physical tree in which an edge is a wire and a vertex is a terminal or a module.

Net classification: A scheme for routing that captures the complex pattern of interrelations between nets due to crosstalk, matching, and mutual sensitivity to noise.

Net (signal set) size: The number of terminals in a net (signal set).

Nested wires: Bent wires that are neighboring, but separate on the same layer.

Node-ended NR edge-set subtree (NN edge-set subtree): An NR edge-set subtree in which each end vertex is a node or a dynamic Steiner point whose moving domain includes a node, and each inner vertex is a Steiner point whose moving domain does not include a node.

No-embedded-branch (NEB) cycle: A cycle generated by connecting two nodes in a binary tree, that does not embed any branches of the binary tree.

NP-complete problem: A decision problem on NP problems that can be transformed from one to another in polynomial time.

NP problem: A decision problem that can be solved by a nondeterministic polynomial-time algorithm.

No-redundancy (NR) edge-set tree: An edge-set tree in which no redundant length can be found by replacing an arbitrary single edge set with a new edge set.

NR edge-set subtree: An edge-set subtree in which no redundant length can be found by replacing an arbitrary single edge set with a new edge set.

Obstacle: A space occupied by routed paths in a routing workspace.

Obstacle-avoiding rectilinear Steiner tree (OARST) problem: A conditional Steiner's problem identical to the Steiner's problem except for the presence of rectilinear obstacles that the segments in the Steiner tree must not intersect.

Obstacle-avoiding short path (OARST) problem: A special case of the OARST problem that considers two-terminal rectilinear interconnections only.

One-and-half layer channel model: A channel model for channel routing, in that one single-layer metal mask and polysilicon bars with a fixed length and distance.

One-block subnet: A subnet whose two adjacent terminals are located on one block.

One-dimensional compaction: A process moving wires and components in one direction to minimize the area among wires and components.

One-dimensional (1-D) Steiner point: A dynamic Steiner point that can be moved in either a horizontal or a vertical direction.

Overlap mode wire layout: A wire layout not in generalized knock-knee mode.

Path: A connection that connects two terminals with a sequence of line segments (physical subedges) and their neighboring clearances.

Permutation: Taking r elements from n ($n \geq r$) distinct elements in order.

Placement: A technique that provides a module configuration for routing by placing modules on a workspace to satisfy one or more objectives.

P problem: A decision problem that can be solved by a deterministic algorithm in polynomial time.

Push aside: A rip-up technique that relocates existing wires to make room for the failed connection while maintaining all existing connections.

Rectilinear distance: A distance measured by a half-perimeter of the rectangle with two vertices (nodes or terminals) as end vertices, often called a Manhattan distance in technical publications.

Rectilinear tree: A tree with every edge length measured by the rectilinear distance.

Redundant length: The positive length difference between an edge-set tree and another edge-set tree generated by replacing a set of edge sets in the previous edge-set tree with a set of new edge sets.

Rip-up rerouting: Moving wires to make room for incomplete wires.

Routable substrate area: The surface area that can be used for conductors, including traces, vias, via-pads, and device lead pads.

Route-based parallelism: An orthogonal parallel decomposition for evaluating potential routes of a segment at the same time.

Routing length: A wire length needed by routing, where specific wires are not considered.

Routing track: A narrow stripe on a workspace used to accommodate a path or part of a path.

Segment-based parallelism: An orthogonal parallel decomposition for routing two-point segments generated from the minimum spanning tree decomposition at the same time.

Segment conflict: A conflict in parallel computing, in that a segment is involved in two grid-conflicts.

Segmented channel routing: A channel routing constrained to assign connections to fixed segments (tracks) whose lengths and positions are predetermined in a channel.

Sequence: A set whose elements are arranged in order.

Shield: An extra connection between two cross-talk connections, on which a proper stable voltage is exerted to reduce the cross-coupling capacitance.

Signal set: A set of signal terminals in a net.

Simulated annealing: An iterative process simulating the thermal equilibrium states, used to approach the objective from an initiation.

Single clique: A clique evolved from one LD clique and its adjacent degenerated edge sets.

Skew: Delay spread between signals in a timing group.

Skin effect: The tendency of a high-frequency current to flow on the surface of a conductor.

Sorted sequence: A set of sequences sorted in either an ascendent or a descendent order.

Spanning tree: A tree without Steiner points.

Standard gate: A logic gate formed by a pair of transistors.

Steiner point: An additional point (vertex) used to connect a set of nodes into a tree.

Steiner's problem: A combinatorial-graph problem for finding a tree connecting a set of nodes with the minimum tree length.

Steiner tree: A tree connecting a set of nodes with additional vertices called Steiner points.

Stem rectangle: The minimum rectangle connecting the projection of the moving domains of two main vertices of a clique when the projection of one moving domain on another is empty; or the rectangle whose two opposite sides are two projections of one moving domain on another.

Straight edge: An edge whose vertices have the same x-coordinate or y-coordinate.

Straight-edge set: An edge set whose vertices have the same x-coordinate or y-coordinate.

Strong rip-up: A rip-up technique that removes a single connection iteratively based on an analysis of the current blocking environment till the failed connection can be routed successfully.

Switchbox problem: A special case of the Steiner's problem, in that all nodes are located on the perimeter of a rectangle.

Symmetric routing: Routing for symmetric interconnections to meet the frequent requirement in a high-speed analog circuit.

Tails: A pair of branches pertaining to two tips in a binary tree.

Terminal: A pin, a joint, or a mount on a module used to electronically connect with other modules.

Time complexity: The time expressed as a function of the size of a decision problem, which is used to measure the computational efficiency of an algorithm for the problem.

Tip: A node with no children in a binary tree.

Top: The vicinity of a grid edge on which the density is the maximum in a topography map (T-map).

Topography map (T-map): A two-dimensional map recording the density of each grid edge.

Tree: A connected graph that connects a set of nodes without cycle.

Two-block subnet: A subnet whose two adjacent terminals are located on two blocks.

Two-dimensional compaction: A process moving wires and components in both x- and y-directions to minimize the area among wires and components.

Two-dimensional (2-D) Steiner point: A dynamic Steiner point that can be moved along both the horizontal and the vertical directions.

Valley: The vicinity of a grid edge on which the density is the minimum in a topography map (T-map).

Vector binary tree: A binary tree in which the swinging direction of the line connecting a father and his grandfather determines the children's location (on the right or the left).

Via: An electrical conductor connecting a net routed on different layers.

Via minimization: A routing process that minimizes the number of vias under a specific condition.

Weight-controlled cost function: A cost function for routing, that contains all the non-idealities to be controlled in the form of a weighted sum, expressing the trade-offs between different simultaneous constraints.

Weight of an edge in channel routing: A function of current density and capacity.

Wire-by-wire parallelism: An orthogonal parallel decomposition for routing different wires at the same time.

Wire density: The signal-wire length per unit area of the substrate.

Wire Layout: The process determining the placement of the routing segments.

Wire segment: A physical line segment in a net.

Wiring: A process assigning wire segments of the wire layout to one of k available layers so that the segments are electronically connected in the right way.

Zone: An area between two neighboring terminals of two segments.

T - #0067 - 101024 - C0 - 234/156/26 [28] - CB - 9780849396229 - Gloss Lamination